Optics and Photonics

Essential Technologies for Our Nation

Committee on Harnessing Light: Capitalizing on Optical Science Trends and Challenges for Future Research

National Materials and Manufacturing Board

Division on Engineering and Physical Sciences

NATIONAL RESEARCH COUNCIL
OF THE NATIONAL ACADEMIES

THE NATIONAL ACADEMIES PRESS
Washington, D.C.
www.nap.edu

THE NATIONAL ACADEMIES PRESS 500 Fifth Street, NW Washington, DC 20001

NOTICE: The project that is the subject of this report was approved by the Governing Board of the National Research Council, whose members are drawn from the councils of the National Academy of Sciences, the National Academy of Engineering, and the Institute of Medicine. The members of the committee responsible for the report were chosen for their special competences and with regard for appropriate balance.

This study was supported by Contract No. ECCS-1041156 between the National Academy of Sciences and the National Science Foundation, and by the following awards: #N66001-10-1-4052 from DARPA-DSO; #N66001-11-1-4091 from DARPA-MTO; #60NANB10D266 from NIST; #W911NF-10-1-0488 from ARO; #DE-DT0002194,TO#16 from DOE-EERE; and #DE-SC0005899 from DOE-BES, as well as support from SPIE, OSA, and the NRC. Any opinions, findings, conclusions, or recommendations expressed in this publication are those of the authors and do not necessarily reflect the views of the organizations or agencies that provided support for the project.

International Standard Book Number-13: 978-0-309-26377-1
International Standard Book Number-10: 0-309-26377-8
Library of Congress Control Number: 2012954592

This report is available in limited quantities from:

National Materials and Manufacturing Board
500 Fifth Street, NW
Washington, DC 20001
nmmb@nas.edu
http://www.nationalacademies.edu/nmmb

Additional copies of this report are available from the National Academies Press, 500 Fifth Street, NW, Keck 360, Washington, DC 20001; (800) 624-6242 or (202) 334-3313; http://www.nap.edu.

Printed in the United States of America

THE NATIONAL ACADEMIES

Advisers to the Nation on Science, Engineering, and Medicine

The **National Academy of Sciences** is a private, nonprofit, self-perpetuating society of distinguished scholars engaged in scientific and engineering research, dedicated to the furtherance of science and technology and to their use for the general welfare. Upon the authority of the charter granted to it by the Congress in 1863, the Academy has a mandate that requires it to advise the federal government on scientific and technical matters. Dr. Ralph J. Cicerone is president of the National Academy of Sciences.

The **National Academy of Engineering** was established in 1964, under the charter of the National Academy of Sciences, as a parallel organization of outstanding engineers. It is autonomous in its administration and in the selection of its members, sharing with the National Academy of Sciences the responsibility for advising the federal government. The National Academy of Engineering also sponsors engineering programs aimed at meeting national needs, encourages education and research, and recognizes the superior achievements of engineers. Dr. Charles M. Vest is president of the National Academy of Engineering.

The **Institute of Medicine** was established in 1970 by the National Academy of Sciences to secure the services of eminent members of appropriate professions in the examination of policy matters pertaining to the health of the public. The Institute acts under the responsibility given to the National Academy of Sciences by its congressional charter to be an adviser to the federal government and, upon its own initiative, to identify issues of medical care, research, and education. Dr. Harvey V. Fineberg is president of the Institute of Medicine.

The **National Research Council** was organized by the National Academy of Sciences in 1916 to associate the broad community of science and technology with the Academy's purposes of furthering knowledge and advising the federal government. Functioning in accordance with general policies determined by the Academy, the Council has become the principal operating agency of both the National Academy of Sciences and the National Academy of Engineering in providing services to the government, the public, and the scientific and engineering communities. The Council is administered jointly by both Academies and the Institute of Medicine. Dr. Ralph J. Cicerone and Dr. Charles M. Vest are chair and vice chair, respectively, of the National Research Council.

www.national-academies.org

COMMITTEE ON HARNESSING LIGHT: CAPITALIZING ON OPTICAL SCIENCE TRENDS AND CHALLENGES FOR FUTURE RESEARCH

PAUL McMANAMON, Exciting Technology, LLC, *Co-Chair*
ALAN E. WILLNER, University of Southern California, *Co-Chair*
ROD C. ALFERNESS, NAE,[1] Alcatel-Lucent (retired), University of California, Santa Barbara
THOMAS M. BAER, Stanford University
JOSEPH BUCK, Boulder Nonlinear Systems, Inc.
MILTON M.T. CHANG, Incubic Management, LLC
CONSTANCE CHANG-HASNAIN, University of California, Berkeley
CHARLES M. FALCO, University of Arizona
ERICA R.H. FUCHS, Carnegie Mellon University
WAGUIH S. ISHAK, Corning Incorporated
PREM KUMAR, Northwestern University
DAVID A.B. MILLER, NAS,[2] NAE, Stanford University
DUNCAN T. MOORE, NAE, University of Rochester
DAVID C. MOWERY, University of California, Berkeley
N. DARIUS SANKEY, Intellectual Ventures
EDWARD WHITE, Edward White Consulting

Staff

DENNIS CHAMOT, Acting Director, National Materials and Manufacturing Board
ERIK B. SVEDBERG, Study Director
HEATHER LOZOWSKI, Financial Associate
RICKY D. WASHINGTON, Administrative Coordinator (until August 2012)
MARIA L. DAHLBERG, Program Associate (until August 2012)
ANN F. LARROW, Program Associate (effective August 2012)
LAURA TOTH, Senior Program Assistant (until February 2012)
PAUL BEATON, Program Officer, STEP[3] (October through December 2011)
CAREY CHEN, Christine Mirzayan Science and Technology Policy Fellow, STEP (October through December 2011)

[1] NAE, National Academy of Engineering.
[2] NAS, National Academy of Sciences.
[3] STEP, Board on Science, Technology, and Economic Policy.

NATIONAL MATERIALS AND MANUFACTURING BOARD

---

[1] NAE, National Academy of Engineering.

Preface

The National Research Council (NRC) undertook the writing of a study on optics and photonics in 1988 (*Photonics: Maintaining Competitiveness in the Information Era*)[1] and then again in 1998 (*Harnessing Light: Optical Science and Engineering for the 21st Century*).[2] Now, after 14 years of dramatic technical advances and economic impact, another study is needed to help guide the nation's strategic thinking in this area. Since 1998 many other countries have developed their own strategic documents and organizations in the area of optics and photonics, and many have cited the U.S. NRC's 1998 *Harnessing Light* study as instrumental in influencing their thinking. The present study, *Optics and Photonics: Essential Technologies for Our Nation*, discusses impacts of the broad field of optics and photonics and makes recommendations for actions and research of strategic benefit to the United States.

To conduct the study, the NRC established the Committee on Harnessing Light: Capitalizing on Optical Science Trends and Challenges for Future Research, a diverse group of academic and corporate experts from across many disciplines critical to optical science and engineering, including materials science, communications, quantum optics, linear and nonlinear optical elements, semiconductor physics, device fabrication, biology, manufacturing, economic policy, and venture capital. The statement of task for this study (given in full in Appendix A) is as follows:

---

[1] National Research Council. 1988. *Photonics: Maintaining Competitiveness in the Information Era.* Washington, D.C.: National Academy Press.

[2] National Research Council. 1998. *Harnessing Light: Optical Science and Engineering for the 21st Century.* Washington, D.C.: National Academy Press.

1. Review updates in the state of the science that have taken place since publication of the National Research Council report *Harnessing Light*;
2. Identify the technological opportunities that have arisen from recent advances in and potential applications of optical science and engineering;
3. Assess the current state of optical science and engineering in the United States and abroad, including trends in private and public research, market needs, examples of translating progress in photonics innovation into competitiveness advantage (including activities by small businesses), workforce needs, manufacturing infrastructure, and the impact of photonics on the national economy;
4. Prioritize a set of research grand-challenge questions to fill identified technological gaps in pursuit of national needs and national competitiveness;
5. Recommend actions for the development and maintenance of global leadership in the photonics-driven industry—including both near-term and long-range goals, likely participants, and responsible agents of change.

It became apparent from the outset that various funding agencies and professional societies that deal with optics and photonics felt a keen need for the NRC to provide an authoritative vision of the field's future. If the field is indeed a key enabling technology that will help drive significant economic growth, then such a study should attempt to make recommendations that can be used to help policy makers and decision makers capitalize on optics and photonics. It was in this spirit that the committee conducted this study.

Several factors, including the following, made the committee's task a challenging one:

1. The field of optics and photonics is extremely broad in terms of the technical science and engineering topics that it encompasses.
2. The field impacts many different market segments, such as energy, medicine, defense, and communications, but as an enabling technology it is not always highlighted in available data about these areas.
3. The field has expanded greatly beyond the United States, such that many other countries have invested heavily in research and development and manufacturing.

Additionally, the area of optics and photonics is typically subsumed as an enabling technology under the heading of other disciplines (e.g., electrical engineering, physics). Therefore, it was challenging to gather data specific to optics and photonics in terms of workforce and economic impact. For example, optics enables common DVD players, but is the economic impact to be gauged by the value of the whole DVD player or just the inexpensive yet high-performance laser that makes the whole system work properly? Similarly, how do we place a value on the fact that the society-transforming Internet could not have grown at such a fast pace, or achieved even close to its current level of performance, without low-loss optical fiber, which by itself is not particularly expensive? The committee grappled with many such questions.

In the course of the study, the committee observed that exciting progress has been made in the field and believes that the future holds much promise. A small anecdotal indication in the popular press of the breadth and depth of the field is that roughly 12 of the 50 best inventions of 2011 listed by *Time* magazine had optics as a key technological part of the invention.[3]

Our entire community owes its sincerest gratitude to the generous sponsors of the study, which include the Army Research Office, the Defense Advanced Research Projects Agency, the Department of Energy, the National Institute of Standards and Technology, the National Research Council, the National Science Foundation, the Optical Society of America, and the International Society for Optics and Photonics (SPIE). Each sponsor was critical to enabling the study to proceed with the necessary resources, and key champion(s) in each of these organizations stepped forward at a crucial time to help out. We also wish to thank the many individuals who helped the committee accomplish its task, including the workshop speakers and study reviewers, and we are extremely grateful to have worked with outstanding committee members.

It was with a deep sense of appreciation that the committee was able to rely on the dedication, professionalism, insight, and good cheer of the NRC staff, primarily Dennis Chamot, Maria Dahlberg, Erik Svedberg, Laura Toth, and Ricky Washington. As the manager of the study, Erik has been a superb and tireless partner, whose keen perspective was invaluable. The committee also extends its thanks to Stephen Merrill, executive director of the National Academies' Board on Science, Technology, and Economic Policy, for engaging his staff during the latter part of this study, especially Paul Beaton, program officer, and Carey Chen, Christine Mirzayan Science and Technology Policy Fellow. In addition, the committee would like to thank Kathie Bailey-Mathae, director of the Board on International Scientific Organizations, for critically helping with the preliminary groundwork leading up to the start of the study.

We sincerely hope that readers of this study find some perspectives that will

[3] Grossman, L., M. Thompson, J. Kluger, A. Park, B. Walsh, C. Suddath, E. Dodds, K. Webley, N. Rawlings, F. Sun, C. Brock-Abraham, and N. Carbone. 2011. Top 50 Inventions. *Time*. Available at http://www.time.com/time/magazine/article/0,9171,2099708,00.html. Accessed October 16, 2012.

help guide future actions, whether such readers are congressional staffers, funding agencies, corporate chief technology officers, or high school students.

Paul McManamon and Alan E. Willner, *Co-Chairs*
Committee on Harnessing Light: Capitalizing on
Optical Science Trends and Challenges for Future Research

Acknowledgments

This report has been reviewed in draft form by individuals chosen for their diverse perspectives and technical expertise, in accordance with procedures approved by the National Research Council's (NRC's) Report Review Committee. The purpose of this independent review is to provide candid and critical comments that will assist the institution in making its published report as sound as possible and to ensure that the report meets institutional standards for objectivity, evidence, and responsiveness to the study charge. The review comments and draft manuscript remain confidential to protect the integrity of the deliberative process. We wish to thank the following individuals for their review of this report:

William B. Bridges (NAS/NAE), California Institute of Technology,
Elsa Garmire (NAE), California Institute of Technology,
James S. Harris (NAE), Stanford University,
Thomas S. Hartwick, Hughes Aircraft Company,
Eric G. Johnson, Clemson University,
Stephen M. Lane, Lawrence Livermore National Laboratory,
E. Phillip Muntz (NAE), University of Southern California, and
Thomas E. Romesser (NAE), Northrop Grumman Aerospace Systems.

Although the reviewers listed above have provided many constructive comments and suggestions, they were not asked to endorse the conclusions or recommendations, nor did they see the final draft of the report before its release. The review of this report was overseen by Peter Banks (NAE), Red Planet Capital

Partners. Appointed by the NRC, he was responsible for making certain that an independent examination of this report was carried out in accordance with institutional procedures and that all review comments were carefully considered. Responsibility for the final content of this report rests entirely with the authoring committee and the institution.

The committee also thanks those who were guest speakers at its meetings and who added to the committee members' understanding of optics and photonics and related issues:

John Ambroseo, Coherent Inc.,
Eugene Arthurs, SPIE,
John Dexheimer, LightWave Advisors, Inc.,
Ed Dowski, Ascentia Imaging,
Julie Eng, Finisar,
Michael Gerhold, U.S. Army Research Office,
Larry Goldberg, National Science Foundation,
Matthew Goodman, Defense Advanced Research Projects Agency,
Linda Horton, Department of Energy,
Kristina Johnson, Consultant,
Christian Jörgens, German Embassy,
Bikash Koley, Google,
Prem Kumar, CLEO,
Minh Le, Department of Energy,
Donn Lee, Facebook,
Robert Leheny, Institute for Defense Analyses,
Frederick J. Leonberger, Eovation Advisors, LLC,
Tingye Li, AT&T Consultant,
Aydogan Ozcan, University of California, Los Angeles,
Mario Paniccia, INTEL,
Kent Rochford, National Institute of Standards and Technology,
Joseph Schmitt, Cardiovascular Division, St. Jude Medical,
Jag Shah, Defense Advanced Research Projects Agency,
Bruce J. Tromberg, University of California, Irvine,
Usha Varshney, National Science Foundation, and
Paul Wehrenberg, Consultant.

Contents

SUMMARY 1

1 INTRODUCTION 13
Motivation for This Study, 15
Enabling Technology, 16
Economic Issues, 17
Global Perspective, 18
Importance of Education, 18
Progress for the Future, 19

2 IMPACT OF PHOTONICS ON THE NATIONAL ECONOMY 20
Introduction, 20
The Economics of Photonics: A Case Study of Lasers, 21
The Economic Impact of the Laser, 22
Funding of Early Laser Development, 23
The Early Laser Market, 24
International Comparison, 25
Conclusions from the Laser Case Study, 27
Estimating the Economic Impact of Photonics—Industry Revenues, Employment, and R&D Investment in the United States, 28
Government and Industrial Sources of R&D Funding in Photonics and Federal Funding of Optics, 32

Changes in Photonics-based Innovations in the United States Since 1980, 37
Venture Capital in Optics, 43
Markets for Technology, Intellectual Property, and U.S. University Technology Licensing, 50
Models of Collaborative R&D and Implications for Photonics Innovation, 52
Semiconductor Manufacturing Technology (SEMATECH), 54
Optoelectronics Industry Development Association (OIDA), 56
National Nanotechnology Initiative, 59
Summary Comments, 60
Proposed National Photonics Initiative, 61
Findings, 62
Recommendations, 63

3 COMMUNICATIONS, INFORMATION PROCESSING, AND DATA STORAGE 64
Introduction, 64
Communications, 65
Information Processing, 69
Data Storage, 72
Impact Example: The Internet, 73
Technical Advances, 75
Communications, 75
Networking, 85
R&D Example Areas, 87
Information Processing, 88
Data Storage, 91
Manufacturing, 92
Communications, 92
Information Processing, 93
Data Storage, 94
Economic Impact, 94
Comparison Between the United States and the Rest of the World, 96
Findings and Conclusions, 97
Recommendations and Grand Challenge Questions, 99

4 DEFENSE AND NATIONAL SECURITY 102
Introduction, 102
Optics and Photonics: Impact on Defense Systems, 104

Technology Overview, 104
Changes Since the *Harnessing Light* Study, 105
Identification of Technological Opportunities from Recent Advances, 108
Manufacturing, 121
U.S. Global Position, 122
Findings and Conclusions, 124
Recommendation and Grand Challenge Questions, 126

5 ENERGY 127
Introduction, 127
Solar Power, 130
Photovoltaic Systems, 133
Concentrated Solar Power Systems, 142
Hybrid Systems, 147
LCOE Outlook for Solar Power Compared to Other Current and Possible Future Fuel Sources, 148
Solid-State Lighting, 150
Findings, 159
Recommendations and Grand Challenge Question, 160

6 HEALTH AND MEDICINE 163
Introduction, 163
Historical Overview of the Impact of Technology on Medicine, 164
Optics and Photonics in Medical Practice Today, 166
Advances in Technology Providing the Opportunity for New Applications of Photonics, 168
Advances in Technology Providing the Opportunity for Future Applications of Photonics, 169
Nucleic Acid Sequence Detection and Mutation Detection, 169
Proteomic Analysis Through Protein and Tissue Arrays, 171
High-Throughput Screening, 171
Flow Cytometry Mass Spectrometry, 174
Ophthalmology, 174
Image-Guided Surgery, 176
Dual Energy CT and Quantitative Image Analysis, 178
Biomedical Optics in Regenerative Medicine, 180
Biomedical Optics in Research, 180
Findings, 181
Recommendations, 183

7 ADVANCED MANUFACTURING 185
Introduction, 185
Production and Innovation in Photonics Technologies:
Three Case Studies, 186
Displays, 186
Solar Cells, 189
Optoelectronic Components for Communications Systems, 194
Similarities and Differences Among the Three Cases, 200
Advanced Manufacturing in Optics, 202
Optical Surfaces, 203
Aspherical Lenses, 204
Fabrication Processes and Equipment, 204
Applications of Photonics in Manufacturing, 205
Photolithography, 206
Lasers in Manufacturing, 211
Additive Manufacturing, 212
Stereolithography, 214
Selective Laser Sintering, 215
Laser Engineered Net Shaping, 217
Photonics and the Future of U.S. Manufacturing, 219
High-Volume Products, 220
Low-Volume Products, 221
The U.S. Manufacturing Workforce, 221
Findings, 223
Recommendations and Grand Challenge Question, 224

8 ADVANCED PHOTONIC MEASUREMENTS AND APPLICATIONS 226
Introduction, 226
Impact of Optics and Photonics on Sensing, Imaging, and Metrology, 227
Technology Overview, 230
Changes Since *Harnessing Light*, 232
Changes in SI Definitions, 232
Development of Attosecond Pulse Trains by Means of the
Generation of High Harmonics, 233
Table-top Availability of Extreme Intensities by Means of
Chirped-Pulse Amplification, 234
Nano-optics and Plasmonics, Negative-Index Materials, and
Transformation Optics, 235
Advances in Controlled Generation of Quantum Light States
and Their Manipulation and Detection, 236

High-Resolution Remote Sensing with Optical Synthetic Aperture Radar, 239
Advances in Adaptive Optical Techniques, 239
Identification of Technological Opportunities from Recent Advances, 239
Manufacturing, 244
U.S. Global Position, 244
Findings, 245
Recommendations and Grand Challenge Question, 246

9 STRATEGIC MATERIALS FOR OPTICS 248
Introduction, 248
Energy Applications, 249
Novel Structures: Sub-Wavelength Optics, Metamaterials, and Photonic Crystals, 250
Technology Challenges of Nanostructured Materials, 254
Optical Materials in the Life Sciences and Synthetic Biology, 257
Findings, 258
Recommendations, 259

10 DISPLAYS 260
Introduction, 260
The Near Future, 262
Overview of Displays, 263
Liquid-Crystal Displays, 263
Touch Displays, 264
OLED Displays, 267
Flexible Displays, 268
Projection Displays, 269
Three-Dimensional Holographic Displays, 269
Display Product Manufacturing, 272
Findings, 272
Recommendations, 272

APPENDIXES

A Statement of Task, with Introductory Information 277
B Acronyms and Abbreviations 280
C Additional Technology Examples 288
D Biographies of Committee Members 331

Summary

Optics and photonics technologies are central to modern life; indeed, UNESCO has recently adopted a resolution declaring 2015 to be the International Year of Light.[1] These technologies enable the manufacture and inspection of all the integrated circuits in every electronic device in use.[2] They give us displays on our smartphones and computing devices, optical fiber that carries the information in the Internet, advanced precision fabrication, and medical diagnostics tools. Optics and photonics technology offers the potential for even greater societal impact over the next few decades. Solar power generation and new efficient lighting, for example, could transform the energy landscape, and new optical capabilities will be essential to supporting the continued exponential growth of the Internet. Optics and photonics technology development and applications have substantially increased across the globe over the past several years. This is an encouraging trend for the world's economy and its people, while at the same time posing a challenge to U.S. leadership in these areas. As described in this study conducted by the National Research Council's (NRC's) Committee on Harnessing Light: Capitalizing on Optical Science Trends and Challenges for Future Research, it is critical that the United States take advantage of these emerging optical technologies for creating new industries and generating job growth.

[1] For more information, see http://www.eps.org/news/106324/.

[2] For example, photolithography is used to create most of the layers in integrated circuits, and cameras inspect the quality afterward.

Each chapter of the present report addresses the developments that have taken place over the 15 years since the publication of the NRC report *Harnessing Light: Optical Science and Engineering for the 21st Century*,[3] technological opportunities that have arisen since then, and the state of the art in the United States and abroad, and recommendations are offered for how to maintain U.S. global leadership.

It is the committee's hope that this study will help policy makers and leaders decide on courses of action that can advance the economy of the United States, provide visionary guidance and support for the future development of optics and photonics technology and applications, and ensure a leadership role for the United States in these areas. Although many unknowns exist in the course of pursuing basic optical science and its transition to engineering and ultimately to products, the rewards can be great. Researchers have achieved some dramatic advances. For example, work in optics and photonics has now provided clocks so stable that they will slip less than 1 second in more than 100 million years. Much more primitive clocks enabled the incredibly useful Global Positioning System (GPS), and it remains to be discovered how these new clock advances can be fully harnessed for the benefit of society. In many ways, the current period might be analogous to the dawn of the laser in 1960, when many of the transforming applications of that extraordinary invention had not yet been contemplated. This is only one example of technology innovation in optics and photonics that can lead to future major applications.

GRAND CHALLENGE QUESTIONS TO FILL TECHNOLOGICAL GAPS

To fill identified technological gaps in pursuit of national needs and national competitiveness, the committee developed five overarching grand challenge questions:

1. How can the U.S. optics and photonics community invent technologies for the next factor-of-100 cost-effective capacity increases in optical networks?

As mentioned in Chapter 3, it is not currently known how to achieve this goal, but the world has experienced a factor-of-100 cost-effective capacity increase every decade thus far, and user demand for this growth is anticipated to continue. Unfortunately, the mechanisms that have enabled the previous gains cannot sustain further increases at that high rate, and so the world will either see increases in capability stagnate or will have to invent new technologies.

[3] National Research Council. 1998. *Harnessing Light: Optical Science and Engineering for the 21st Century.* Washington, D.C.: National Academy Press.

2. How can the U.S. optics and photonics community develop a seamless integration of photonics and electronics components as a mainstream platform for low-cost fabrication and packaging of systems on a chip for communications, sensing, medical, energy, and defense applications?

In concert with meeting the fifth grand challenge, achieving this grand challenge would make it possible to stay on a Moore's law-like path of exponential performance growth. The seamless integration of optics and photonics at the chip level has the potential to significantly increase speed and capacity for many applications that currently use only electronics, or that integrate electronics and photonics at a larger component level. Chip-level integration will reduce weight and increase speed while reducing cost, thus opening up a large set of future possibilities as devices become further miniaturized.

3. How can the U.S. military develop the required optical technologies to support platforms capable of wide-area surveillance, object identification and improved image resolution, high-bandwidth free-space communication, laser strike, and defense against missiles?

Optics and photonics technologies used synergistically for a laser strike fighter or a high-altitude platform can provide comprehensive knowledge over an area, the communications links to download that information, an ability to strike targets at the speed of light, and the ability to robustly defend against missile attack. Clearly this technological opportunity could act as a focal point for several of the areas in optics and photonics (such as camera development, high-powered lasers, free-space communication, and many more) in which the United States must be a leader in order to maintain national security.

4. How can U.S. energy stakeholders achieve cost parity across the nation's electric grid for solar power versus new fossil-fuel-powered electric plants by the year 2020?

The impact on U.S. and world economies from being able to answer this question would be substantial. Imagine what could be done with a renewable energy source, with minimal environmental impact, that is more cost-effective than nonrenewable alternatives. Although this is an ambitious goal, the committee poses it as a grand challenge question, something requiring an extra effort to achieve. Today, it is not known how to achieve this cost parity with current solar cell technologies.

5. How can the U.S. optics and photonics community develop optical sources and imaging tools to support an order of magnitude or more of increased resolution in manufacturing?

Meeting this grand challenge could facilitate a decrease in design rules for lithography, as well as providing the ability to do closed-loop, automated manufacturing of optical elements in three dimensions. Extreme ultraviolet (EUV) is a challenging technology to develop, but it is needed in order to meet future lithography needs. The next step beyond EUV is to move to soft x rays. Also, the limitations in three-dimensional resolution on laser sintering for three-dimensional manufacturing are based on the wavelength of the lasers used. Shorter wavelengths will move the state of the art to allow more precise additive manufacturing that could eventually lead to three-dimensional printing of optical elements.

The committee believes that these five grand challenges are the top priorities in their respective application areas, and that because of their diverse nature, further prioritization among them is not advisable. These grand challenge questions are discussed in the main text immediately after the first key recommendation that supports the challenge and are drawn from the findings and recommendations throughout the report. They are discussed in the chapter in which they first appear, and occasionally in succeeding chapters.

REPORT CONTENT AND KEY RECOMMENDATIONS

This report is divided into chapters based on application areas, with crosscutting chapters addressing the impact of photonics on the national economy, advanced manufacturing, and strategic materials. Following an introductory Chapter 1, Chapter 2 discusses the impacts of photonics technologies on the U.S. economy.

Chapters 3 through 10 each cover a particular area of technological application. As mentioned, the discussion of each application area typically begins with a review of updates in the state of the science since the publication of the NRC's report *Harnessing Light*, as well as the technological opportunities that have arisen from recent advances in and potential applications of optical science and engineering. Included are recommended actions for the development and maintenance of global leadership in the photonics-driven industry, including both near-term and long-range goals, likely participants, and responsible agents of change. As relevant to their respective topics, the chapters assess the current state of optical science and engineering in the United States and abroad, including trends in private and public research, market needs, examples of translating progress in photonics innovation into global competitive advantage (including activities by small businesses), workforce needs, manufacturing infrastructure, and the impact of photonics on the national economy.

Following is a chapter-by-chapter overview of the content of Chapters 2 through 10, including the key recommendations from each.

Chapter 2: Impact of Photonics on the National Economy

Chapter 2 considers the economic impact of optics and photonics on the nation and the world. This chapter uses a case study of lasers to discuss the conceptual challenges of developing estimates of the economic impact of photonics innovation. It also addresses the problems associated with using company-level data to provide indicators of the economic significance of the "photonics sector" within the U.S. economy. Additionally, this chapter discusses the ways in which the changing structure of the innovation process within photonics reflects broader shifts in the sources of innovation within the U.S. economy. The chapter also considers the results of recent experiments in public-private and inter-firm research and development (R&D) collaboration in other high-technology areas for the photonics sector. Possibly the most important finding of the committee in this area is related to the pervasive nature of optics and photonics as an enabling technology.

> **Key Recommendation:** The committee recommends that the federal government develop an integrated initiative in photonics (similar in many respects to the National Nanotechnology Initiative) that seeks to bring together academic, industrial, and government researchers, managers, and policy makers to develop a more integrated approach to managing industrial and government photonics R&D spending and related investments.

This recommendation is based on the committee's judgment that the photonics field is experiencing rapid technical progress and rapidly expanding applications that span a growing range of technologies, markets, and industries. Indeed, in spite of the maturity of some of the constituent elements of photonics (e.g., optics), the committee believes that the field as a whole is likely to experience a period of growth in opportunities and applications that more nearly resembles what might be expected of a vibrantly young technology. But the sheer breadth of these applications and technologies has impeded the formulation by both government and industry of coherent strategies for technology development and deployment.

A national photonics initiative would identify critical technical priorities for long-term federal R&D funding. In addition to offering a basis for coordinating federal spending across agencies, such an initiative could provide matching funds for industry-led research consortia (of users, producers, and material and equipment suppliers) focused on specific applications, such as those described in Chapter 3 of this report. In light of near-term pressures to limit the growth of or even reduce federal R&D spending, the committee believes that a coordinated initiative in photonics is especially important.

The committee assesses as deplorable the state of data collection and analysis of photonics R&D spending, photonics employment, and sales. The development of better historical and current data collection and analysis is another task for which a national photonics initiative is well suited.

Key Recommendation: The committee recommends that the proposed national photonics initiative spearhead a collaborative effort to improve the collection and reporting of R&D and economic data on the optics and photonics sector, including the development of a set of North American Industry Classification System (NAICS) codes that cover photonics; the collection of data on employment, output, and privately funded R&D in photonics; and the reporting of federal photonics-related R&D investment for all federal agencies and programs.

It is essential that an initiative such as the proposed national photonics initiative be supported by coordinated measurement of the inputs and outputs in the sector such that national policy in the area can be informed by the technical and economic realities on the ground in the nation.

Chapter 3: Communications, Information Processing, and Data Storage

Chapter 3 considers communications, information processing, and data storage. The Internet's growth has fundamentally changed how business is done and how people interact. Photonics has been a key enabler allowing this communication revolution to occur. The committee anticipates that this revolution will continue, with additional demands driving significant increases in bandwidth and an even heavier reliance on the Internet. So far there has been a factor-of-100 increase in capacity each decade. However, there exists a technology wall inhibiting achievement of the next factor-of-100 growth.

Key Recommendation: The U.S. government and private industry, in combination with academia, need to invent technologies for the next factor-of-100 cost-effective capacity increase in long-haul, metropolitan, and local-area optical networks.

The optics and photonics community needs to inform funding agencies, and information and entertainment providers, about the looming roadblock that will interfere with meeting the growing needs for network capacity and flexibility. There is a need to champion collaborative efforts, including consortia of companies, to find new technology—transmission, amplification, and switching—to carry and route at least another factor-of-100 capacity in information over the next 10 years.

> **Key Recommendation:** The U.S. government, and specifically the Department of Defense (DOD), should strive toward harmonizing optics with silicon-based electronics to provide a new, readily accessible and usable, integrated electronics and optics platform.

They should also support and sustain U.S. technology transition toward low-cost, high-volume circuits and systems that utilize the best of optics and electronics in order to enable integrated systems to seamlessly provide solutions in communications, information processing, biomedical, sensing, defense, and security applications. Government funding agencies, the Department of Defense, and possibly a consortium of companies requiring these technologies should work together to implement this recommendation. This technology is one approach to assist in accomplishing the first key recommendation in Chapter 3 concerning the factor-of-100 increase in Internet capability.

> **Key Recommendation:** The U.S. government and private industry should position the United States as a leader in the optical technology for the global data center business.

Optical connections within and between data centers will be increasingly important in allowing data centers to scale in capacity. The committee believes that strong partnering between users, content providers, and network providers, as well as between businesses, government, and university researchers, is needed for ensuring that the necessary optical technology is generated, which will support continued U.S. leadership in the data center business.

Chapter 4: Defense and National Security

In Chapter 4, the committee discusses defense and national security. It is becoming increasingly clear that sensor systems are the next "battleground" for dominance in intelligence, surveillance, and reconnaissance. Comprehensive knowledge across an area will be a great defense advantage, along with the ability to communicate information at high bandwidths and from mobile platforms. Laser weapon attack can provide a significant advantage to U.S. forces. Defense against missile attacks, especially ballistic missiles, is another significant security need. Optical systems can provide synergistic capability in all these areas.

Key Recommendation: The U.S. defense and intelligence agencies should fund the development of optical technologies to support future optical systems capable of wide-area surveillance, exquisite long-range object identification, high-bandwidth free-space laser communication, "speed-of-light" laser strike, and defense against both missile seekers and ballistic missiles. Practical application for these purposes would require the deployment of low-cost platforms supporting long dwell times.

These combined functions will leverage the advances that have been made in high-powered lasers, multi-function sensors, optical aperture scaling, and algorithms that exploit new sensor capabilities, by bringing the developments together synergistically. These areas have been pursued primarily as separate technical fields, but it is recommended that they be pursued together to gain synergy. One method of maintaining this coordination could include reviewing the coordination efforts among agencies on a regular basis.

Chapter 5: Energy

Chapter 5 deals with optics and photonics in the energy area. Both the generation of energy and the efficient use of energy are discussed in terms of critical national needs. Photonics can provide renewable solar energy, while solid-state lighting can help reduce the overall need for energy used for lighting.

Key Recommendation: The Department of Energy (DOE) should develop a plan for grid parity across the United States by 2020.

"Grid parity" is defined here as the situation in which any power source is no more expensive to use than power from the electric grid. Solar power electric plants should be as cheap, without subsidies, as alternatives. It is understood that this will be more difficult in New England than in the southwestern United States, but the DOE should strive for grid parity in both locations.

Even though significant progress is being made toward reducing the cost of solar energy, it is important that the United States bring the cost of solar energy down to the price of other current alternatives without subsidy and maintains a significant U.S. role in developing and manufacturing solar energy alternatives. There is a need not only for affordable renewable energy but also for creating jobs in the United States. A focus in this area can contribute to both. Lowering the cost of solar cell technology will involve both technology and manufacturing advances.

Solid-state lighting can also contribute to energy security in the United States.

Key Recommendation: The DOE should strongly encourage the development of highly efficient light-emitting diodes (LEDs) for general-purpose lighting and other applications.

For example, the DOE could move aggressively toward its 21st-century lightbulb, with greater than 150 lm/W, a color rendering index greater than 90, and a color temperature of approximately 2800 K. Since one major company has already published results meeting the technical requirements for the 21st-century lightbulb, the DOE should consider releasing this competition in 2012. Major progress is being made in solid-state lighting, which has such advantages over current lighting alternatives as less wasted heat generation and fast turn-on time. The United States needs to exploit the current expertise in solid-state lighting to bring this technology to maturity and to market.

Chapter 6: Health and Medicine

Chapter 6 discusses the application of optics and photonics to health and medicine. Photonics plays a major role in many health-related areas. Medical imaging, which is widely used and is still a rapidly developing area, is key to many health-related needs, both for gaining understanding of the status of a patient and for guiding and implementing corrective procedures. Lasers are used in various corrective procedures in addition to those for the eye. There is still great potential for further application of optics and photonics in medicine.

Key Recommendation: The U.S. optics and photonics community should develop new instrumentation to allow simultaneous measurement of all immune-system cell types in a blood sample. Many health issues could be addressed by an improved knowledge of the immune system, which represents one of the major areas requiring better understanding.

Key Recommendation: New approaches, or dramatic improvements in existing methods and instruments, should be developed by industry and academia to increase the rate at which new pharmaceuticals can be safely developed and proved effective. Developing these approaches will require investment by the government and the private sector in optical methods integrated with high-speed sample-handling robotics, methods for evaluating the molecular makeup of microscopic samples, and increased sensitivity and specificity for detecting antibodies, enzymes, and important cell phenotypes.

Chapter 7: Advanced Manufacturing

Chapter 7 addresses the field of advanced manufacturing and the way in which it relates to optics and photonics. Advanced manufacturing is critical for the economic well-being of the United States. While there are issues concerning the ability of the United States to compete successfully in high-volume, low-cost manufacturing, it is likely that the United States can continue to be a strong competitor in lower-volume, high-end manufacturing. Additive manufacturing has the potential to allow the production of parts near the end user no matter where the design is done. Thus, if the end user is in the United States, it is there that the printing or manufacturing would occur. Optical approaches, such as laser sintering, are very important approaches to three-dimensional printing.

> **Key Recommendation:** The United States should aggressively develop additive manufacturing technology and implementation.

Current developments in the area of lower-volume, high-end manufacturing include, for example, three-dimensional printing, also called additive manufacturing. With continued improvements in the tolerance and surface finish, additive manufacturing has the potential for substantial growth. The technology also has the potential to allow three-dimensional printing near the end user no matter where the design is done.

> **Key Recommendation:** The U.S. government, in concert with industry and academia, should develop soft x-ray light sources and imaging for lithography and three-dimensional manufacturing.

Advances in table-top sources for soft x rays will have a profound impact on lithography and optically based manufacturing. Therefore, investment in these fields should increase to capture intellectual property and maintain a leadership role for these applications.

Chapter 8: Advanced Photonic Measurements and Applications

Chapter 8 discusses sensing, imaging, and metrology in relation to optics and photonics. Sensing, imaging, and metrology have made significant progress since the publication of the NRC's *Harnessing Light* in 1998.[4] Notable developments include having at least one Nobel Prize awarded for developing dramatic increases in

---

[4] National Research Council. 1998. *Harnessing Light: Optical Science and Engineering for the 21st Century.* Washington, D.C.: National Academy Press.

the precision of time measurement.[5] Single-photon detectors have been developed, but at this time they are only available with a dead time after detection, not allowing single-photon sensitivity for detecting all incoming photons. Extreme nonlinear optics has made significant progress, providing the potential for soft x-ray sources and imaging. Entangled photons and squeezed states are new areas for R&D in the optics and photonics field, allowing sensing options never previously considered.

> **Key Recommendation:** The United States should develop the technology for generating light beams whose photonic structure has been prearranged to yield better performance in applications than is possible with ordinary laser light.

Prearranged photonic structures in this context include generation of light with specified quantum states in a given spatiotemporal region, such as squeezed states with greater than 20-dB measured squeezing in one field quadrature, Fock states of more than 10 photons, and states of one and only one photon or two and only two entangled photons with greater than 99 percent probability. These capabilities should be developed with the capacity to detect light with over 99 percent efficiency and with photon-number resolution in various bands of the optical spectrum. The developed devices should operate at room temperature and be compatible with speeds prevalent in state-of-the-art sensing, imaging, and metrology systems. U.S. funding agencies should give high priority to funding research and development—at universities and in national laboratories where such research is carried out—in this fundamental field to position the U.S. science and technology base at the forefront of applications development in sensing, imaging, and metrology. It is believed that this field, if successfully developed, can transfer significant technology to products for decades to come.

> **Key Recommendation:** Small U.S. companies should be encouraged and supported by the government to address market opportunities for applying research advances to niche markets while exploiting high-volume consumer components. These markets can lead to significant expansion of U.S.-based jobs while capitalizing on U.S.-based research.

Chapter 9: Strategic Materials for Optics

Chapter 9 deals with strategic materials for optics. The main developments in materials for optics and photonics are the emergence of metamaterials and the

[5] For example, the 2005 Nobel Prize in physics. More information can be found at http://www.nobelprize.org/nobel_prizes/physics/laureates/2005/. Accessed August 2, 2012.

realization of how vulnerable the United States is to the need for certain critical materials. At this time, some of those materials are available only from China.

> **Recommendation:** The U.S. R&D community should increase its leadership role in the development of nanostructured materials with designable and tailorable optical material properties, as well as process control for uniformity of production of these materials.

Chapter 10: Displays

Chapter 10 addresses display technology. The major current display industry is based on technologies invented primarily in the United States, but this industry's manufacturing operations are located mostly overseas. Labor costs were a consideration, but other factors such as the availability of capital were significant in creating this situation. However, the United States is still dominant in many of the newer display technologies, and it still has an opportunity to maintain a presence in those newer markets as they develop.

> **Recommendation:** U.S. private companies and the Department of Defense should ensure a leadership role by funding R&D related to new materials for flexible, low-power, holographic and three-dimensional display technologies.

CONCLUDING COMMENTS

In reviewing the technologies considered here, a number of potential future opportunities have come to light that allow one to imagine changes to daily life: for example, electronic imaging devices implantable in the eye which can restore sight to the blind; cost-effective, laser-based, three-dimensional desktop printing of many different types of objects; the generation, detection, and manipulation of single photons in the same way as is done with single electrons, and doing it all on a photonic integrated circuit; the use of optics as interconnects between integrated circuit chips, with dramatic increases in power efficiency and speed; the unfurling of a flexible display on a smartphone or the watching of holographic images at home; and the ability of mobile lasers to neutralize threats from afar with high accuracy and speed. These are just a few interesting examples of potential changes that can occur as a result of the enabling technologies considered in this study.

1

Introduction

Optics and photonics are technical enablers for many areas of the economy, and dramatic technical advances have had a major impact on daily life. For example, in the last decade, advances in optical fiber communications have permitted a nearly 100-fold increase in the amount of information that can be transmitted from place to place, enabling a society-transforming Internet to thrive. As noted in the introduction to Charles Kao's 2009 Nobel Prize lecture on his work in optical fiber communications, "the work has fundamentally transformed the way we live our daily lives."[1] Indeed, optical fiber communications have enabled what Thomas Friedman has called a "flat world."[2] Without optics, the Internet as we know it would not exist.

The phrase "optics and photonics" is used throughout this study to capture light's dual nature (1) as a propagating wave, like a radio wave, but with a frequency that is now a million times higher than that of a radio wave; and (2) as a collection of traveling particles called photons, with potential as a transformative field similar in impact to electronics. Further proof that optics and photonics are technical enablers can be seen in the laser. A laser provides a source of light that can be (1) coherent, meaning that a group of photons can act as a single unit; and

---

[1] Kao, C.K. "Sand from Centuries Past: Send Future Voices Fast." Nobel Lecture. 2009. Available at http://www.nobelprize.org/nobel_prizes/physics/laureates/2009/kao_lecture.pdf. Accessed July 30, 2012.

[2] Friedman, T.L. 2005. *The World Is Flat.* New York, N.Y.: Farrar, Straus, and Giroux.

(2) monochromatic, meaning that the photons can have a well-defined single color. Today we can see how these effects are used in many areas. With light:

- High amounts of energy can be precisely directed with low loss.
- Many different properties of waves (i.e., degrees of freedom such as amplitude, frequency, phase, polarization, and direction) can be accurately manipulated.
- Waves can be coherently processed to have high directionality, speed, and dynamic range.

BOX 1.1
Optics, Electro-optics, Optoelectronics, and Photonics: Definitions and the Emergence of a Field

Optics—the science that deals with the generation and propagation of light—can be traced to 17th-century ideas of Descartes concerning transmission of light through the aether, Snell's law of refraction, and Fermat's principle of least time. These ideas were subsequently built upon through the 19th century by Hooke (interference of light and wave theory of light), Boyle (interference of light), Grimaldi (diffraction), Huygens (light polarization), Newton (corpuscular theory), Young (interference), Fresnel (diffraction), Rayleigh, Kirchhoff, and, of course, Maxwell (electromagnetic fields). The end of the 19th century marked the close of the era of classical optics and the start of quantum optics. In 1900, Max Planck's introduction of energy quanta marked the first steps toward quantum theory and an early understanding of atoms and molecules. With the demonstration in 1960 of the first laser, many of the fundamental and seemingly disconnected principles of optics established by Einstein, Bose, Wood, and many others were focused and drawn together.

"Electro-optics" and "optoelectronics" are both terms describing subfields of optics involving the interaction between light and electrical fields. Although John Kerr, who discovered in 1875 that the refractive index of materials changes in response to an electrical field, could arguably be regarded as the inaugurator of the field of electro-optics, the term "electro-optics" first gained popularity in the literature in the early 1960s. By 1964 authors from RAND could be found publishing from a group called the Electro-Optical Group. In 1965 the Quantum Electronics Council of the Institute of Electrical and Electronics Engineers (IEEE) was formed from IEEE's Electronic Devices Group and Microwave Theory and Techniques Group; in 1977 became an IEEE society; and in 1985 took the name Lasers and Electro-Optics Society, thus legitimizing the use of the name in the professional field.

The exact origins and limits of the term "optoelectronics" are difficult to pin down. Some claim that optoelectronics is a subfield of electro-optics involving the study and application of electronic devices that source, detect, and control light. Colloquially, the term "optoelectronics" is most commonly used to refer to the quantum mechanical effects of light on semiconductor materials, sometimes in the presence of an electrical field. Semiconductors started to assume serious importance in optics in 1953, when McKay and McAfee demonstrated electron mul-

MOTIVATION FOR THIS STUDY

Although the fields of optics and photonics have developed gradually (Box 1.1), important changes have occurred over the past several years that merit study and related action:

1. The science and engineering of light have enabled dramatic technical advances.
2. Globalization of manufacturing and innovation has accelerated.

tiplication in silicon and germanium p-n junctions, and Neumann indicated separately in a letter to a colleague that that one could obtain radiation amplification by stimulated emission in semiconductors. Japan's Optoelectronics Industry and Technology Development Association was established in 1980, and the U.S. counterpart is the Optoelectronics Industry Development Association.

As used in its present sense, the term "photonics" appeared as "la photonique" in a 1973 article by French physicist Pierre Aigrain. The term began to be seen in print in English around 1981 in press releases, annual reports of Bell Laboratories, and internal publications of Hughes Aircraft Corporation and in the more general press. In 1982, the trade magazine *Optical Spectra* changed its name to *Photonics Spectra*, and in 1995 the International Society for Optics and Photonics (SPIE) debuted *Photonics West*, arguably one of the largest conferences in optics and photonics. Sternberg defines "photonics" as the "engineering applications of light," involving the use of light to detect, transmit, store, and process information; to capture and display images; and to generate energy. However, in the professional literature, "photonics" is used almost synonymously with the term "optics," referring equally to both science and applications. The term "photonics" continues to gain popularity today. In 2006 Nature Publishing Group established the journal *Nature Photonics*, and in 2008 the Lasers and Electro-Optics Society became the IEEE Photonics Society.

SOURCES:

Brown, R.G.W., and E.R. Pike. 1995. A history of optical and optoelectronic physics in the twentieth century. In *Twentieth Century Physics, Vol. III*, L.M. Brown, A. Pais, and B. Pippard, eds. Bristol, U.K., and Philadelphia, Pa.: Institute of Physics Publishing; New York, N.Y.: American Institute of Physics Press.

IEEE Global History Network. 2012. "IEEE Photonics Society History." Available at http://www.ieeeghn.org/wiki/index.php/IEEE_Photonics_Society_History. Accessed August 1, 2012.

Nature Publishing Group. 2006. "Nature Publishing Group Announces the Launch of Nature Photonics." Available at http://www.nature.com/press_releases/Nature_Photonics_launches.pdf. Accessed August 1, 2012.

SPIE. 2011. "History of the Society." Available at http://spie.org/x1160.xml. Accessed August 3, 2012.

Sternberg, E. 1992. *Photonic Technology and Industrial Policy: U.S. Responses to Technological Change*. Albany, N.Y.: State University of New York Press.

3. Optics and photonics have become established as enabling technologies for a multitude of industries that are vital to our nation's future.

Accordingly, the National Research Council's Committee on Harnessing Light: Capitalizing on Optical Science Trends and Challenges for Future Research undertook a new study to examine the current state of the art and economic impact of optics and photonics technologies, with an eye toward ensuring that optics and photonics continue to enable a vibrant and secure future for U.S. society.

ENABLING TECHNOLOGY

Optics and photonics, an enabling technology with widespread impact, exhibits the characteristics of a general-purpose technology, that is, a technology in which advances foster innovations across a broad spectrum of applications in a diverse array of economic sectors. Improvements in those sectors in turn increase the demand for the technology itself, which makes it worthwhile to invest further in improving the technology, thus sustaining growth for the economy as a whole. The transistor and integrated circuit are good examples of general-purpose technologies. The importance of photonics as an enabling technology since 1998 can be highlighted by a few examples:

- A cell phone can enable video chats and perform an Internet search, with optics and photonics playing a key part. The most obvious contribution of optics is the high-resolution display and the camera. In addition, the cell phone uses a wireless radio connection to a local cell tower, and the signal is converted to an optical data stream for transmission along a fiber-optic network. An Internet search conducted on the phone will be directed over these fibers to a data center, and in a given data center clusters of co-located computers talk to each other through high-capacity optical cables. There can be more than 1 million lasers involved in the signaling.
- People are surrounded by objects whose manufacture was enabled by highly accurate directed-energy light. For example, nearly every microprocessor has been fabricated using optical lithographic techniques, and in nearly all advanced manufacturing, high-power lasers are used for cutting and welding.
- Optics is rapidly changing medical imaging, making it possible not only to see with higher resolution inside the body but also to distinguish between subtle differences in biological material. Swallowed capsules can travel through the body and send images back to a doctor for diagnosis. Today, the relatively young field of optical coherence tomography has the poten-

tial to save thousands of lives annually[3] by providing dramatically better images for early detection of disease. Optical spectroscopic techniques can provide valuable information from blood and tissue samples that is critical in early detection and prevention of health problems, and eye, dental, and brain surgery now uses focused lasers for ablating, cutting, vaporizing, and suturing.

- In World War II, only a small fraction of the bombs dropped from airplanes hit their target. "Smart" bombs debuted in Vietnam. Although the Thanh Hoa Bridge withstood 871 sorties by conventional bombs and 11 U.S. planes were lost, the bridge was destroyed with four sorties and no losses the first time smart bombs were used. In Iraq and Afghanistan, smart bombs are the norm.[4] The critical advance is accurate targeting using laser designators and laser-guided munitions. Moreover, situational awareness of the battlefield and of enemy terrain provides information for targeting. Imaging systems using LIDAR (light detection and ranging), such as HALOE, can provide wide-area three-dimensional imaging. Even wider-area passive sensors such as ARGUS-IS can provide highly detailed mapping of a country in days as opposed to months.

Additional examples of optics and photonics as enabling technologies are discussed in subsequent chapters and also in Appendix C.

ECONOMIC ISSUES

From an economic standpoint, an enabling technology like optics and photonics tends to be commercialized outside the industry, and profits can be generated by companies that do not consider themselves a part of the photonics industry. These companies are more inclined to invest in previously validated applications for which photonics can but does not necessarily provide the sole technology solution, rather than to invest in photonics in particular. Since 2000, the photonics industry has tended to receive little interest from the investment community and little financial analyst coverage, and start-up companies in photonics can have difficulty acquiring seed capital.[5]

However, a large fraction of the major companies in the United States rely on

[3] Center for Integration of Medicine and Innovative Technology. Capabilities brochure. Available at http://www.cimit.org/images/media_center/CapabilitiesBrochure.pdf. Accessed July 30, 2012.

[4] Air University Review. 1987. "The Decisive Use of Air Power?" Available at http://www.airpower.maxwell.af.mil/airchronicles/aureview/1987/werrell.html. Accessed July 30, 2012.

[5] This subject is addressed further in Chapter 2.

photonics-enabled technologies to be competitive in the marketplace.[6] To move forward in general, having an optics and photonics technology roadmap that focuses on meeting needs in specific market applications and that is synergistic with business and marketing trends could help to improve business development, profitability, and growth.

GLOBAL PERSPECTIVE

In considering actions for global leadership in the photonics industry, the committee took note of several important points. For example, although many key optics and photonics innovations occurred in the United States, including in display and communications technologies, the multibillion-dollar display industry has moved mostly to Southeast Asia, with a negligible fraction of display production remaining in the United States. Furthermore, whereas the United States for decades led the manufacture of telecommunications equipment, China went from having no company in 1998 in the top 10 largest telecommunications companies in the world to having three such companies in 2011. A similar scenario exists for Chinese companies that specialize in selling optical components and subsystems. By contrast, data centers continue to be located overwhelmingly in the United States, possibly because the United States has the most effective communications infrastructure at the moment.

A theme evident in several of the presentations made to the committee was that innovation will remain critical to ensuring a U.S. leadership position in optics and photonics. The United States has acclaimed educational institutions and a creative, entrepreneurial corporate spirit. According to the U.S. Patent and Trademark Office, the number of patents granted to the United States in 2010 in the field of optics was more than 50 percent greater than that granted to the next-nearest country. Yet, according to the records of the Optical Society of America, the number of research papers submitted to its journals in 2010 by scholars from the Pacific Rim countries exceeded the number of papers submitted by North American authors.[7]

IMPORTANCE OF EDUCATION

Education plays a critically important role in ensuring a vibrant future for the United States in the fields of optics and photonics. Today, the United States has

[6] See, for example, the National Center for Optics and Photonics Education's (OPTEC's) *Photonics: An Enabling Technology*, for fields that are important. Available at http://www.op-tec.org/pdf/Enabling_Technology_9NOV2011.pdf. Accessed July 30, 2012.

[7] Cao, J. 2012. A new journal in optics and photonics—*Light: Science and Applications*. Editorial. *Light: Science and Applications* 1:Online. Available at http://65.199.186.23/lsa/journal/v1/n3/full/lsa20123a.html. Accessed July 26, 2012.

many outstanding universities that educate students from around the world in the classroom and in research laboratories. Over the past several years, many institutions outside the United States have also invested heavily in excellent educational facilities. Because education is inextricably linked to innovation in optics and photonics, the committee underscores the importance to the nation of maintaining a strong U.S. educational infrastructure in optics and photonics. Although the present study does not focus on education, it does mention specific examples that might benefit from action, including the training of skilled technicians as well as ensuring that an adequate numbers of citizens can be hired by the defense industry. The committee concluded that improvements in technical education are needed to increase the quality of skilled blue-collar workers in optics and photonics.

PROGRESS FOR THE FUTURE

Although many of the innovations in optics and photonics (i.e., the science and engineering of optical waves and photons) have occurred in the United States, U.S. leadership is far from secure. The committee has heard compelling arguments that, if the United States does not act with strategic vision, future scientific advances and economic benefits might be led by others.

It is the committee's hope that this study will help policy makers and leaders decide on courses of action that can advance the future of optics and photonics; promote a greener, healthier, and more productive society; and ensure a leadership position for the United States in the face of increasing foreign competition.

In general, the committee's recommendations call for improved management of U.S. public and private research and development resources, emphasizing the need for public policy that encourages adoption of a portfolio approach to investing in the wide and diverse opportunities now presented by optics and photonics.

2

Impact of Photonics on the National Economy

INTRODUCTION

The vast diversity of applications enabled by photonics poses both economic promise and policy challenges. On the one hand, technical advances in fundamental principles of photonics may have broad impacts in many applications and economic sectors. On the other hand, this diversity means that monitoring public and private investment, employment, output, and other economic aspects of photonics is difficult. Photonics is a broad technology rather than an industry, and the economic data assembled by U.S. government agencies do not support a straightforward assessment of the "economic impact" of photonics. For example, there are no North American Industry Classification System (NAICS) codes that enable the tracking of revenue, employment, and industrial research and development (R&D) spending in photonics-related fields, and we lack data on government R&D spending in photonics. The absence of such information reduces the visibility of photonics within the industrial community and impedes the development of more coherent public policies to support the development of this constellation of technologies and applications.

This chapter takes the following form: First, a case study of lasers is used to introduce the field of photonics, and the conceptual challenges of developing estimates of the economic impact of photonics innovations are discussed. Next, company-level data are presented, and the challenges associated with using such data to provide indicators of the economic significance of the "photonics sector" within the U.S. economy are addressed. Next is a discussion of sources of R&D

investment within photonics, including government and company funding of R&D, followed by an examination of the ways in which the changing structure of the innovation process within photonics (including sources of R&D funding) reflects broader shifts in the sources of innovation within the U.S. economy. That section motivates the subsequent discussions of the role of venture-capital finance in photonics innovation, the role of university licensing, and the implications of offshore growth in the production of optics and photonics products for innovation in the field. This discussion of the changing structure of innovation finance and performance in the United States leads to the next section, which considers the implications of recent experiments in public-private and inter-firm R&D collaboration in other high-technology sectors for the photonics sector. Finally, conclusions and recommendations are presented.

THE ECONOMICS OF PHOTONICS: A CASE STUDY OF LASERS

The laser is a central technology within photonics, and a brief history of its development and expanding applications provides some insights into the economic effects of the much broader field of photonics, as well as underscoring the difficulties of measuring the economic impact of such a diverse field. First demonstrated in 1960 by Theodore Maiman of Hughes Aircraft, the laser built on fundamental research on microwave technology by Charles Townes and Arthur Schawlow at Columbia University and Bell Labs, respectively. The laser exhibits many of the characteristics of a "general-purpose technology"[1] (other examples include information technology [IT], steam power, and electrical power), in that laser technology itself has been transformed by a series of important innovations, with numerous new types of lasers developed over the past 50 years. Innovations in lasers have broadened the applications of this technology, many of which have produced dramatic improvements in the performance of technologies incorporat-

[1] Rosenberg, N., and M. Trajtenberg. 2004. A general-purpose technology at work: The Corliss steam engine in the late-nineteenth-century United States. *Journal of Economic History* 64:61-99. In this paper, Rosenberg and Trajtenberg highlight four characteristics of a "general-purpose technology" (GPT): "first, it is a technology characterized by general applicability, that is, by the fact that it performs some generic function that is vital to the functioning of a large number of using products or production systems. Second, GPTs exhibit a great deal of technological dynamism: continuous innovational efforts increase over time the efficiency with which the generic function is performed, benefiting existing users, and prompting further sectors to adopt the improved GPT. Third, GPTs exhibit 'innovational complementarities' with the application sectors, in the sense that technical advances in the GPT make it more profitable for its users to innovate and improve their own technologies. Thus, technical advance in the GPT fosters or makes possible advances across a broad spectrum of application sectors. Improvements in those sectors increase in turn the demand for the GPT itself, which makes it worthwhile to further invest in improving it, thus closing up a positive loop that may result in faster, sustained growth for the economy as a whole" (p. 65).

ing lasers (e.g., fiber-optic communications). Over the course of the 50 years since its invention, the laser has been used in applications ranging from communications to welding to surgery.

The Economic Impact of the Laser

One measure of the economic impact of the laser is provided by Baer and Schlachter's 2010 study for the Office of Science and Technology Policy (OSTP),[2] which compiled data on the size of three economic sectors in which lasers have found important applications. Baer and Schlachter listed these as follows:

- Transportation (production of transport equipment, etc.), estimated to account for $1 trillion in output during 2009-2010;
- The biomedical sector ($2.5 trillion); and
- Telecommunications, e-commerce, information technology ($4 trillion).

The value of lasers deployed in each of these three sectors was respectively estimated at $1.3 billion ($CO_2$ and fiber), $400 million (solid-state and excimer lasers), and $3.2 billion (diode and fiber lasers).

It is important to distinguish between the role of lasers as "enabling" the growth of these three sectors and the role of this technology as "indispensable" to these sectors, because the distinction is central to analyses of the economic impact of any new technology. The fundamental question that arises in this context is, What would have happened in the absence of the laser? That is, what if substitutes had been employed to realize some if not all of the benefits associated with the laser's applications in these sectors? What would have been the cost (both in terms of higher prices and reduced functionality) associated with using non-laser substitutes? In some areas (e.g., surgery, some fields of optical communication), substitutes might well have been unavailable or would have performed so poorly as to render them useless. In other fields such as welding, however, substitutes for lasers that presented fewer cost and performance penalties might well have appeared. In some cases, substitutes for lasers might well have improved their performance and reliability over time.

In the case of the laser as with most major innovations, the data and the methodology necessary to conduct counterfactual thought experiments of this sort are lacking, which makes it difficult to develop credible estimates of economic impact. These analytic challenges are no less significant in assessing the impacts of other

[2] Baer, T., and F. Schlachter. 2010. *Lasers in Science and Industry: A Report to OSTP on the Contribution of Lasers to American Jobs and the American Economy.* Available at http://www.laserfest.org/lasers/baer-schlachter.pdf. Accessed June 25, 2012.

photonics technologies currently in use, and they are truly forbidding where one seeks to predict the economic impact of future applications that have only begun to emerge.

Nonetheless, it seems clear that the laser has been adopted in a diverse array of applications, some of which have underpinned the growth of entirely new methods for the transmission of information.[3] Equally important is the way in which continued innovation in laser technology has enabled and complemented innovation in technologies using lasers. This mutual enhancement further extends the adoption of these applications as performance improves and costs decline. Moreover, the appearance of new applications and markets for lasers has created strong incentives for further investment in innovation in lasers. All of this feedback and self-reinforcing dynamics are classic features of general-purpose technologies. Lasers are one example of such a technology within the field of photonics.

Funding of Early Laser Development

The development of laser technology shares a number of characteristics with other postwar U.S. innovations, in fields ranging from information technology to biotechnology. Like these other technologies, much of the research (especially the fundamental research) that underpinned the laser and its predecessor, the maser, relied on federal funding. Similar to the experience with IT, much of this federal R&D funding was motivated by the national security applications of lasers during a period of high geopolitical tension.[4] Industry funded a considerable amount of laser-related R&D, much of which focused on development and applications, but

[3] Interestingly, optical communication was the only foreseen application of the laser in 1958. See, for example, Sette, D. 1965. Laser applications to communication. *Zeitschrift für angewandte Mathematik und Physik ZAMP* 16(1):156-169.

[4] Bromberg, J.L. 1991. *The Laser in America, 1950-1970.* Cambridge, Mass.: MIT Press. In this study, Bromberg emphasizes another characteristic of federally and industrially financed R&D in the field of lasers: the extent of linkage among research and researchers in U.S. industry, federal laboratories, and academia during the 1945-1980 period: "Academic scientists were linked to industrial scientists through the consultancies that universities held in large and small firms, through the industrial sponsorship of university fellowships, and through the placement of university graduates and postdoctoral fellows in industry. They were linked by joint projects, of which a major example here is the Townes-Schawlow paper of [*sic*] optical masers, and through sabbaticals that academics took in industry and industrial scientists took in universities. Academic scientists were linked with the Department of Defense R&D groups, and with other government agencies through tours of duty in research organizations such as the Institute for Defense Analyses, through work at DoD-funded laboratories such as the Columbia Radiation Laboratory or the MIT Research Laboratory for Electronics, and through government study groups and consultancies. They were also linked by the fact that so much of their research was supported by the Department of Defense and NASA" (p. 224). Similar linkages among industry, government, and military research characterized the early years of development of the U.S. computer and semiconductor industries, in contrast to their European and Japanese counterparts.

much of this R&D investment (particularly in the early years of the laser's development) was motivated by the prospect of significant federal procurement contracts for military applications of lasers.

The early work in the 1950s of Townes at Columbia University on masers, for example, was financed in large part by the Joint Services Electronics Program, a multi-service military R&D program that sought to sustain after 1945 the wartime research activities of the Massachusetts Institute of Technology (MIT) and Columbia Radiation Laboratories, both established during World War II. Military funding supported early work on masers and lasers at RCA, Stanford University, and Hughes Aircraft. R&D related to lasers at the National Bureau of Standards also was closely overseen by military representatives. By 1960, according to Bromberg,[5] the Department of Defense (DOD) was investing roughly $1.5 million (1960 dollars) in extramural R&D on lasers, an amount that rose rapidly after Maiman's demonstration of the ruby laser at Hughes; by 1962, according to Bromberg's estimates, the DOD was spending roughly $12 million on laser-related R&D, one-half of the total U.S. R&D investment in the technology. In 1963, total DOD R&D investment, including intramural projects, approached $24 million, which increased to just over $30 million in the late 1970s.[6] Another tabulation of military R&D investment estimates total military laser-related R&D spending through 1978 at more than $1.6 billion (all amounts in nominal dollars).[7]

The Early Laser Market

The military also was a major source of demand in the early laser industry, although its share of the market declined over time as civilian applications and markets grew rapidly. According to Bromberg, the DOD share of the laser market fell from 63 percent in 1969 to 55 percent in 1971.[8] Although the DOD dominated the government market for lasers, other federal agencies also were important purchasers, and Seidel estimates that total government purchases of lasers amounted to nearly 56 percent of the total market in 1975, increasing to slightly more than 60 percent by 1978.[9] Commercial laser sales grew from $1.985 billion in 1983 to

---

[5] Bromberg, J.L. 1991. *The Laser in America, 1950-1970.* Cambridge, Mass.: MIT Press.

[6] Koizumi, K. 2008. *AAAS Report XXXIII: Research & Development FY 2009,* Chapter 5. Available at http://www.aaas.org/spp/rd/09pch5.htm. Accessed July 30, 2012.

[7] Seidel, R. 1987. From glow to flow: A history of military laser research and development. *Historical Studies in the Physical and Biological Sciences* 18:111-147.

[8] Bromberg, J.L. 1991. *The Laser in America, 1950-1970.*

[9] Seidel, R. 1987. From glow to flow: A history of military laser research and development. *Historical Studies in the Physical and Biological Sciences* 18:111-147.

$2.285 billion in 1984, according to DeMaria;[10] government sales in these same 2 years amounted to $1.23 billion and $1.3 billion, respectively. The government share of laser sales almost certainly has continued to decline in more recent years.

The dominance of the early laser market by the military services had important implications for the development of the embryonic laser industry. In contrast to the military services of other NATO member nations, U.S. military procurement officials rarely excluded new firms from procurement competitions, although in many cases these firms had to arrange for a "second source" of their products to avoid supply interruptions. The prospect of military procurement contracts therefore attracted new firms to enter the laser industry and underlaid a growth in the total number of firms working with laser development. The number of new firms in the industry also grew rapidly because of the growth of new laser applications in diverse civilian markets, as well as the growth of new types of laser technologies. Clearly, the military contracts sped up the laser development. The appearance of the diode-pumped solid-state laser in 1988, however, may have triggered the exit from the industry of a large number of firms, and the number of active firms fell to 87 by 2007, during a period of rapidly increasing sales for the industry as a whole.

International Comparison

Although data allowing for a comparison of the structure of the laser industries of the United States and other nations are not readily available, it is likely that the number of independent producers of lasers in other nations exhibited less dramatic growth and decline. Assuming that this characterization of the laser industries of the United States and other nations is accurate, the differences reflected the prominent role of government demand for lasers in the United States, as well as the important role of U.S. venture capital in financing new-firm entrants into the laser industry.

The origin of U.S. and Japanese scientific publications appearing in *Applied Physics Letters* from 1960 through 2009 on the topic of semiconductor lasers was analyzed by Shimzu (2011);[11] data in the study by Shimzu suggest a contrast in the sources of leading-edge laser R&D during this period. Established U.S. firms in the areas of electronics, IT, and communications dominated semiconductor-laser publications during 1960-1964, accounting for more than 80 percent of publications of U.S. origin. These firms' share of publications dropped sharply after 1964, to 30 to

[10] DeMaria, A.J. 1987. "Lasers in Modern Industries." In *Lasers: Invention to Application*, J.H. Ausubel and H.D. Langford, eds. Report of the National Academy of Engineering. Washington, D.C.: National Academy Press.

[11] Shimuzu, H. 2011. Scientific breakthroughs and networks in the case of semiconductor laser technology in the U.S. and Japan, 1960s-2000s. *Australian Economic History Review* 51:71-96.

35 percent during 1965-1974, before increasing again to 61 percent during 1975-1979 and 46 percent during 1980-1984. By 2005-2009, however, the established-firm share of U.S. scientific publications in semiconductor lasers had dropped to less than 5 percent. Start-up firms, which contributed no semiconductor-laser publications during 1960-1980, accounted for more than 10 percent of publications during 1985-1989 and 9.25 percent during 2005-2009. U.S. university-based researchers accounted for the majority of U.S. semiconductor-laser publications throughout the 1965-2009 period, as their share grew from slightly more than 57 percent in 1965-1969 to almost 85 percent during 2005-2009. (See Figure 2.1.)

The data in the Shimuzu study on publications of Japanese origin in semiconductor-laser research in *Applied Physics Letters*[12] indicate a minimal role for start-up firms as sources of research. Although all papers published in this prestigious journal are reviewed by scientific peers, the burden of translation into English may well introduce some bias into this comparison—papers of Japanese origin effectively have to clear a higher "quality threshold" to appear in this journal. This potential source of bias should be kept in mind in comparing Japanese and U.S. publications. Established Japanese corporations, which accounted for no scientific papers during 1960-1969 (see Figure 2.1), contribute a declining share of scientific papers of Japanese origin in semiconductor lasers, although their share declined somewhat less significantly, from 75 percent during 1970-1974 to nearly 40 percent during 2005-2009. Japanese start-up firms, however, played almost no role as a source of scientific publications, appearing only after 2000 in Shimuzu's data, with a share of 0.74 percent during 2000-2004 and 2.94 percent during 2005-2009. Japanese universities, which accounted for less than 30 percent of papers of Japanese origin in this journal and field before 1990, by 2005-2009 contributed more than 55 percent. (See Figure 2.1.)

Although covering only one area of laser technology and limited to one scientific journal, the data analyzed by Shimuzu clearly indicate that new-entrant firms in the United States accounted for a much larger share of scientific activity (as represented by publications) in semiconductor lasers than was true of Japanese start-ups, whereas established Japanese firms have maintained a more prominent role as sources of scientific publications into the 21st century than have U.S. established firms in electronics, communications, or IT. The role of university researchers as sources of published scientific research, however, appears to have grown significantly in both nations, albeit more dramatically in the United States than in Japan.

[12] Ibid.

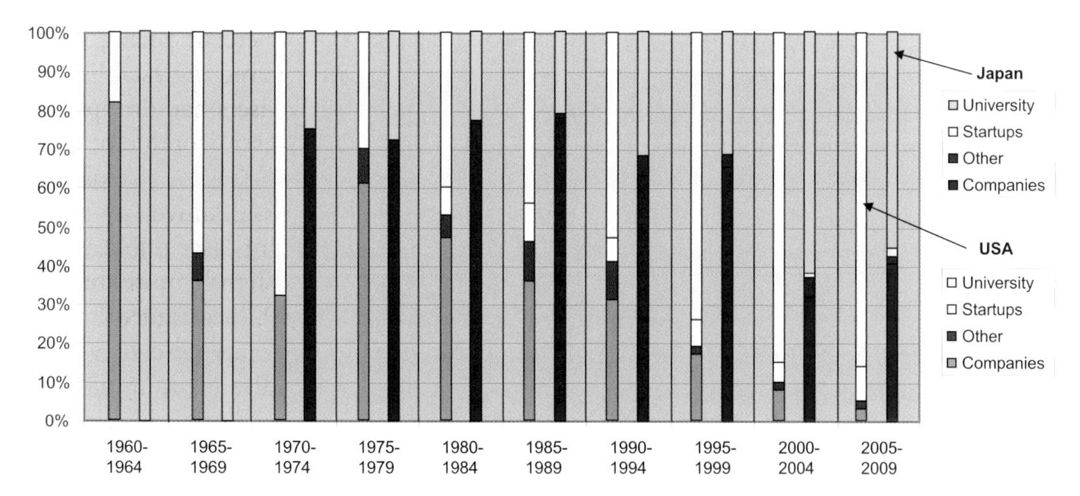

FIGURE 2.1 Comparison between the United States and Japan with respect to different sectors' contributions to scientific publications that appeared in *Applied Physics Letters* from 1960 through 2009 on the topic of semiconductor lasers. (In the set of two bars for each time period, U.S. data are on the left, and Japanese data are on the right.) SOURCE: Based on data in Shimuzu, H. 2011. Scientific breakthroughs and networks in the case of semiconductor laser technology in the U.S. and Japan, 1960s-2000s. *Australian Economic History Review* 51:71-96.

Conclusions from the Laser Case Study

This brief overview of the development of laser technology and the U.S. laser industry highlights several issues that are relevant to overall photonics technology. The difficulty of measuring the "economic impact" of lasers reflects the need for any such assessment to rely on assumptions about the availability or timing of the appearance of substitute technologies. These difficulties are more serious for predictions of the economic impact of technologies currently under development. Such predictions rely on guesses about the nature of substitutes and markets, as well as predictions concerning the pace and timing of the adoption of new technologies. The laser's development also highlights several of the features of general-purpose technologies, namely, their widespread adoption, driven in many cases by continued innovation and improvement in the focal technology, as well as the ways in which users of the technology in adopting sectors contribute to new applications that rely in part on incremental improvements in the technology. In the view of the committee, many other technologies in the field of photonics share these characteristics with lasers.

The development of laser technology and the laser industry in the United States also displays some contrasts with the experiences of other nations, particularly in the important direct and indirect role played by the federal government in the early stages of the technology's development. Federal R&D and procurement spending, much of which was derived from military sources, influenced both the pace of development of laser technology and the structure of the laser industry, revealed most plainly in the contrasts between U.S. and Japanese scientific publications in laser technology. Moreover, the high levels of mobility of researchers, funding, and ideas among industry, government, and academia were important to the dynamism of the U.S. laser industry in its early years, with few formal policies geared toward "technology transfer" between government or university laboratories and industry such as those in place today. Although military R&D spending continues to account for roughly 50 percent of total federal R&D spending (which now accounts for roughly one-third of total national R&D investment, down significantly from its share during the period of laser-technology development),[13] the share of long-term research within the military R&D budget has been under severe pressure in recent years, and congressional restorations of executive branch cuts in this spending have often taken the form of earmarks. Moreover, as the laser industry matures and nonmilitary markets exert much greater influence over the evolution of applications for this technology, the ability of military R&D to guide broad technological advances in this field has declined.

ESTIMATING THE ECONOMIC IMPACT OF PHOTONICS—INDUSTRY REVENUES, EMPLOYMENT, AND R&D INVESTMENT IN THE UNITED STATES

This section employs estimates of revenues, employment, and R&D investment for a sample of firms that are active in the field of photonics to illustrate the breadth of photonics-based industrial activity in the U.S. economy.

The data were provided by the International Society for Optics and Photonics (SPIE) and the Optical Society of America (OSA) and include 336 unique (avoiding double counting of corporations that appear on more than one membership list) corporate members in 2011; 1,009 unique companies that had exhibited at one of the two trade shows in 2011; and 1,785 unique companies listed as employers of

[13] National Science Foundation. 2012. Chapter 4, "R&D: National Trends and International Comparison—Highlights." In *Science and Engineering Indicators 2012*. Available at http://www.nsf.gov/statistics/seind12/c4/c4h.htm#s6. Accessed July 30, 2012.

professional societies' individual members in 2011.[14] Table 2.1 lists the number of publicly traded and privately held companies in each of these three groups. Note that although the companies listed *within* each of the three groups described above appear only once, there is overlap among companies appearing on each of the three lists (i.e., a single firm may be listed as a member of one or the other society, as well as an exhibitor and an employer of society members). Aggregating all unique companies across the three groups produces a list of 2,442 unique U.S. companies active in some way within photonics, 285 of which are public and 2,157 of which are privately held, as shown in Figure 2.2. As a point of comparison, there were approximately 5.9 million "employer" firms (firms with payroll) in the United States in 2008, and approximately 17,000 publicly traded companies.[15] Thus the present study's count of companies across the three lists comprises approximately 0.04 percent of all U.S. employer firms and 1.7 percent of all U.S. publicly traded companies.

Data on revenues, employment, and R&D spending in 2009 and 2010 for 282 of the 285 publicly traded companies that are listed as members, employers, or exhibitors can be seen in Table 2.2.[16] The total revenues associated with these 282 public companies in 2010 amounted to $3.085 trillion, they invested $166 billion on R&D (amounting to 5.4 percent of revenues), and employed 7.4 million individuals. As a comparison point, "employer" firms in the United States in 2008 created an aggregate of $29.7 trillion in revenues and employed an aggregate of 120

[14] NAICS or other industry-specific public databases on economic activity in photonics do not currently exist. In an attempt to create a rough estimate of economic activity in photonics, the committee collected three types of information with help from the two largest professional societies in photonics: SPIE and the Optical Society of America (OSA). This information included (1) a list of all U.S.-headquartered member companies for each society, (2) a list of U.S.-headquartered exhibiting companies at the largest trade conference for each society, and (3) a list of all U.S.-headquartered companies associated with individual members of the professional society. The information provided by these societies covers only 2011. In the analysis of this information, the list of member companies was considered to be a rough estimate of companies with strong participation in optics and photonics in 2011, the list of exhibiting companies as a rough estimate of companies selling products involving photonics in 2011, and the list of companies associated with individual members of the professional society a rough estimate of companies with some activities in photonics in 2011. This list of firms also served as the basis for compiling estimates of economic activity during 2010 for the subset of firms for which data were available (see text). It is important to emphasize that each of these estimates is *very* rough, and it is plausible that some photonics-specialist firms are not captured by these estimates, while other firms for which photonics represents a small share of overall revenues, employment, or R&D investment may be included.

[15] According to the U.S. Census Bureau. 2008. "Statistics about Business Size." Available at http://www.census.gov/econ/susb/introduction.html. Accessed June 25, 2012.

[16] Data from Standard & Poor's Compustat. Available at http://www.compustat.com/. Accessed June 25, 2012.

TABLE 2.1 Number of Unique Companies in 2011 That Were Corporate Members, Participated in One of the Two Largest Trade Shows, or Were Associated with Individual Members Across the Two Largest Professional Societies in Photonics

	Public		Private		Total
By Type	No.	%	No.	%	No.
Corporate members	45	13	291	87	336
Exhibited at trade shows	107	11	902	89	1,009
Employed professional society members	243	14	1,542	86	1,785

SOURCE: Data contributed by SPIE and the Optical Society of America, compiled by Carey Chen, Board on Science, Technology, and Economic Policy of the National Academies.

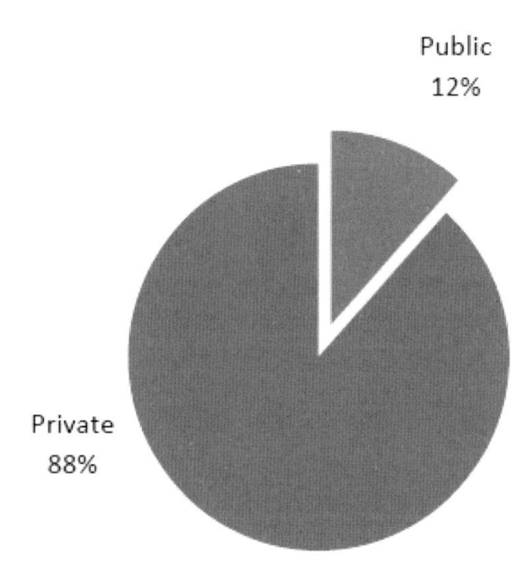

FIGURE 2.2 Percentage in 2011 of public versus private companies across the 2,442 unique companies recorded within the SPIE and OSA databases. SOURCE: Data contributed by SPIE and OSA, and subsequently collated by Carey Chen, Board on Science, Technology, and Economic Policy of the National Academies.

million individuals.[17] Thus, the public firms listed as active in photonics accounted for approximately 10 percent of U.S.-based employer firms' revenues and 6 percent of U.S.-based employer firms' aggregate employment in 2010.

Data were also used from Dun and Bradstreet to estimate revenue and R&D expenditures for the publicly traded and privately held firms listed as corporate

[17] Public company listings contributed by SPIE and the Optical Society of America. Revenue, employee, and research and development (R&D) expenditure data subsequently collected from Compustat.

TABLE 2.2 Revenues, Number of Employees, and R&D Expenditures from 282 Unique Public Companies in 2009 and 2010

	2009	2010
Revenue ($ millions)	2,741,289	3,085,292
No. of employees (000s)	7,159	7,415
R&D expenditures ($ millions)	151,104	166,603
R&D: % of revenue	5.5%	5.4%

SOURCE: Public company listings contributed by SPIE and the Optical Society of America. Revenue, employee, and research and development (R&D) expenditure data subsequently collected from Compustat. Compiled by Carey Chen, Board on Science, Technology, and Economic Policy of the National Academies.

TABLE 2.3 Revenue and Number of Employees for the 336 Unique Companies (in 2010) That Had Corporate Members with at Least One of the Two Professional Societies

Company Type	No.	Revenue ($ millions)	No. of Employees (000s)	R&D Expenditures ($ millions)
Public	45	501,551	1,495	27,455
Private	291	1,394	10	Not available
TOTAL	336	502,945	1,506	

SOURCE: Company listings contributed by SPIE and the Optical Society of America. Revenue, employee, and research and development (R&D) expenditure data subsequently collected for public companies from Compustat and for private companies from Dunn and Bradstreet. Compiled by Carey Chen, Board on Science, Technology, and Economic Policy of the National Academies.

members of SPIE or OSA, on the assumption that photonics sales and innovation-related activities are likely to be much more significant within these firms than within those listed as exhibitors or employers of professional society members. This group of public and private corporate member companies was responsible for $503 billion in revenues in 2010 (roughly one-sixth of the aggregate revenues for the more comprehensive list of firms summarized in Table 2.1) and employed 1.5 million individuals (slightly more than one-tenth of the employment associated with the more comprehensive list of firms). Table 2.3 reports total revenues and R&D investment for the publicly traded and privately held firms within this population of "photonics specialists."[18] Clearly, the firms that can be defined as "photonics specialists" account for a much smaller share of overall U.S. employment and industrial revenues.

[18] R&D expenditures were not available from Dun and Bradstreet for privately held companies.

These data suggest that a small (although non-negligible) proportion of total U.S. firms is accounted for by those U.S.-headquartered firms with sufficient activity in photonics to be active as a member, an exhibitor, or an employer of a professional-society member in the database of one of the two largest U.S. photonics-related professional societies. The aggregate revenues associated with these firms, however, represent a relatively large proportion of aggregate U.S. employer firm revenues (10 percent) and employment (6 percent). This study interprets these data as indicative of the pervasiveness of photonics innovation and technology within this economy. Data for firms identified here as specialists in photonics suggest that these firms account for a much smaller share of total U.S. industrial revenues and employment, although they are still a significant source of economic activity.

Like the data from the Baer and Schlachter study for OSTP on lasers cited above,[19] these data convey some sense of the breadth of photonics-based industrial activity within this economy. While not directly measuring the economic impact of photonics within the U.S. economy in 2011, they do reflect the general-purpose nature of photonics. This field of technology influences innovation and employment across a broad swath of the economy. Once again, the estimates used in this analysis underscore gaps in existing public databases and the need for better measurement and tracking of photonics-related R&D, employment, and industrial activity to enable better understanding of the full economic impact of so pervasive a technology.

GOVERNMENT AND INDUSTRIAL SOURCES OF R&D FUNDING IN PHOTONICS AND FEDERAL FUNDING OF OPTICS[20]

One of the only previous attempts to estimate overall U.S. R&D investment in the field of photonics and to compare R&D investment in optoelectronics among Western Europe, Japan, and the United States is the 1992 study by Sternberg,[21] which covers only 1981-1986. Sternberg in turn relies on an unpublished study for the National Institute of Standards and Technology (NIST) by Tassey,[22] which

[19] Baer and Schlachter. 2010. *Lasers in Science and Industry.*

[20] This section uses rough estimates of agency-level R&D spending in areas related to optics and photonics to discuss overall trends in U.S. R&D investment in the field of photonics. To address the lack of regularly tracked data or recent published studies on U.S. R&D investment in photonics, the committee requested data on all optics- and photonics-related programs from all government agencies identified as potentially supporting R&D in these areas. The results of this data-collection effort are discussed in the text.

[21] Sternberg, E. 1992. *Photonic Technology and Industrial Policy.* Albany, N.Y.: State University of New York Press.

[22] Tassey, G. 1985. *Technology and Economic Assessment of Optoelectronics.* Planning report. Gaithersburg, Md.: National Bureau of Standards.

Sternberg claims includes only nondefense R&D spending for the United States. (Sternberg does not discuss whether or not Tassey's study omits defense-related R&D spending in Japan or Western Europe.) Sternberg's data indicate that U.S. R&D investment from industry and government sources grew from $69 million in 1981 to $339 million in 1986, while Japanese investment grew from $112 million to $344 million, and European investment grew from $30 million to $165 million. As noted above, Sternberg claims that Tassey's data, which form the basis for these comparative estimates, omit U.S. defense-related government R&D spending, which he estimates to be as much as $230 million in 1986. If Sternberg's revision of Tassey's estimates is credible, U.S. R&D investment in optoelectronics as of the middle of the 1980s greatly exceeded the combined investments of Europe and Japan. In a separate calculation for fiscal year (FY) 1990, Sternberg estimates that "Science and Technology Funding" from DOD sources[23] for photonics amounted to $655 million, which exceeds his estimate of U.S. photonics R&D funding from all sources for 1986.

A 1996 study on R&D spending in optoelectronics alone found that Japanese firms spent much more than U.S. firms on optoelectronics R&D during 1989-1993.[24] But the study also showed that the U.S. government spent significantly more on optoelectronics R&D during this period than the Japanese government spent, investing more in R&D in 1990 alone than the Japanese government had spent in 15 years of government support. (See Tables 2.4 and 2.5.)

Since the publication of Sternberg's monograph in 1992, studies in Canada and Europe have attempted to estimate public funding of photonics. The Canadian Photonics Consortium estimated Canadian funding of photonics from public sources in 2008 to have been approximately $136 million,[25,26] roughly two-thirds of the estimated $219.7 million that the U.S. government invested in R&D in optoelectronics alone in 1993 (see Table 2.4). The European Union (EU) Framework 7 Programme invested €165 million (U.S. $210 million at average international ex-

---

[23] It remains unclear whether or not this funding includes development work.

[24] Japanese Technology Evaluation Center (JTEC). 1996. *Optoelectronics in Japan and the United States.* Report of the Loyola University Maryland's International Technology Research Institute, Baltimore, Md.

[25] Photonics21. 2010. "Lighting the Way Ahead, Photonics21 Strategic Research Agenda." Dusseldorf, Germany: European Technology Platform. Available at http://www.photonics21.org/download/SRA_2010.pdf. Accessed June 25, 2012.

[26] Canadian Photonics Consortium. 2008. "Photonics: Making Light Work for Canada, A Survey by the Canadian Photonics Consortium." Available at http://www.photonics.ca/Making%20Light%20Work%20for%20Canada_2008.pdf. Accessed June 25, 2012.

TABLE 2.4 Optoelectronics R&D Spending by U.S. Firms, 1989-1993 ($ millions)

Source of Funding	1989	1990	1991	1992	1993
In-house	533.5	598.9	651.3	615.7	619.8
Federal government	189.7	200.1	204.8	178.6	219.7
Customer	8.7	11.6	15.3	17.1	20.4
Joint venture	0.5	2.5	7.8	0.4	5.2
Other	3.2	2.2	3.5	3	2.6
TOTAL	735.6	815.3	882.7	814.8	867.7

SOURCE: Japanese Technology Evaluation Center (JTEC). 1996. *Optoelectronics in Japan and the United States.* Report of the Loyola University Maryland's International Technology Research Institute. Baltimore, Md.: International Technology Research Institute. Available at http://www.wtec.org/loyola/opto/toc.htm. Accessed June 25, 2012.

TABLE 2.5 U.S. Government-Funded Optoelectronics R&D, by Funding Organization, 1989-1993 ($ million)

Source of Funding	1989	1990	1991	1992	1993
DOD	122	120	116.6	107.8	145.6
ARPA	4.5	6.7	9	22.5	21.7
Armed Forces (Army, Navy, Air Force)	48.1	55.3	57.5	33.3	38.1
NASA	10.7	12.7	15	8	5.9
DOE/National Lab	2.4	1.6	1.5	0.7	0.3
NIH	0.6	0.6	1.1	1.2	1
NSF	0.3	0.4	0.4	0.3	0.3
NIST	—	—	—	0.9	3.1
Other	1.4	2	2.2	2.4	3.6
TOTAL	190	199.3	203.3	177.1	219.6

NOTE: Acronyms are defined in Appendix B of this report. SOURCE: Japanese Technology Evaluation Center (JTEC). 1996. *Optoelectronics in Japan and the United States.* Report of the Loyola University Maryland's International Technology Research Institute. Baltimore, Md.: International Technology Research Institute. Available at http://www.wtec.org/loyola/opto/toc.htm. Accessed June 25, 2012.

change rates for the year 2010) in photonics in 2010 as part of its information and communications technology work program (EU Framework 7,[27] IRS[28]).[29]

---

[27] European Commission. 2010. "Information and Communication Technologies Work Programme 2011-12." Seventh Framework Programme (FP7). Available at http://cordis.europa.eu/fp7/ict/. Accessed June 25, 2012.

[28] Internal Revenue Service. 2010. "Internal Revenue Service Yearly Average Currency Exchange Rates." Available at http://www.irs.gov/businesses/small/international/article/0,,id=206089,00.html. Accessed June 25, 2012.

[29] Photonics is also funded through other EU mechanisms and work programs, and so this is not a complete representation (e.g., medical imaging would be in the "Health" work program). In addition, individual countries, including both France and Germany, have individual programs focused on photonics. Nonetheless, the EU 2010 investment in R&D in information and communications

No recent studies have attempted to estimate the scale or agency sources of U.S. government R&D support for photonics. As part of the committee's data-collection efforts, each agency was given a one-page description of this study, including a brief description of what the committee included in its definition of optics and photonics, along with examples of optics and photonics technologies and applications. The responding agencies were as follows: within the Department of Defense, the Air Force Office of Scientific Research (AFOSR), Army Research Office (ARO), Office of Naval Research (ONR), High Energy Laser-Joint Technology Office (HEL JTO), and Defense Advanced Research Projects Agency (DARPA); Department of Homeland Security (DHS); Department of Energy (DOE); National Institutes of Health (NIH); NIST; and the National Science Foundation (NSF). Although all of the agencies and programs contacted made a good-faith effort to respond, serious gaps nonetheless remained in the committee's estimates of total federal R&D support for photonics during FY 2006-FY 2010. The committee obtained a total estimate of more than $53 billion for federal photonics-related R&D during this period. Nearly $45 billion of this total is based on DOD R&D investments that appear to involve photonics in some fashion but cannot be verified as limited solely or even primarily to photonics. Similarly, the estimated $4.4 billion in photonics R&D attributed to NIH includes investments in other technological fields. However, much of the NIST R&D investment in photonics is omitted from the estimates. Because of these complications, the committee recommends that an estimate of $53 billion for federal R&D investments in photonics during FY 2006-FY 2010 be interpreted more realistically as an upper bound on a "true" total that may well be anywhere from $25 billion to $55 billion.

Within individual federal programs, more reliable and precisely defined estimates of photonics-related R&D investment were obtained. Many of these agency- or program-specific investments in photonics R&D have grown during FY 2006-2010. Within the DOD, for example, DARPA R&D funding in optics and photonics has almost doubled during this period, rising to $486 million by FY 2010 (see Figure 2.3).

The second-largest funder of optics and photonics R&D, and the largest civilian funder, is NIH, which accounts for 80 percent of the reported nondefense photonics R&D in Table 2.5. Here again, the funding of optics and photonics technologies appears to have grown during the last decade. Figure 2.4 shows aggregate annual funding of optics and photonics by NIH based on a search for all proposals granted between 2000 and 2011 that included the words "optics," "photonics," "opto," or "laser" in their abstract, project title, or project terms. While the data for 2011 ex-

---

technology-related photonics still amounts to little more than the estimated U.S. government investment in R&D in optoelectronics alone in 1993 (Table 2.4).

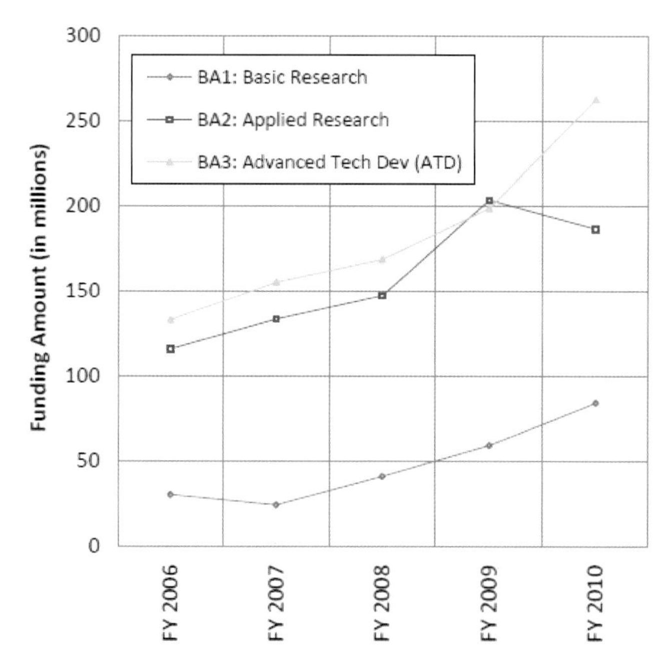

FIGURE 2.3 DARPA funding in optics and photonics. SOURCE: Based on data collected from the Defense Advanced Research Projects Agency (DARPA) Research and Development, Test and Evaluation, and Defense-Wide Budgets Database (http://www.darpa.mil/NewsEvents/Budget.aspx), compiled by Carey Chen, Board on Science, Technology, and Economic Policy of the National Academies.

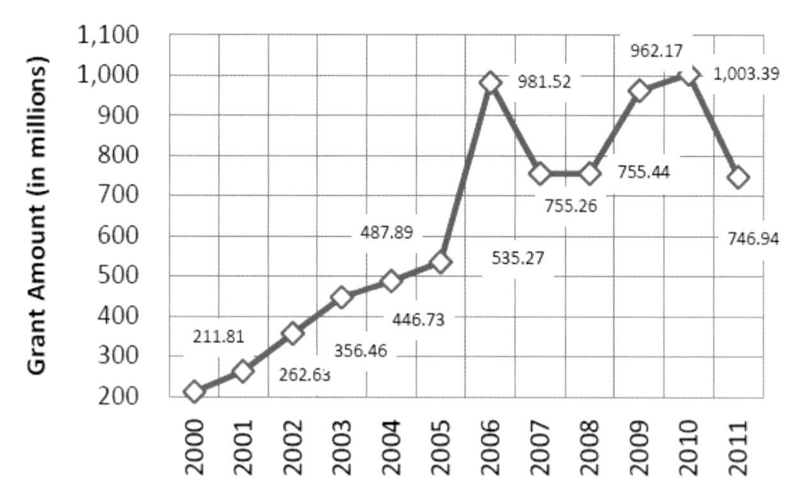

FIGURE 2.4 Funding by the National Institutes of Health (NIH) between 2000 and 2011 in optics and photonics based on a keyword search in the NIH RePORTer Database. SOURCE: Data collected from the National Institutes of Health RePORTer Database on October 10, 2011, compiled by Carey Chen, Board on Science, Technology, and Economic Policy of the National Academies.

tend only through October 10, NIH personnel reported that the full-year funding level for 2011 is likely to be lower than that for 2010.[30]

According to the data provided by NIH RePORTer search, the National Cancer Institute (NCI) has provided the most funding, at 15.8 percent of the NIH total, followed by the National Eye Institute (NEI) at 11.8 percent and the National Heart, Lung, and Blood Institute (NHLBI) at 9.9 percent.

Other data from the federal agencies also suggest that federal photonics R&D spending has grown in recent years. For example, Figure 2.5 compiles the total funding during 1977-2011 associated with grants either fully or partially funded by the NSF Electronics, Photonics, and Magnetic Devices (EPMD) Division.

Finally, data on DOE funding of solar energy and photovoltaics R&D (Figure 2.6) suggest similar upward trends in the last decade, although a longer time series suggests that federal funding in solar energy may well have been equally substantial in the late 1970s. Although federal R&D funding in this field during the 1980s and 1990s was lower than that of the Japanese and German governments, federal R&D funding has grown substantially in recent years, such that it once again exceeds these other governments' investments (see Figure 2.7).

In summary, the data made available by federal agencies for this committee's attempt at a complete estimate of federal photonics-related R&D investment suggest that the DOD dominates funding of optics and photonics, with NIH being the second-largest contributor.[31] Federal R&D spending in optics and photonics also appears to have grown during the last decade.

CHANGES IN PHOTONICS-BASED INNOVATION IN THE UNITED STATES SINCE 1980

This section discusses several aspects of the changing structure of the public and private R&D institutions and investments that have underpinned innovation in photonics and other technologies in the United States since 1980. These structural changes have reduced the role of large industrial firms as performers of R&D and have increased the importance of smaller firms, many of which are funded through venture capital, and at least some of which rely on university-licensed intellectual

[30] It is important to note that the search tool uses two different coding methods, one pre-2008 (each institute's judgment of how a grant should be coded) and one post-2008 (an automated trans-NIH coding system called RCDC). It does not appear that this difference in coding should affect the rising trend observed from 2000 to 2006, as this is before the change in 2008.

[31] It should be noted that, although every effort was made to be comprehensive in this estimate, the data provided were incomplete. The extraordinary dearth of data on the scale and sources of federal R&D funding in optics and photonics (as is the case in assessing the economic significance of photonics) leaves one "flying blind." In the committee's judgment, the creation of more reliable and comprehensive data on federal R&D spending in this field is essential to the formulation of a more effective and coherent public policy.

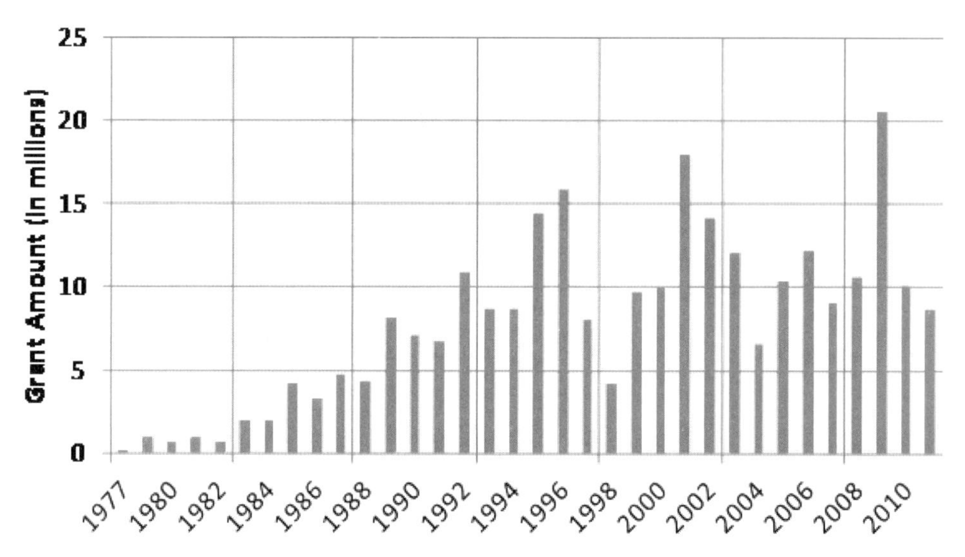

FIGURE 2.5 National Science Foundation (NSF) Electronics, Photonics, and Magnetic Devices (EPMD) Division funding in optics and photonics. NOTE: To show the trend in underlying sources, two outliers (a 4-year grant of $4.8 million in total awarded in 2002 and a 9-year grant of $6.3 million in total awarded in 1994) were removed from the analysis. In addition, three cooperative agreements (totaling $12.9 million, $13.8 million, and $17.4 million, respectively, the first awarded in 1999 and the last two both awarded in 2003) were excluded from the above plot. The first and last were agreements for the Nanoscale Science and Engineering Centers (NSECs), and the middle one was a renewal proposal for the National Nanofabrication Users Network. SOURCE: Data collected from the NSF Electronics, Photonics, and Magnetic Devices (EPMD) Division database on October 31, 2011, compiled by Carey Chen, Board on Science, Technology, and Economic Policy of the National Academies.

property. The role of venture capital in photonics innovation is discussed in the next section, and potential approaches to inter-firm and public-private collaboration in technology development are examined in a subsequent section. Chapter 7, "Advanced Manufacturing," discusses another important structural change in U.S. R&D and innovation during this period—the movement of much of the production activity in photonics and other high-technology industries to foreign locations.

Since 1980, the mix of private and public R&D investment in United States has shifted from roughly 50/50 federal/private to roughly 30/70 (Figure 2.8). Moreover, within the defense-related R&D budget, the share of "research" (6.1-6.3) has declined somewhat (see Figure 2.9). To the extent that U.S. photonics technology development benefited heavily from defense-related R&D and procurement spending, the decline in the share of total R&D accounted for by defense may have reduced federal support for photonics.[32] As the case study of lasers discussed above

[32] As noted above, the committee lacked sufficient data to reach strong conclusions about federal agency-level R&D spending in photonics since 1980.

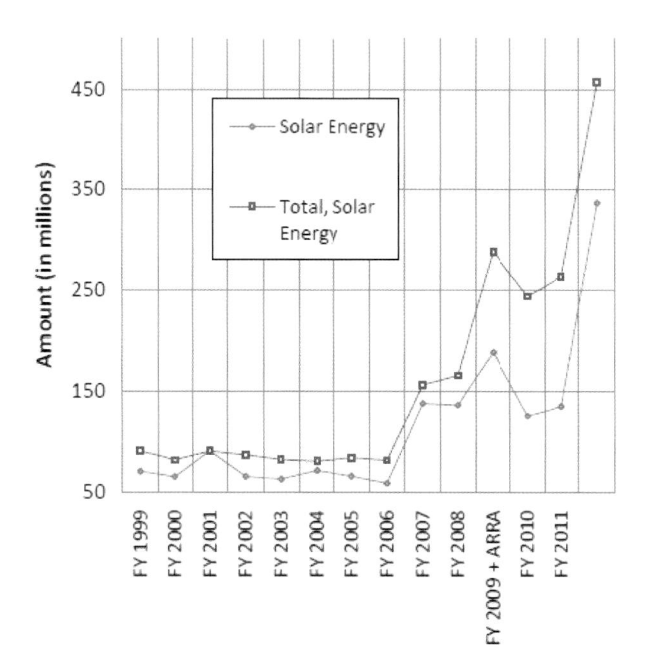

FIGURE 2.6 Department of Energy (DOE) funding in solar energy and photovoltaics R&D. SOURCE: Data provided by the Department of Energy on October 18, 2011, compiled by Carey Chen, Board on Science, Technology, and Economic Policy of the National Academies.

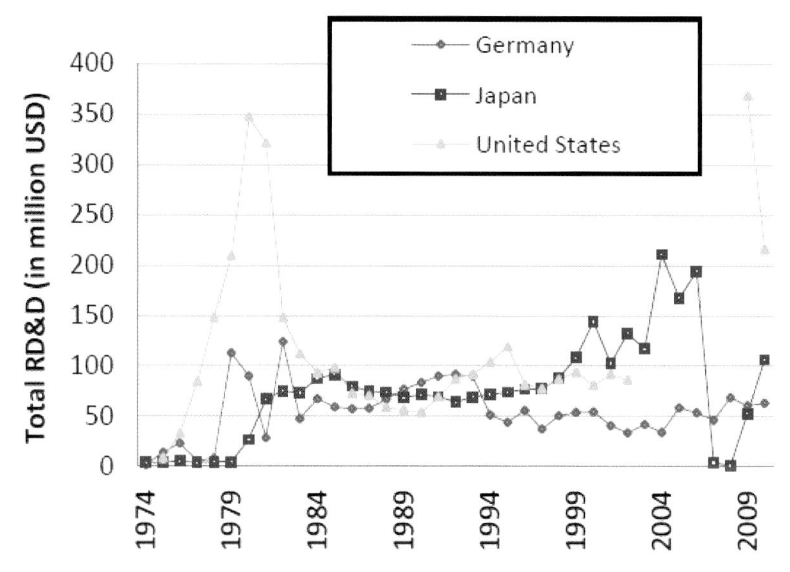

FIGURE 2.7 Federal research, development and demonstration (RD&D) budget in photovoltaics (at 2010 prices and exchange rates). SOURCE: International Energy Agency (IEA) Online Statistics © OECD/IEA 2012. Reprinted with permission.

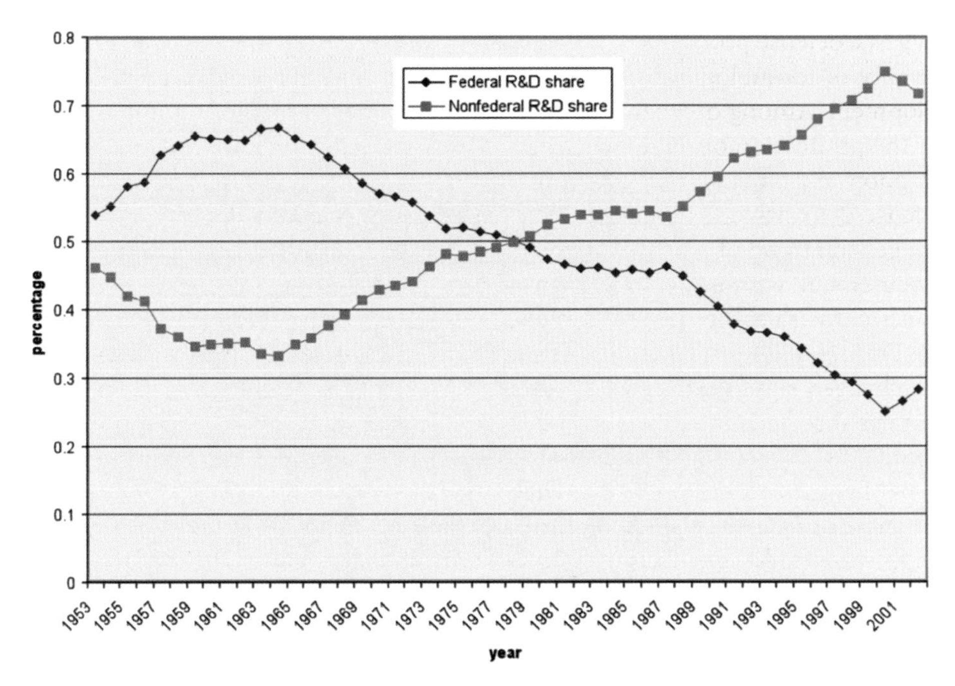

FIGURE 2.8 Federally and non-federally funded research and development (R&D) between 1953 and 2002. SOURCE: National Science Foundation (NSF), *Science and Engineering Indicators*, various years.

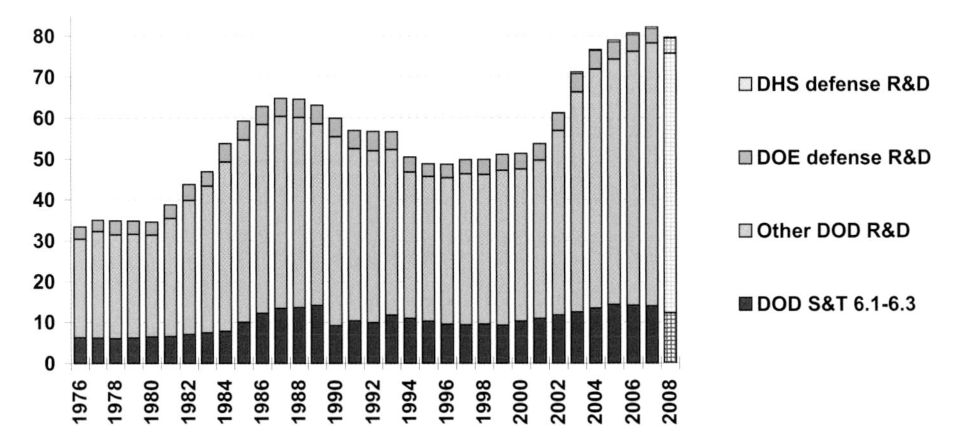

FIGURE 2.9 Trends in defense R&D, FY 1976-2008 (House) in billions of constant FY 2007 dollars. NOTE: Contribution from Department of Homeland Security (DHS) defense R&D (shown at the top of years 2003-2008) can be difficult to discern in lower-resolution formats. SOURCE: American Association for the Advancement of Science (AAAS) analyses of R&D in *AAAS Reports VIII-XXXII*. FY 2008 figures are the latest AAAS estimates of FY 2008 appropriations; 2007 figures include enacted supplementals; R&D includes conduct of R&D and R&D facilities. DOD S&T figures are not strictly comparable for all years because of changing definitions. JULY '07 REVISED © 2007 AAAS. Reprinted with permission.

suggests, defense procurement demand was an especially powerful impetus to the advance of technologies in photonics-related fields in the early years of their development. Among other things, defense procurement provided an initial market for the production of many new entrant firms. It is plausible that the U.S. market for photonics technologies now is dominated (in terms of total revenues) by non-defense demand to a much greater extent than in previous decades.[33] Certainly, this general trend of a declining role in the scope and influence of defense-related procurement within the evolution of a technology-intensive sector is broadly consistent with earlier trends in information technology and semiconductors.[34]

During 1984-2001, the structure of U.S. industrial R&D performance also has changed significantly (see Figure 2.10). The share of overall industrial R&D performance accounted for by the largest U.S. firms has declined significantly, from 60 percent in 1984 to less than 40 percent by 2001. The share of the smallest firms (fewer than 500 employees) has more than doubled during this period, and the share of industrial R&D performance accounted for by firms with 500-9,999 employees also has grown significantly. Some of this growth in the share of smaller firms may reflect alliances with larger firms and other types of outsourcing. Nonetheless, the role of the largest U.S. firms in industrial R&D has declined significantly during this period. The committee's studies in this chapter and in Chapter 7 suggest that trends in the structure of photonics-specific industrial R&D are likely to resemble these overall trends. These trends also may reflect growth in the role of inter-firm "markets for technology," as smaller, specialized firms develop new technologies for sale or license to larger firms that seek to commercialize or incorporate these technologies into their products. Another factor in the growth of U.S. technology licensing activity has been the increased presence of U.S. universities in patenting and licensing faculty research discoveries, a trend that was encouraged and legitimized by the passage of the Bayh-Dole Act of 1980.[35] Such "markets for technology" also have benefited from another significant development during 1984-2001 in the United States—the shift in policy to stronger patent-holder rights.

Finally, non-U.S. markets have increased their share of overall global demand for many high-technology products, and offshore sites for the production and R&D activities of many U.S. high-technology firms also have assumed greater importance, as noted in Chapter 7. In most high-technology industries, U.S. firms' offshore R&D facilities tend to focus on product development more than on basic

---

[33] The committee lacked data on the precise role of defense procurement in the development of the overall field of photonics or on the trends in defense procurement spending over time.

[34] Mowery, D.C. 2011. "Federal Policy and the Development of Semiconductors, Computer Hardware, and Computer Software: A Policy Model for Climate Change R&D?" In *Accelerating Energy Innovation*, R.M. Henderson and R.G. Newell, eds. Chicago, Ill.: University of Chicago Press.

[35] Mowery, D.C., R.R. Nelson, B.N. Sampat, and A.A. Ziedonis. 2004. *Ivory Tower and Industrial Innovation: University-Industry Technology Transfer Before and After the Bayh-Dole Act.* Stanford, Calif.: Stanford University Press.

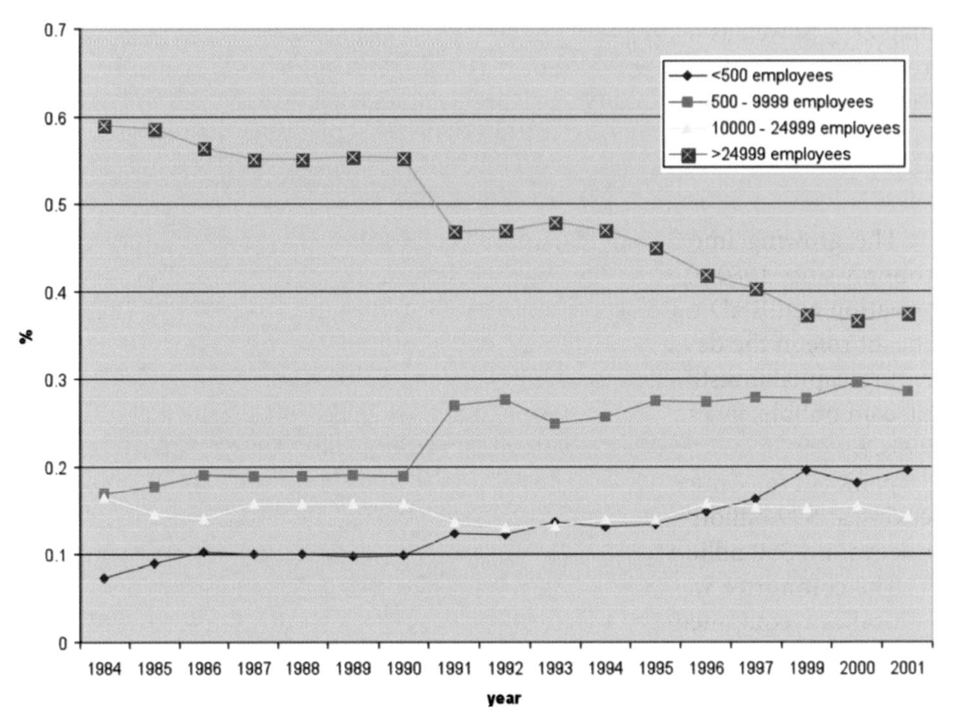

FIGURE 2.10 Firm-size class shares of industry-performed research and development (non-federally funded) between 1984 and 2001. SOURCE: NSF, *Science and Engineering Indicators,* various years.

research. Their growth (and especially the growth of U.S. firms' R&D activities in industrializing economies) reflects the increased sophistication of non-U.S. application environments (reflecting, among many other factors, greater broadband deployment in many non-U.S. economies). The United Nations Conference on Trade and Development (UNCTAD) *World Investment Report 2005* noted that multi-national firms from the United States, Europe, and Japan all had increased the share of their offshore R&D activities located in the People's Republic of China, Singapore, Malaysia, and South Korea. According to the UNCTAD report, the developing-economy share of U.S. multi-national firms' foreign R&D investment grew from 7.6 percent in 1994 to 13.5 percent by 2002.[36] The case studies in

[36] Similarly, German firms increased the share of their foreign R&D investment in developing and transitional (including new Eastern European EU member) economies from 2.7 percent to 7.2 percent between 1995 and 2003. United Nations Conference on Trade and Development (UNCTAD). 2005. *World Investment Report 2005: Transnational Corporations and Internationalization of R&D.* New York, N.Y.: UNCTAD.

Chapter 7, as well as the experiences of the committee, suggest that both offshore production and at least some offshore R&D investment by both U.S. and non-U.S. firms in photonics have grown considerably since 1990.

VENTURE CAPITAL IN OPTICS

The growing importance of smaller firms as R&D performers in the U.S. economy since 1980 partly reflects expanded venture-capital funding for new-firm formation and R&D in high-technology sectors. Venture capital has played a significant role in the development of the optics and photonics industry. While total venture-capital investment in the United States may be lower today than during the dot-com bubble, investments continue to be significantly higher than in the early 1990s. As shown in Figure 2.11, in 1995 the total capital invested in venture-backed companies in the United States was $3.7 billion. This investment grew dramatically, peaking at $99 billion in 2000. It then fell to a low in 2003, increasing to a minor peak around $30 billion in 2007.

The committee was able to use data on venture-capital investments in communications, equipment and services, and telecommunications, to provide insights

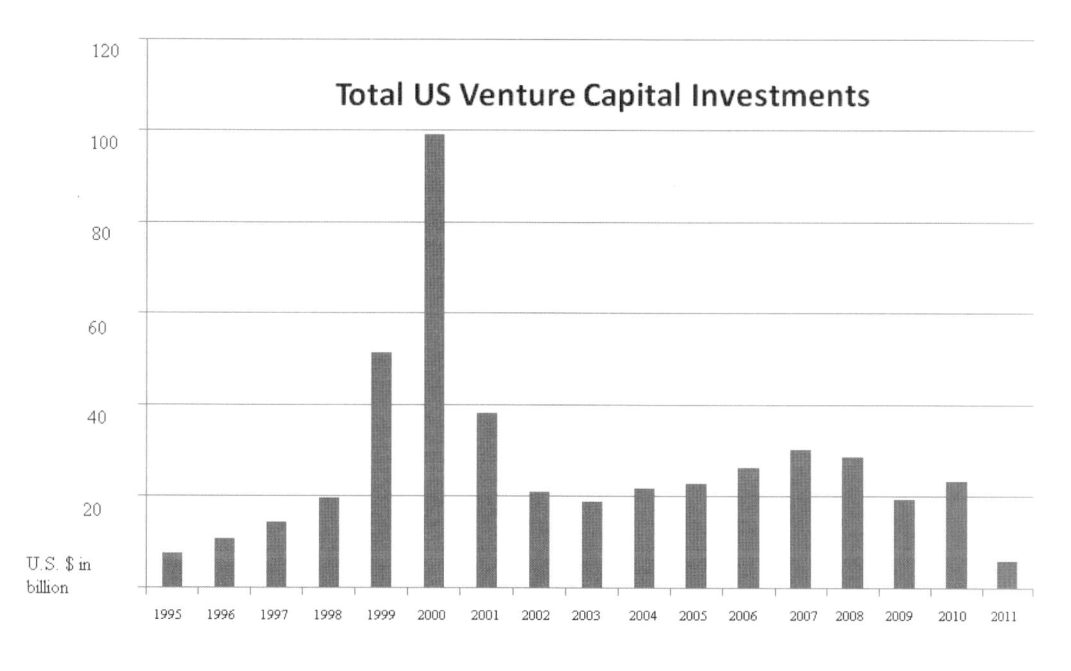

FIGURE 2.11 Total U.S. venture-capital investments, 1995-2010. SOURCE: PricewaterhouseCoopers/National Venture Capital Association MoneyTree™ Report, Data: Thomson Reuters. Reprinted with permission.

into the investments made in optics and photonics companies.[37] The communications, equipment and services, and telecommunications sectors were among the leading market applications for optics and photonics technologies between 1993 and 2000. Nonetheless, venture-capital investment data for these sectors omit photonics-based applications in solar, biotechnology, and many defense applications.

Figures 2.12 and 2.13 report the estimated total venture-capital investment in network and telecommunications services during 1995-2011, as well as the share of total venture-capital investment in the United States directed toward these sectors. At the peak in 2000, optical investments in network and telecommunications services were estimated to be close to $8 billion, which represented roughly 28 percent of the total venture-capital dollars invested in network and telecommunications services. Between 1998 and 2000, the percent of dollars invested in optics and photonics companies in network and telecommunications services grew from 15 percent to 20 percent of total venture dollars invested in network and telecommunications services to a peak of 28 percent. In 2010, optical and photonics investments in communications, equipment and services, and telecommunications fell from a total of $8 billion to less than $2 billion.

Categorizing optical and photonics investments is much more difficult to do today than in 2000. Between 1993 and 2000, most of the optics investments were in the areas of communications, reflecting growth in fiber-optic networks for long-haul communications. In 1995 the primary technologies that represented optics investments included lasers, fiber optics, optical electronics, imaging instruments, and optical components. The incredible growth of the World Wide Web, the passage of the Telecommunications Act of 1996, and the new ability of individuals to invest in public stocks through online services that allow the trading of stocks at an unprecedented rate expanded demand for optical technologies that were at the center of the infrastructure needed to build the new information highway.

By 1998 an optical communications-centered view of public companies had developed within the U.S. investment community. Investment banking analysts began to follow level-three services/infrastructure systems companies (e.g., Ciena, ADC, and Nortel), components/subsystems companies (e.g., Uniphase, JDS, SDL, Inc., and LightPath), and supporting technologies companies (e.g., Newport, Aeroflex, and VerTel). By 2000, 200 to 300 optical components companies received nearly $10 billion in venture financing, 8 new firms had initial public offerings (IPOs), and 10 more filed for IPOs.[38] However, the financing of so many new entrants led to intense competition for market share in a market that soon faced slower growth. In 2001, more than 100 optical companies sought financing and

[37] Analysts have never collected data for photonics-specific venture-capital funding.

[38] Chang, M. 2010. What is the state of investment in photonics? *Laser Focus World* 46(5):33.

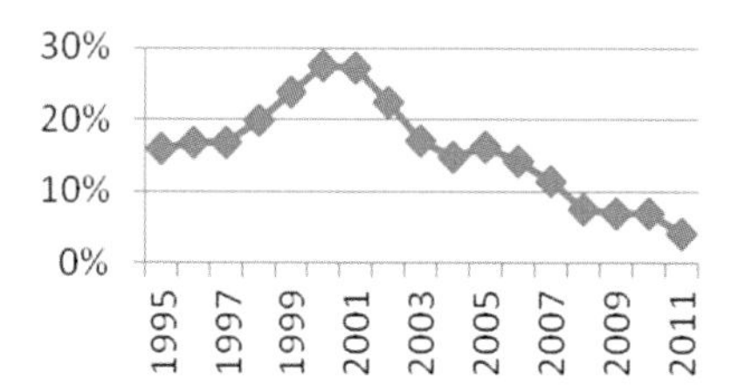

FIGURE 2.12 Network and telecommunications percent of total annual venture-capital investments, 1995-2011. SOURCE: PricewaterhouseCoopers/National Venture Capital Association MoneyTree™ Report, Data: Thomson Reuters. Reprinted with permission.

more than 30 companies planned to file for IPOs. The optimistic growth forecasts for overall telecommunications were not realized, resulting in a crash in the market that led venture capitalists to direct their investments away from optical networking companies.

During the past decade, the investment community's expected returns from the traditional optics and photonics sectors declined sharply. Figure 2.14 shows the change in internal rate of return (IRR) expectations from communications companies compared with IRRs in the 1990s that were greater than 100 percent. From Figure 2.14, one can see that communications companies, defined here to include significant optics and photonics, had a mean IRR of greater than 100 percent for those funded in 1993 through 1998, with the highest return occurring for

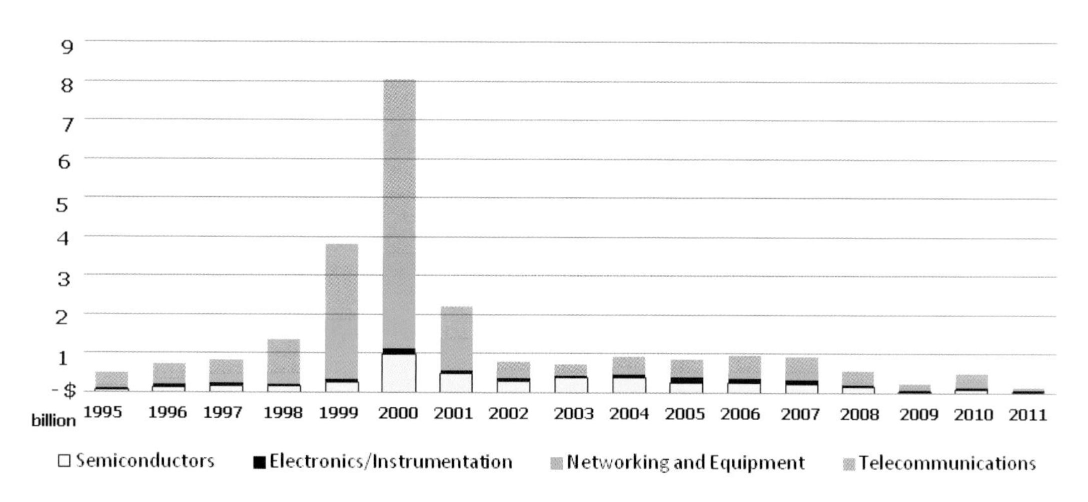

FIGURE 2.13 Trends (in U.S.$ billions) in U.S. venture series A rounds venture investments with proxy-relation to optics and photonics, 1995-2011. SOURCE: PricewaterhouseCoopers/National Venture Capital Association MoneyTree™ Report, Data: Thomson Reuters; courtesy of John Dexheimer, LightWave Advisors, Inc. Reprinted with permission.

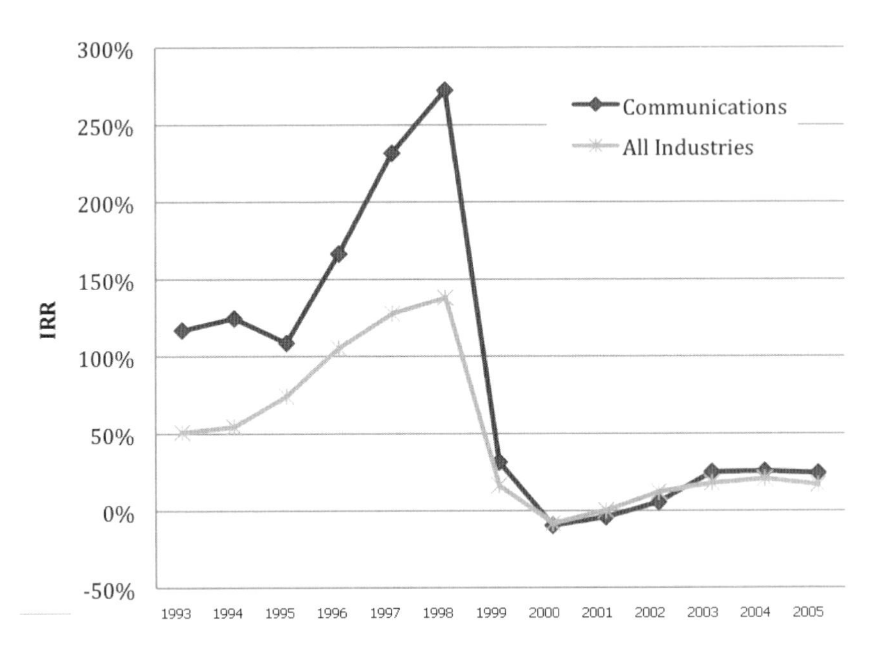

FIGURE 2.14 U.S. venture internal rate of return (IRR) on vintage year companies, 1993-2005. (Shown is the calculated mean return on investments in terms of IRRs for companies in a given year known as the vintage year.) SOURCE: Data collected by Cambridge Associates. Reprinted with permission.

those companies funded in 1998 (a mean IRR of 272 percent). The mean IRR for communications companies with exits funded in 1999 dropped to 31 percent and was negative for those companies funded between 2000 and 2001. It can also be seen that the mean IRRs for all venture-backed companies with a successful exit performed worse than the communications sector during the late 1990s, which drove the demand for venture capital-investments during the 1999-2001 period. For companies funded in those years, the mean IRR was negative.

The expectations of the venture industry for specific sectors also are affected by trends in the overall stock market. Figure 2.15 shows the performance of the OEM [original equipment manufacturers] Capital Photonics Index relative to the S&P 500, revealing that the S&P 500 outperformed the photonics sector during the early part of 2011. If the traditional photonics industry fails to outperform the S&P 500, it is unlikely that the venture industry will shift its investment focus away from social networking and software companies, which currently account for a large share of venture-capital investments, back to photonics. Accordingly, the U.S. venture-capital industry may play a smaller role as a source of investment in photonics innovation in the telecommunications sector, for the near future at least.

Further support for this forecast of the modest near-term effects of venture-

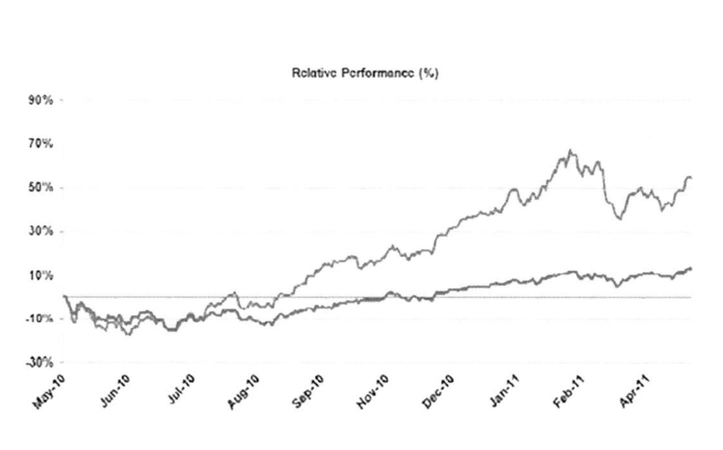

Advanced Photonix	Infinera Corp.
Agilent	IPG Photonics
Andor Technology	JDS Uniphase
AXT Inc.	KLA Tencor
Cognex Corp.	LighPath Tech
Coherent Inc.	Newport Corp.
CyberOptics Corp.	Oclaro, Inc.
Dynasil Corp.	Ophir Optronics
EXFO Inc	PPGI Photonics
Finisar Corp.	Rolfin-Sinar
Gooch & Housego	Rudolph Tech
Halma plc	Zygo Corp.
II-VI Inc.	

FIGURE 2.15 Relative performance of OEM Capital Photonics Index (lower red line) versus S&P 500 (upper blue line), May 2010 through April 2011. SOURCE: Courtesy of OEM Capital, Inc., as presented by John Dexheimer to the National Research Council, August 24, 2011. Reprinted with permission.

capital funding on photonics innovation in telecommunications applications is provided by Figure 2.16, which shows the results of a venture-capital survey conducted by Deloitte and the National Venture Capital Association (NVCA) to determine the anticipated investments of venture capital over the next 5 years in the telecommunications and new media/social networking markets.

The optics and photonics companies that survived the downturn in the telecommunications equipment market have diversified into the biotechnology, energy, imaging, and defense applications. Since the venture-capital industry associations do not track the photonics industry as a separate sector, the challenge again becomes one of determining a correct metric to ascertain the role of venture financing in the contemporary optics and photonics industry. The bursting of the telecommunications bubble in March 2000 also means that investment analysts now do not track firms in these fields closely. Today only 3 investment analysts track the company JDSU (a company that produces optoelectronic components for telecommunications and data communications), compared to more than 40 in 2000. By contrast, First Solar has 30 analysts and Cree has 50, which may be interpreted as an indicator that more private investment has been directed to solar, lighting, and display technologies. Other markets in which optical technologies are used as a key enabler that have received attention from the venture-capital community include biopharmaceutical tools, environmental sensing devices, and medical devices.

With communications no longer a driving factor for many venture capitalists

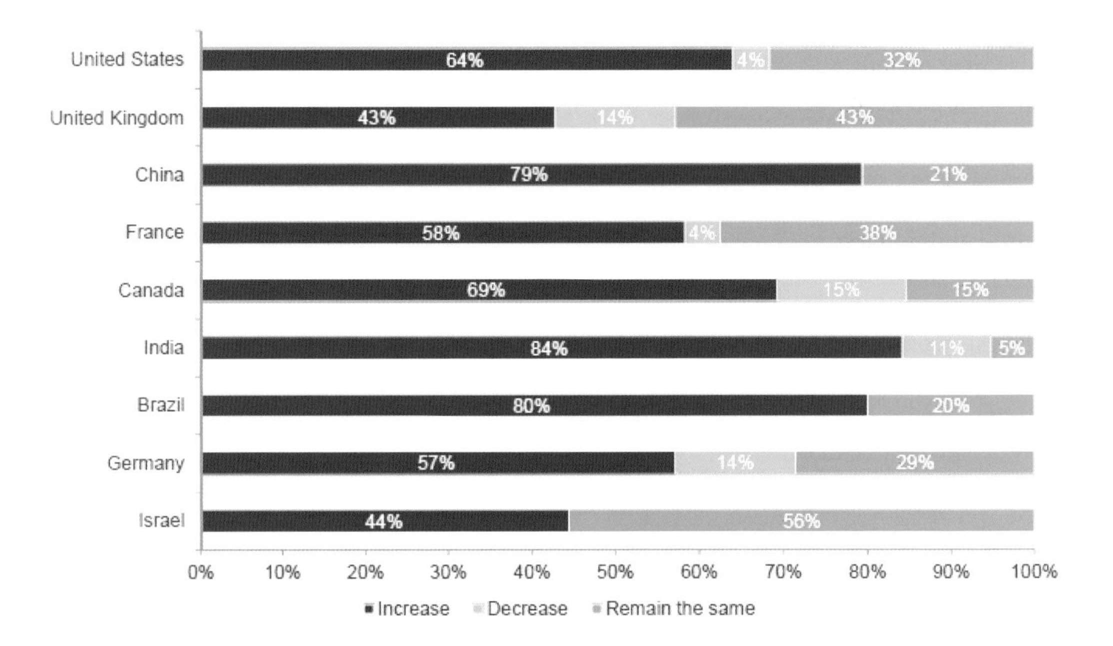

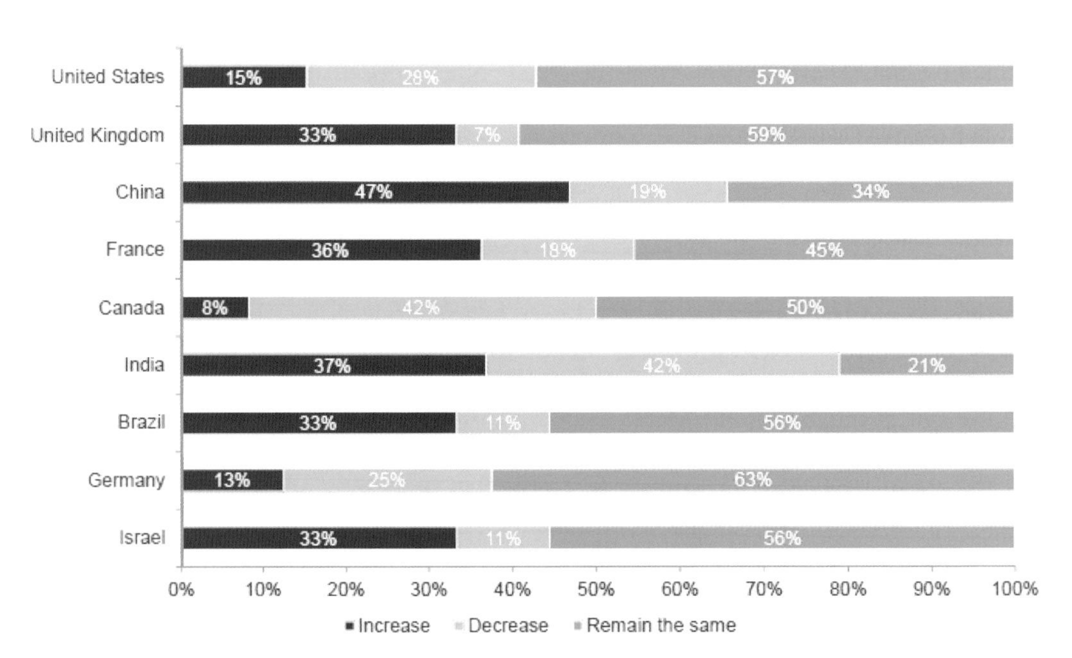

FIGURE 2.16 Anticipated investment levels in terms of total capital over the next 5 years—new media/social networking (*top*) and telecommunications (*bottom*). SOURCE: 2011 Global Venture Capital Survey, June, by Deloitte; National Venture Capital Association—Next 5 Year Venture Capital Forecast of Allocations. Reprinted with permission.

and optics now spread across many applications, the optics and photonics community may benefit significantly from an initiative to regain the attention of the venture-capital community and to reduce reliance on government funding. (See Box 2.1.) In addition, for policy makers to have the ability to track and measure the impact of optics and photonics, the optics industry needs to encourage organizations like the NVCA to refine its economic and financial data collection to track venture-capital investments in optics and photonics across the diversity of markets in which optics and photonics can be enabling technologies. If returns from optics investments can be better quantified, venture capitalists may be more likely to track and invest in optics and photonics as they did for the brief period in the mid-1990s.

BOX 2.1
A Note on Government Funding of Small and Medium-Sized Businesses

As suggested in this chapter in the earlier section entitled "Government and Industrial Sources of R&D Funding in Photonics and Federal Funding of Optics," in addition to venture capital and angel funding, government funding can contribute to the financing of small firms that are research and development performers. Major government programs funding small and medium-sized enterprises include the Small Business Innovation Research (SBIR) Program (which channels funds through multiple federal agencies, including the Department of Defense [DOD], Department of Energy [DOE], National Institutes of Health [NIH], and National Science Foundation [NSF]; the related Small Business Technology Transfer [STTR] Program; and varying programs at the National Institute of Standards and Technology [NIST], including the earlier Advanced Technology Program [ATP] and the more recent Technology Innovation Program [TIP]).

In 2010, 11 federal agencies together provided more than $2 billion in SBIR funding. Individual programs in the DOD (including the Defense Advanced Research Projects Agency [DARPA]), DOE, NIH, and NSF also can target early technology development and commercialization by small firms outside of the SBIR initiatives. An estimated 20 to 25 percent of funding for early-stage technology development comes from the federal government. Nevertheless, small technology-based firms (500 or fewer employees) received only 4.3 percent of extramural government R&D dollars in 2005 (in contrast to medium and large firms, which received 50.3 percent). Of this 4.3 percent, 2.5 percent (or 58 percent of the total) came from SBIR and STTR funds.

NOTE: The committee was unable to collect per-agency numbers on SBIR (and other small business-oriented) funding of optics and photonics. Nevertheless, it is likely that SBIR funding and other programs (including NIST's ATP and TIP, as well as some DARPA initiatives) have played an important role in funding early-stage technology development in photonics.

SOURCES: Branscomb, L.M., and P.E. Auerwwald. 2002. *Between Invention and Innovation: An Analysis of Funding for Early-Stage Technology Development.* Washington, D.C.: National Institute of Standards and Technology; National Research Council. 2008. *An Assessment of the SBIR Program.* Washington, D.C.: The National Academies Press.

MARKETS FOR TECHNOLOGY, INTELLECTUAL PROPERTY, AND U.S. UNIVERSITY TECHNOLOGY LICENSING

As is discussed above in this chapter, structural change in the U.S. R&D system has expanded the importance of licensing transactions involving intellectual property, and the committee believes that these transactions play an important role in photonics innovation in particular. Further evidence of the increased importance of licensing transactions and patented intellectual property in general is provided by the America Invents Act (Public Law No. 112-29), which was signed into law on September 16, 2011. This act, which represented the first comprehensive overhaul of U.S. patent policy in decades, was intended in part to improve the quality of patents granted by the U.S. Patent Office and included steps to further harmonize U.S. patent policy with that of other nations.

As the discussion of lasers pointed out, U.S. universities have long been an important source of ideas and discoveries in the broad field of photonics. In many ways the role of U.S. universities in the U.S. photonics industry appears to be more significant than is true of university research in the photonics industries of other nations. The Bayh-Dole Act of 1980 (Public Law No. 96-517) was passed with broad bipartisan support in order to catalyze the commercialization by U.S. firms of U.S. universities' research advances, and in the wake of the act's passage, U.S. university patenting has grown. Many if not all U.S. research universities have established campus offices of technology licensing to oversee the patenting and licensing to industry of research advances.[39] In the case of the photonics industry, universities have clearly over the last four decades been playing an expanding role in early-stage R&D. For example, there has been an increase in the percent of overall publications in optoelectronics with at least one academic author over the four decades between 1967 and 2007.[40]

The expanded licensing activities of U.S. universities have also attracted considerable criticism from at least some sectors of U.S. industry, notably firms in information technology. R. Stanley Williams of Hewlett Packard, a firm with a long history of close research collaboration with U.S. universities (and a firm active in photonics research and innovation), stated in testimony before the U.S. Senate Commerce Committee's Subcommittee on Science, Technology and Space:

> Largely as a result of the lack of federal funding for research, American Universities have become extremely aggressive in their attempts to raise funding from large corporations. . . . Large U.S. based corporations have become so disheartened and

[39] Although lacking precise data, the committee believes that university-licensed intellectual property has been an important source of innovation in the U.S. photonics industry.

[40] Doutriaux, T. 2009. "The Resiliency of the Innovation Ecosystem: The Impact of Offshoring on Firms versus Individual Technology Trajectories." Work toward a Master's Thesis. Advisor: E. Fuchs. Pittsburgh, Pa.: Carnegie Mellon University.

> disgusted with the situation they are now working with foreign universities, especially the elite institutions in France, Russia and China, which are more than willing to offer extremely favorable intellectual property terms. (September 17, 2002)[41]

These remarks from a leading industrial research manager suggest that for at least some U.S. firms, U.S. universities' patent and licensing policies have become an impediment to collaboration rather than a facilitator of such collaboration, which remains essential to innovation in a U.S. economy that faces challenges from foreign nations with increased technological capabilities. In some cases, frictions between university licensing professionals and U.S firms reflect an unrealistic assessment by university personnel of the financial returns associated with "driving a hard bargain" in licensing terms for a single patent. In December 2005, in response to this criticism and other industry statements of dissatisfaction, four large IT firms (Cisco, Hewlett Packard, IBM, and Intel) and six universities (Carnegie Mellon University; Rensselaer Polytechnic Institute; University of California, Berkeley; Stanford University; University of Illinois at Urbana-Champaign; and University of Texas at Austin) agreed on a "statement of principles" for collaborative research on open-source software that emphasizes the liberal dissemination of the results of collaborative work funded by industrial firms.[42]

In 2007, a group of technology-licensing managers that included representatives from Stanford University; the California Institute of Technology (Caltech); the University of California, Berkeley; and other leading U.S. research universities as well as the Association of American Medical Colleges, issued a list titled "In the Public Interest: Nine Points to Consider in Licensing University Technology" that emphasized the importance of ensuring access to universities' intellectual property in the public interest.[43] Finally, the National Research Council (NRC) convened a committee of experts and practitioners from industry and academia to consider best practices in university technology licensing, and issued a report titled

---

[41] American Society of Mechanical Engineers (ASME). 2002. Statement available at http://www.memagazine.org/contents/current/webonly/webex319.html. Accessed April 2, 2005.

[42] The "Open Collaboration Principles" cover "just one type of formal collaboration that can be used when appropriate and will co-exist with other models, such as sponsored research, consortia and other types of university/industry collaborations, where the results are intended to be proprietary or publicly disseminated." According to the principles, "The intellectual property created in the collaboration [between industry and academic researchers] must be made available for commercial and academic use by every member of the public free of charge for use in open source software, software related industry standards, software interoperability and other publicly available programs as may be agreed to by the collaborating parties." Ewing Marion Kauffman Foundation. 2006. Available at http://www.kauffman.org/pdf/open_collaboration_principles_12_05.pdf. Accessed October 17, 2012.

[43] Association of University Technology Managers. 2007. "In the Public Interest: Nine Points to Consider in Licensing University Technology." White paper. Available at http://www.autm.net/Nine_Points_to_Consider.htm. Accessed July 25, 2012.

Managing University Intellectual Property in the Public Interest, in 2011.[44] Notable among the conclusions of the report was that "[university] patenting and licensing practices should not be predicated on the goal of raising significant revenue for the institution. The likelihood of success is small, the probability of disappointed expectations high, and the risk of distorting and narrowing dissemination efforts is great" (p. 5).

The committee believes that the "Nine Points" document cited above, and the conclusions of the NRC's 2011 report on U.S. university technology licensing, provide valuable guidelines for U.S. universities' management of their photonics-related intellectual property. The committee supports the conclusions of these expert groups.

MODELS OF COLLABORATIVE R&D AND IMPLICATIONS FOR PHOTONICS INNOVATION

As is noted above, the structure of the U.S R&D system has changed since 1980. As the share of federal R&D funding has declined, so also the role of large-firm R&D laboratories has decreased in significance, the influence of defense-related procurement within maturing high-technology sectors such as lasers (and, the committee believes, other photonics technologies) has declined, and offshore R&D has grown in importance. What do these trends imply for the structure of federal R&D in photonics and the ability of such R&D investments to produce significant economic payoffs for U.S. taxpayers? In other high-technology sectors, ranging from nanotechnology to semiconductors, one policy that has proven useful is public-private collaboration in R&D.[45]

It is widely accepted in economics that "in the absence of policy intervention, the social rate of return to R&D expenditure exceeds the private rate, leading to a socially suboptimal rate of investment in R&D" (p. 22).[46] This market failure raises important questions for how public policy should seek to foster increased investment in R&D. In addition to public funding of R&D, one useful policy tool to internalize the externalities (e.g., that the social benefits exceed the private-firm

[44] National Research Council. 2011. *Managing University Intellectual Property in the Public Interest.* Washington, D.C.: The National Academies Press.

[45] An essential first step in developing such a policy is a better accounting of the current federal investment in photonics R&D.

[46] Jaffe, A. 2002. Building programme evaluation into the design of public research-support programmes. *Oxford Review of Economic Policy* 18(1):22-34.

benefits) of R&D is public-private partnerships or research.[47,48] Past research has found a positive impact of Japanese consortia and of ATP-funded U.S. government-industry joint ventures on the research productivity of participants in the technological areas targeted by the consortia.[49,50,51] Indeed, in addition to the support provided by the government's funding, research consortia can play an important role in supporting network formation, thus increasing knowledge flows among participants,[52,53,54,55] and supporting skills[56,57] and the creation of new industries.[58]

In addition to considering research consortia, this section looks at several less widely researched models of coordinated technology development. As Bergh discusses in his paper "Manufacturing Infrastructure for Optoelectronics,"[59] it considers three models for the coordination of technology development for a shorter or longer term and with more or less government funding. The first model, SEMATECH, is a not-for-profit research consortium established in 1987 to provide a research facility in which member companies could improve their semiconductor manufacturing process technology. The second, the Optoelectronics Industry Development Association (OIDA), is a not-for-profit partnership of North American suppliers and users of optoelectronic components, established in 1991 to improve the competitiveness of the North American optoelectronics industry with public

---

[47] Spence, A.M. 1984. Cost reduction, competition, and industry performance. *Econometrica* 52(1): 101-121.

[48] Katz, M.L. 1986. An analysis of cooperative research and development. *RAND Journal of Economics* 17(4):527-543.

[49] Branstetter, L., and M. Sakakibara. 1998. Japanese research consortia: A microeconometric analysis of industrial policy. *Journal of Industrial Economics* 46(2):207-233.

[50] Branstetter, L., and M. Sakakibara. 2002. When do research consortia work well and why? Evidence from Japanese panel data. *American Economic Review* 92(1):143-159.

[51] Sakakibara, M. 2003. Knowledge sharing in cooperative research and development. *Managerial and Decision Economics* 24:117-132.

[52] Tripsas, M., S. Schrader, and M. Sobrero. 1995. Discouraging opportunistic behavior in collaborative R&D: A new role for government. *Research Policy* 24:367-389.

[53] McEvily, B., and A. Zaheer. 1999. Bridging ties: A source of firm heterogeneity in competitive capabilities. *Strategic Management Journal* 20:1133-1156.

[54] Whitford, J. 2005. *The New Old Economy: Networks, Institutions, and the Organizational Transformation of American Manufacturing.* Oxford, U.K.: Oxford University Press.

[55] Fuchs, E. 2010. Rethinking the role of the state in technology development: DARPA and the case for embedded network governance. *Research Policy* 39:1133-1147.

[56] McEvily, B., and A. Zaheer. 1999. Bridging ties: A source of firm heterogeneity in competitive capabilities. *Strategic Management Journal* 20:1133-1156.

[57] Whitford, J. 2005. *The New Old Economy.*

[58] Fuchs, E. 2010. Rethinking the role of the state in technology development: DARPA and the case for embedded network governance. *Research Policy* 39:1133-1147.

[59] Bergh, A. 1996. Manufacturing infrastructure for optoelectronics. *Lasers and Electro-Optics Society Annual Meeting Conference Proceedings.* IEEE. Lasers and Electro-Optics Society LEOS-96. November 18-21, 1996.

funding from various agencies as well as private membership funding. The third, the National Nanotechnology Initiative (NNI), is one of the largest federal interagency R&D programs; established in 2000, today it coordinates funding from 25 federal departments and agencies for nanotechnology research and development. An effective industry coalition would take time and resources to develop and therefore would need staunch commitment by stakeholders. One goal would be to create a collective voice that is knowledgeable and credible to center activities in photonics, to provide a positive influence to the industry, and to keep government agencies and the public informed. The examples below provide insights into how these goals have been achieved through institutions in the United States historically. The model provided by the German Fraunhofer Institutes is discussed in Box 2.2.

Semiconductor Manufacturing Technology (SEMATECH)

As noted above, SEMATECH was established in 1987 to provide a research facility in which member companies could develop next-generation manufac-

BOX 2.2
Fraunhofer Institutes

The committee heard reports about the unique and successful photonics activities of the German Fraunhofer Institutes. These institutes represent a novel approach to fostering leading-edge research by a combination of universities, companies, and government. In this approach (1) the government provides an overarching forum and core funding, (2) the industry provides a healthy percentage of the funding but is focused on key areas in which the industry has interest, and (3) the universities provide the intellectual workforce to achieve technological advances. Although it is debated whether this model would be effective in the U.S. institutional structure, this combination has been successful in providing leadership, maintaining interest, and producing impressive technical advances within the German context.

Even if the full model is not transferable, aspects of these highly respected Fraunhofer Institutes might be valuable models for the United States, given that these institutes have been playing a pivotal role in technology commercialization and enabling Germany to have a leadership position in photonics. For example, this model was instrumental in the recent formation of the €360 million High-Tech Foundation Fund. This fund includes industry and government participation to provide seed capital for start-up companies to commercialize technologies that are spun out of the Fraunhofer Institutes. This fund also enables industry leaders to steer critically needed seed capital to worthy photonics start-up companies without requiring government agencies to make the selections.

SOURCES: Fraunhofer. 2012. Available at http://www.fraunhofer.de/en/html. Accessed June 26, 2012. See also German Center for Research and Innovation. 2012. Available at http://www.germaninnovation.org/about-us. Accessed June 26, 2012.

turing technology.[60,61] SEMATECH sought to support horizontal collaboration among U.S. semiconductor producers on the development of process technology. This initial focus, however, proved to be infeasible because of the importance of firm-specific process expertise for the competitive advantage of individual semiconductor manufacturers.[62,63] As a consequence, SEMATECH altered its research agenda to a vertical collaboration model that sought to improve the technological capabilities of U.S. suppliers of semiconductor manufacturing equipment.[64] This shift in its research agenda was associated with a shift in SEMATECH's intellectual property policies—from licensing research results to member firms on an exclusive basis for 2 years, to member firms' receiving priority in ordering and receiving new models of equipment resulting from SEMATECH-funded research.[65] By the mid-1990s, SEMATECH's interactions with equipment and material suppliers fell into four main categories: joint development projects, equipment improvement projects, provision of technology "roadmaps," and expanded communication between suppliers and member firms.[66]

Although SEMATECH's role in the improvement of the U.S. industry's competitiveness may be difficult to prove, SEMATECH met most of its revised objectives in the development of process technology, the supply of manufacturing equipment, and collaboration between manufacturers, suppliers, and research centers.[67] Some research suggests that SEMATECH reduced the duplication of member R&D spending [68,69] and that economic returns to member companies outweighed their membership costs.[70] Several lessons for the design and structure of public-private consortia can be drawn from the experience of SEMATECH. First, SEMATECH focused primarily on short-term research, with 80 percent of all R&D efforts

[60] Grindley, P., D. Mowery, and B. Silverman. 1994. SEMATECH and collaborative research: Lessons in the design of high-technology consortia. *Journal of Policy Analysis and Management* 13(4):723-758.

[61] Link, A., D. Teece, and W. Finan. 1996. Estimating the benefits from collaboration: The case of SEMATECH. *Review of Industrial Organization* 11:737-751.

[62] Grindley et al. 1994. SEMATECH and collaborative research.

[63] Carayannis, E., and J. Alexander. 2004. Strategy, structure, and performance issues of precompetitive R&D consortia: Insights and lessons learned from SEMATECH. *IEEE Transactions on Engineering Management* 51(2):226-232.

[64] Grindley et al. 1994. SEMATECH and collaborative research.

[65] Grindley et al. 1994. SEMATECH and collaborative research.

[66] Grindley et al. 1994. SEMATECH and collaborative research.

[67] Grindley et al. 1994. SEMATECH and collaborative research.

[68] Irwin, D., and P. Klenow. 1996. High-tech R&D subsidies: Estimating the effects of SEMATECH. *Journal of International Economics* 40:323-344.

[69] Irwin, D., and P. Klenow. 1996. SEMATECH: Purpose and performance. *Proceedings of the National Academy of Sciences* 93:12739-12742.

[70] Link, A., D. Teece, and W. Finan. 1996. Estimating the benefits from collaboration: The case of SEMATECH. *Review of Industrial Organization* 11:737-751.

focused on outcomes within 1 to 3 years.[71] Second, SEMATECH was organized originally by industry, its operations were led and its decisions directed by industry, and it retained substantial support in the form of funding from industry.[72] Third, SEMATECH operated with a fairly centralized organizational structure rather than as an umbrella consortium of independent projects (although this characterization is less true for the equipment projects), which facilitated adjustment of its research agenda and operations in response to the changing needs of the industry.[73,74] Fourth, SEMATECH drew top executives from member firms into organizational decisions and management.[75,76] Fifth, SEMATECH was a consortium of established companies with underlying strengths in product and process technology.[77]

Optoelectronics Industry Development Association (OIDA)

Similar to the situation in the semiconductor industry, the value of a photonics community coalition is apparent in providing leadership to help interface with industry and government on policy matters, as well as in informing the general public and the investment community on current matters. However, previous attempts to form photonics industry trade associations have had limited success, possibly because these organizations did not receive sufficiently broad industry participation. For example, the Laser Electro-Optics Manufacturers Association and OIDA were composed largely of photonics technology manufacturers and tended not to have support from the applications for those outcomes from industry. The case of OIDA is further discussed here.

In 1988, a National Research Council study entitled *Photonics: Maintaining Competitiveness in the Information Era* recommended the formation of "an industry association that could help organize consortia to conduct cooperative research and address technical problems and policy issues beyond the scope of any one

---

[71] It is important to point out that the R&D activities of SEMATECH were complemented by two other initiatives. SEMI/SEMATECH represented the U.S. semiconductor equipment producers within SEMATECH, and operated with a modest funding base contributed by the members of the Semiconductor Equipment Manufacturing Industry Association (SEMI); and the Semiconductor Research Corporation (SRC), which enlisted the members of SEMATECH and other U.S. semiconductor firms, supported long-term R&D at U.S. universities.

[72] Grindley et al. 1994. SEMATECH and collaborative research.

[73] Browning, L., J. Beyer, and J. Shelter. 1995. Building cooperation in a competitive industry: SEMATECH and the semiconductor industry. *Academy of Management Journal* 38(1):113-151.

[74] Grindley et al. 1994. SEMATECH and collaborative research.

[75] Browning et al. 1995. Building cooperation in a competitive industry.

[76] Grindley et al. 1994. SEMATECH and collaborative research.

[77] Grindley et al. 1994. SEMATECH and collaborative research.

organization."[78] In 1991, OIDA was founded as a North American partnership of suppliers and users of optoelectronics components to improve the competitiveness of the North American optoelectronics industry.[79] Early on, OIDA undertook an Optoelectronic Technology Roadmap Program, intended to identify the critical paths for the development of enabling optoelectronic technologies. This roadmap exercise concluded in 1996 that mastering volume manufacturing was essential to its members' ability to reduce costs and improve competitiveness.[80]

The U.S. optoelectronics industry and OIDA have a long-standing association with NIST as well as with the Department of Defense. In October 1997, at the request of OIDA leaders, NIST organized a photonics manufacturing competition within the Advanced Technology Program (ATP) that led to the funding by ATP of 10 proposals from industry.[81] In 1992, DARPA began a series of programs, which continued through 2009, to promote the development of new optoelectronics technologies, including direct funding for OIDA workshops and operating expenses. In 1994, NIST's Optoelectronics Division was founded "to provide the optoelectronics industry and its suppliers and customers with comprehensive and technically advanced measurement capabilities, standards, and traceability to those standards."[82] As indicated by NIST,[83] the division's mission was to maintain close contact with the optoelectronics industry through major industry associations, including the Optoelectronics Industry Development Association, and to represent NIST at the major domestic and international standards organizations in optoelectronics such as the Telecommunications Industry Association and the American National Standards Institute. This division is now part of the new Quantum Electronics and Photonics Division.[84]

OIDA remains active in technology roadmapping and in the improvement of member-firm capabilities in high-volume manufacturing, including support for such "R&D infrastructure" as an optoelectronics foundry.[85] Although the found-

---

[78] National Research Council. 1988. *Photonics: Maintaining Competitiveness in the Information Era.* Washington, D.C.: National Academy Press.

[79] Bergh, A. 1996. Manufacturing Infrastructure for Optoelectronics.

[80] Bergh, A. 1996. Manufacturing Infrastructure for Optoelectronics.

[81] Kammer, R., Director, National Institute of Standards and Technology. 1998. Prepared Remarks. Optoelectronics Industry Development Association. Washington, D.C., October 2.

[82] National Research Council. 1999. *An Assessment of the National Institute of Standards and Technology Measurement and Standards Laboratories: Fiscal Year 1999.* Washington, D.C.: National Academy Press.

[83] More information is available from the NIST Physical Measurement Laboratory at http://www.nist.gov/pml/. Accessed August 6, 2012.

[84] Quantum Electronics and Photonics Division. Physical Measurement Laboratory, National Institute of Standards and Technology (NIST), website. Available at http://www.nist.gov/pml/div686/. Accessed July 25, 2012.

[85] These efforts have produced few measureable outcomes.

ing membership of OIDA included larger companies, such as AT&T, Bellcore, Corning, IBM, 3M, Hewlett-Packard Company, and Motorola, the number of member companies has declined over the course of OIDA's history. Further, in contrast to SEMATECH membership, many OIDA member firms are smaller, with less developed technologies, and membership fees for all are low compared to SEMATECH—in the low thousands to tens of thousands. Along with its modest industry funding, a key challenge for OIDA is that of enlisting the participation of senior industry executives in managerial positions. Perhaps most importantly, in contrast to SEMATECH, OIDA has lacked a clear R&D agenda. This consortium's inability to develop such a focus reflects the diversity of applications characteristic of photonic semiconductors, which complicates agreement among member firms on technological goals. Inasmuch as photonics manufacturing technology is less mature than semiconductor process technologies, the optoelectronics industry might be better served by a university-government-private partnership that focuses more intensively on early-stage R&D. The Semiconductor Research Corporation (SRC) is an interesting contrast to SEMATECH in this respect. (See Box 2.3.)

BOX 2.3
Semiconductor Research Corporation

The Semiconductor Research Corporation (SRC) is another semiconductor research and development consortium whose structure contrasts with that of SEMATECH. In 1982, the Semiconductor Industry Association launched the Semiconductor Research Association as a cooperative research organization to "enhance basic research in semiconductor related disciplines" by funding "long-term, pre-competitive research in semiconductor technology at U.S. universities." In contrast to SEMATECH—which was created in 1987 out of an SRC initiative—SRC uses horizontal collaborations between member firms and academic researchers to define and fund long-term technology developments central to the survival and success of the semiconductor industry.

In addition to SRC's receiving funding from member firms, SRC program directors also seek matching funds from federal, state, and local governments. Federal-level collaborators have included the U.S. Army Research Office, the Defense Advanced Research Projects Agency, the National Institute of Standards and Technology, and the National Science Foundation.

In 2000, the SRC board committed to globalizing the SRC membership and research base. To date, however, little empirical research exists on the history, processes, or successes of SRC. SRC is viewed by several of its member companies as an ongoing success and warrants further study as an interesting model for research consortia focused on long-term pre-competitive research.

SOURCE: Semiconductor Research Corporation. 2012. "About Semiconductor Research Corporation." Available at http://www.src.org/about/. Accessed August 3, 2012.

National Nanotechnology Initiative

In September 1998, the Interagency Working Group on Nanotechnology (IWGN) was formed within the National Science and Technology Council of the Office of Science and Technology Policy.[86] As described in the National Research Council's 2002 report *Small Wonders, Endless Frontiers: A Review of the National Nanotechnology Initiative* (from which material in this paragraph is drawn substantially and, in some cases, extracted), this group formalized the operations of a set of staff members from several agencies that in November 1996 had begun to meet regularly to discuss their plans and programs in nanoscale science and technology. In August 1999, IWGN's plan for an initiative in nanoscale science and technology was approved by the President's Council of Advisors on Science and Technology (PCAST) and OSTP, and in its 2001 budget submission to Congress, the Clinton administration raised nanoscale science and technology to a federal initiative, referring to it as the National Nanotechnology Initiative (NNI). The National Science and Technology Council (a cabinet-level committee with membership drawn from federal agencies across the government) formed the Nanoscale Science, Engineering, and Technology (NSET) Subcommittee to focus on NNI activities. The National Nanotechnology Coordination Office, established in 2001, provides technical guidance and administrative support to the NSET Subcommittee, facilitates multiagency planning, conducts activities and workshops, and prepares information and reports. The NRC also provides feedback to NNI through its triennial review of NNI;[87] such a review is currently ongoing.[88]

Today NNI is one of the largest federal interagency R&D programs, coordinating funding for nanotechnology research and development among 25 participating federal departments and agencies. Its federal funding grew from $225 million in FY 1999 to $464 million in 2001[89] and an estimated $1.639 billion in 2010.[90,91] As

---

[86] National Research Council. 2002. *Small Wonders, Endless Frontiers: A Review of the National Nanotechnology Initiative.* Washington, D.C.: National Academy Press.

[87] National Research Council. 2006. *A Matter of Size: Triennial Review of the National Nanotechnology Initiative.* Washington, D.C.: The National Academies Press.

[88] See National Research Council. 2012. "Interim Report for the Triennial Review of the National Nanotechnology Initiative, Phase II." Prepublication copy. Washington, D.C.: The National Academies Press. Available at http://www.nap.edu/catalog.php?record_id=13517. Accessed October 23, 2012.

[89] National Research Council. 2002. *Small Wonders, Endless Frontiers.*

[90] National Research Council. 2002. *Small Wonders, Endless Frontiers.*

[91] Office of Science and Technology Policy (OSTP): A Decade of Investments in Innovation Coordinated Through the National Nanotechnology Initiative. National Nanotechnology Initiative Investments by Agency from FY 2001 through FY 2010. OSTP's "High Value Data Sets." The White House. Available at http://www.whitehouse.gov/administration/eop/ostp/library/highvalue. Accessed December 26, 2011.

stated on the NNI webpage,[92] NNI has four goals: (1) maintain a world-class research and development program aimed at realizing the full potential of nanotechnology; (2) facilitate the transfer of new technologies into products for economic growth, jobs, and other public benefit; (3) develop educational resources, a skilled workforce, and the supporting infrastructure and tools to advance nanotechnology; and (4) support responsible development of nanotechnology.

NNI has facilitated several developments to enhance dialogue and coordination among nanoscale R&D programs at federal agencies; these include working groups, an infrastructure network involving an integrated partnership of user facilities at 13 campuses across the United States, centers to support the development of tools for fabrication and analysis at the nanoscale, and NNI-industry consultative boards to facilitate networking among industry, government, and academic researchers, analyze policy impacts at the state level, and support programmatic and budget redirection within agencies. In contrast to either OIDA or SEMATECH, NNI is focused more intensively on priority setting and support for more fundamental, long-term research in this emerging technology.

Given the diversity of applications characteristic of photonics and the relative immaturity both of much of the science and much of the industry, the National Nanotechnology Initiative may provide an interesting model for the increased coordination and tracking of long-term funding of research in photonics. Since the writing of the 1998 NRC report *Harnessing Light: Optical Sciences and Engineering for the 21st Century*,[93] there has been an explosion of new applications for photonics. Indeed, in spite of the maturity of some of the constituent elements of photonics (e.g., optics), the committee believes that photonics as a whole is likely to experience a period of growth in opportunities and applications that more nearly resembles what might be expected of a vibrantly young technology.

SUMMARY COMMENTS

The preceding overview of some recent experiments in collaborative R&D makes apparent several implications for similar efforts in photonics. First, industry participation and leadership, both intellectual and financial, are essential. Second, such an industry commitment to collaborative R&D may be more difficult in a sector that spans a diverse array of applications and is populated mainly by new, relatively small firms. Finally, consortia (such as SEMATECH) in which industry plays a major role in establishing and funding the R&D agenda may not be well suited for supporting long-term research. An interesting exception may be the model

[92] National Nanotechnology Initiative. Available at http://www.nano.gov/. Accessed August 6, 2012.

[93] National Research Council. 1998. *Harnessing Light: Optical Science and Engineering for the 21st Century.* Washington, D.C.: National Academy Press.

presented by the Semiconductor Research Corporation (see Box 2.3), which has supported work by academic researchers in the United States. Nonetheless, given that SRC programs tend to focus on developing technologies to achieve specific end goals that involve stakeholders from a single industry, the long-term precompetitive research agenda supported by the SRC may be insufficient by itself to deal with the diversity of applications that are the focus of R&D within the field of photonics.

Rather than endorsing any single structure for the support of R&D that involves collaboration among industry, government, and academic researchers, the committee believes that a higher-level venue for discussion and assessment of R&D priorities is needed. Any such structure could include among its activities R&D consortia of the type represented by SRC, or SEMATECH, or still other models for collaborative R&D. A research consortium is more likely to succeed in a focused application of the optics and photonics landscape. In the absence of some coordinating initiative, it will prove difficult to develop an effective strategy for public and private R&D investment that seeks to support longer-term R&D and to translate innovation into economic opportunities for U.S. firms and employees across the diversity of emerging photonics applications. Accordingly, the committee's judgment is that the time is overdue for a federal initiative in photonics that seeks to engage industry, academic, and government researchers and policy makers in the design and oversight of R&D and related programs that include federal as well as industry funding.

Proposed National Photonics Initiative

A national photonics initiative would coordinate agency-level investment in photonics-related R&D and could provide partial support for other technology-development initiatives, including R&D consortia funded by federal and industry sources. The committee believes that a number of experiments in coordinating across industry-government-academia in different fields of R&D and technology development are warranted. As pointed out in the next chapter, "Communications, Information Processing, and Data Storage," one application area that may be particularly ripe for such a public-private consortium is large-scale data communications and storage, now the focus of an initiative overseen by OSTP.[94] Finally, as with NNI, a national photonics initiative would assume responsibility for developing, coordinating, and measuring federal funding and national outputs (such as economic indicators) in photonics, to help inform national policy.

---

[94] Office of Science and Technology Policy. 2012. "Big Data Press Release." Available at http://www.whitehouse.gov/sites/default/files/microsites/ostp/big_data_press_release_final_2.pdf. Accessed July 30, 2012.

FINDINGS

Key Finding: Photonics is a key enabling technology with broad applications in numerous sectors of the U.S. economy. The diversity of applications associated with photonics technologies makes it difficult to quantify accurately the economic impacts of photonics in the past and even more difficult to predict the future economic and employment impacts of photonics.

Key Finding: Given the diversity of its applications and the enabling character of photonics technology, data on photonics industry output, employment, and firm-financed R&D investment are not currently reported by U.S. government statistical agencies, further complicating analysis of this technology's economic impact and prospects. Although the 1998 National Research Council study *Harnessing Light: Optical Science and Engineering for the 21st Century* reached a similar conclusion and recommended that members of the photonics community be involved in the next round of Standard Industrial Classification (SIC) or North American Industry Classification System (NAICS) development, no such action was taken by federal statistical agencies.

Finding: Another significant gap in the economic data on photonics is a lack of systematic collection or reporting by the federal government of its significant investment in photonics R&D. As a result, the most basic data are lacking for estimating the overall federal R&D investment in this technology field or the allocation of federal photonics R&D investments among different fields and applications.

Finding: The private organizations that monitor U.S. venture-capital investment trends also do not collect information on the full spectrum of photonics-related venture-capital investments. Changes in the structure of the U.S. R&D and innovation systems mean that the importance of venture-capital funding for the formation of new firms in photonics, as well as for these firms' investments in R&D and technology commercialization, has grown; thus these gaps in data on venture-capital investment hamper the ability to monitor innovation in photonics.

Finding: Many of the important early U.S. innovations in photonics relied on R&D performed in large industrial laboratories and benefited as well from defense-related R&D and procurement spending. The structure of the R&D and innovation processes in photonics, similar to other U.S. high-technology industries, appears to have changed somewhat, with universities, smaller firms, and venture-capital finance playing more prominent roles. These changes in the structure of R&D funding and performance within photonics increase the potential importance of inter-firm collaboration and public-private collaboration in photonics innovation.

RECOMMENDATIONS

Key Recommendation: The committee recommends that the federal government develop an integrated initiative in photonics (similar in many respects to the National Nanotechnology Initiative) that seeks to bring together academic, industrial, and government researchers, managers, and policy makers to develop a more integrated approach to managing industrial and government photonics R&D spending and related investments.

This recommendation is based on the committee's judgment that the photonics field is experiencing rapid technical progress and rapidly expanding applications that span a growing range of technologies, markets, and industries. Indeed, in spite of the maturity of some of the constituent elements of photonics (e.g., optics), the committee believes that the field as a whole is likely to experience a period of growth in opportunities and applications that more nearly resembles what might be expected of a vibrantly young technology. But the sheer breadth of these applications and technologies has impeded the formulation by both government and industry of coherent strategies for technology development and deployment.

A national photonics initiative would identify critical technical priorities for long-term federal R&D funding. In addition to offering a basis for coordinating federal spending across agencies, such an initiative could provide matching funds for industry-led research consortia (of users, producers, and material and equipment suppliers) focused on specific applications, such as those described in Chapter 3 of this report. In light of near-term pressures to limit the growth of or even reduce federal R&D spending, the committee believes that a coordinated initiative in photonics is especially important.

The committee assesses as deplorable the state of data collection and analysis of photonics R&D spending, photonics employment, and sales. The development of better historical and current data collection and analysis is another task for which a national photonics initiative is well suited.

Key Recommendation: The committee recommends that the proposed national photonics initiative spearhead a collaborative effort to improve the collection and reporting of R&D and economic data on the optics and photonics sector, including the development of a set of North American Industry Classification System (NAICS) codes that cover photonics; the collection of data on employment, output, and privately funded R&D in photonics; and the reporting of federal photonics-related R&D investment for all federal agencies and programs.

It is essential that an initiative such as the proposed national photonics initiative be supported by coordinated measurement of the inputs and outputs in the sector such that national policy in the area can be informed by the technical and economic realities on the ground in the nation.

3

Communications, Information Processing, and Data Storage

INTRODUCTION

Optics has become the way by which most information is sent over nearly all the distance that it travels. The remarkable growth of networks and the Internet over the past decade has been enabled by previous generations of optical technology. Optics is, furthermore, the only technology with the physical headroom to keep up with this exponentially growing demand for communicating information. The exceptionally high carrier frequency allows transmission of high bandwidths. Satisfying that future demand will, however, require the continued research and development (R&D) of new technology for optics and continued innovation in its interface with electronics. The use of optics will not be restricted to the traditional market of long-distance telecommunication. Increasingly optics will be used for ever-shorter distances, possibly providing few-millimeter or shorter links between the silicon chips themselves. One key driver for such shorter distances will be to attain sufficient density of communications. A second key driver will be to control the growing power dissipation that is due to switching/transmission elements and the environmental impact of information processing. This chapter summarizes the developments in optics over the last 15 years (i.e., since the publication of the National Research Council's [NRC's] *Harnessing Light*[1]) in information commu-

[1] National Research Council. 1998. *Harnessing Light: Optical Science and Engineering for the 21st Century.* Washington, D.C.: National Academy Press.

nications, processing, and storage and indicates key directions for the future for optics in these fields.

Communications

Optical communications networks provide the underlying high-capacity, ubiquitous connectivity that underpins the global Internet. Figure 3.1 characterizes the growth of communication and computing between 1986 and 2007, based on a broad collection of data.[2] Approximately around the year 2000, Internet traffic took over from voice telephone as the single largest communication format for information. Now Internet traffic dominates completely. All of the long-distance communications on the Internet are over optical fiber.

Major advances in transmission techniques and technologies have allowed network providers to provide extremely cost-effective network upgrades that have kept pace with the extraordinary appetite for broadband Internet services. That growth, as exemplified in Figure 3.1, has driven network bandwidth demands by a factor of 100 over the last 10 years. That increase has been enabled by realizing the full potential of wavelength division multiplexing (WDM) that has resulted in fibers carrying as many as 100 separate wavelengths. In addition, the capacity per wavelength in commercially deployed terrestrial networks has increased from a maximum of 10 gigabits per second (Gb/s) per wavelength when the first edition of *Harnessing Light* was published in 1998, to 100 Gb/s today. As a result, per fiber transmission capacities in terrestrial systems today as high as 5-10 terabits per second (Tb/s) are possible.[3] Transoceanic capacities have lagged somewhat behind terrestrial values because the long amplifier-only distances and the desire to extend the amplifier spacing have made upgrading to per wavelength capacities above 10 Gb/s problematic. Nevertheless, transoceanic per fiber capacities of approximately 1 Tb/s are typical. For the future there are expectations that this growth will continue as more video content calls for bandwidth and that there is a need for another factor-of-100 growth in the coming 10 years as well.

Major advances have also been achieved in both cost-effectively managing the large capacity in today's WDM optical networks and in leveraging the value proposition of optical amplifiers to provide multi-wavelength amplification over network mesh and ring architectures. Reconfigurable, wavelength-routed networks—in which wavelength-defined units of capacity can be added, dropped, or switched

---

[2] Hilbert, M., and P. Lopez. 2011. The world's technological capacity to store, communicate, and compute information. *Science* 332(6025):60-65.

[3] Alferness, R.C. 2008. "Optical Communications—A View to the Future." 24th European Conference on Optical Communications, Brussels, Belgium, September 22. Available at http://ieeexplore.ieee.org/xpls/abs_all.jsp?arnumber=4729111&tag=1. Accessed June 26, 2012.

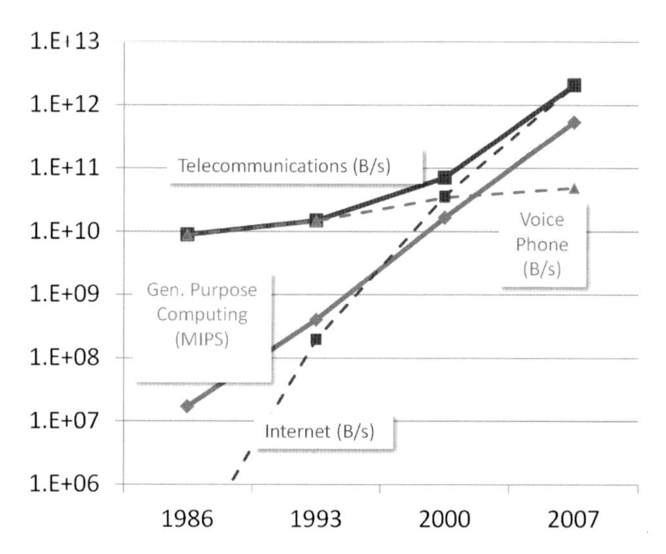

FIGURE 3.1 Traffic In bytes per second (B/s) (1 byte = 8 bits) on the Internet, on voice telephone, and overall, 1986-2007. Also shown is the capacity to compute in general-purpose machines, expressed in millions of instructions per second (MIPS). SOURCE: MIPS graph based on data extracted from Hilbert, M., and P. Lopez. 2011. The world's technological capacity to store, communicate, and compute information. *Science* 332(6025):60-65.

from one fiber route to another fiber route directly in the optical domain without the need for conversion to electronics—are now heavily deployed in long-haul terrestrial networks as well as metropolitan networks. Wavelength-routed networks provide cost-effective solutions because they allow data on wavelengths passing through a node at a multi-route network node to remain in the optical domain and benefit from the cost-effective multi-wavelength amplification enabled by optical amplifiers, rather than needing to be individually electronically regenerated. The large increase in capacity demand has ensured that a prerequisite for the economic viability of such networks—namely, that the capacity demand between any two node pairs on the network be at least as large as that which can be carried by a single wavelength—is met.

WDM optical networks require reconfigurable optical add/drop multiplexers (ROADMs) to, under network electrical control, drop or add wavelength channels at a node and to switch wavelength channels from one fiber route to another. ROADMs are key enablers that have evolved significantly in their functionality, providing increasing levels of flexibility, and in their capacity, or number of fiber ports and wavelengths per fiber, over the last decade. Further progress in these network elements and their enabling technologies will be essential to addressing the growing demand for capacity.

Ultimately, networks are no better than the access capacity that they provide to the end user, whether that customer is a business or a residence. Increasingly that access is through an optical link. The last decade has seen significant increase in the deployment of fiber in the access network, initially to the curb, but increasingly also directly to the business or home.

The United States is not a world leader in fiber to the home or business. It ranks roughly at number 11 among the countries in the world, with approximately 7 percent penetration. The penetration of fiber to the home in the United States is roughly 5 percent.[4]

Passive optical networks (PONs) are the primary broadband optical delivery architecture, providing the shared bandwidth of a fiber to multiple users (16 to 64 users). Initially, systems provided shared bandwidth of 2.5 Gb/s. New systems operating at a total bandwidth of 10 Gb/s are becoming available. Just as wavelength multiplexing has provided cost-effective bandwidth enhancement in long-haul and metropolitan networks, as capacity demand in the access network increases to enable new broadband services, it is expected that WDM will be employed for capacity expansion. A critical requirement will be robust, low-cost WDM optical components to operate in the outside plant. Research in this area at a somewhat modest level is ongoing; an example of U.S.-funded research in this area is the National Science Foundation (NSF) Engineering Research Center for Integrated Access Networks.[5]

In spite of the dramatic achievements that optics has brought to communication networks over the last decade (or, perhaps, because of them), the demand for higher bandwidth, both in the fixed and the mobile domains, continues to grow rapidly. Comparing projections from a number of sources, it seems conservative to suggest that network capacity demand will grow at the rate of at least a factor of 100 over the next 10 years, approximately following the recent historical trend shown in Figure 3.1.[6] Ubiquitous video is the key driver. Increasingly that video is two-way as more end users upload video to sharing websites. Mobile video is also growing at an extremely rapid rate—92 percent compound annual growth rate according to a Cisco report[7]—that puts large bandwidth demand on the backhaul

[4] Montagne, R. 2010. "Understanding the Digital World." Presented at the FTTH Council Europe Conference. Available at http://www.ftthcouncil.eu/documents/Reports/Market_Data_December_2010.pdf. Accessed June 26, 2012.

[5] More information is available at the website of the Center for Integrated Access Networks at www.cian-erc.org. Accessed June 26, 2012.

[6] Korotky, S.K., R.-J., Essiambre, and R.W. Tkach. 2010. Expectations of optical network traffic gain afforded by bit rate adaptive transmission. *Bell Labs Technical Journal* 1(4):285-296.

[7] CISCO. 2012. "Cisco Visual Networking Index: Forecast and Methodology, 2010-2015." White paper. Available at http://www.cisco.com/en/US/solutions/collateral/ns341/ns525/ns537/ns705/ns827/white_paper_c11-481360.pdf. Accessed June 26, 2012.

network, which is increasingly requiring optical links. One can expect also that increasingly mobile backhaul will not be done by point-to-point links as is done today, but rather will be part of an extended-access network which serves end points that include wireless base stations, small enterprises, curb drops, and homes.

Broadband mobility will require ubiquitous, cost-effective backhaul, resulting in the convergence of wire-line and wireless networks that are currently often separate. The nature of traffic in an Internet- and video-driven network could suggest different overall network architecture, with merging between the metropolitan and access networks. Integration of data centers into the network is key today and will be even more so in the future, especially as cloud services become more pervasive. Already in 2003, the NRC's Committee on Coping with Increasing Demands on Government Data Centers stated:

> Recommendation: Data centers should aggressively adopt newer, more "bleeding edge" technical approaches where there might be significant return on investment. This should be done carefully to minimize the inevitable failures that will occur along the way. Even with the failures, the committee believes the cost savings and improvements for end users will be substantial when compared to the methods practiced today.[8]

Just as capacity demand has required the advantages of optics to be pushed further to the edges of the network, data centers will increasingly depend on optics for interconnection, and eventually for reconfigurability and switching. A future of "cloud services" in which most if not all digital services are performed by shared resources in the network is looking increasingly attractive. This new paradigm, which would depend on ubiquitous, instant, and highly reliable access to the network, could place demands on the network equivalent to those of the transition from voice to data in the late 1990s. The advent of WDM, made cost-effective with the optical amplifier, has enabled the cost-effective scaling of optical point-to-point transmission systems to meet the exponential demand growth. ROADMs and optical cross-connects have made it possible to leverage that advantage to multi-node ring and mesh networks. As a result, optical networks that underpin the Internet have been able to keep pace with exploding demand over the last 10 years. However, realistically, the several-orders-of-magnitude capacity increase resulting from many wavelengths will not be duplicated by simply adding even more wavelengths, because of limited optical amplifier bandwidth and fiber power limits in the fiber to mitigate transmission impairment.[9]

---

[8] National Research Council. 2003. *Government Data Centers: Meeting Increasing Demands.* Washington, D.C.: The National Academies Press.

[9] Chraplyvy, A. 2009. "The Coming Capacity Crunch." Plenary paper. 35th European Conference on Optical Communications, Vienna, Austria, September 21. Available at http://ieeexplore.ieee.org/xpls/abs_all.jsp?arnumber=5287305&tag=1. Accessed June 26, 2012.

The optical communication industry is highly challenged to find the key new technologies, architectures, and techniques that will enable cost-effective growth in capacity to keep up with the capacity demand that video, including mobile video, will make on the network. Achieving that goal by simply increasing the number of wavelengths, or by increasing the bit rate per wavelength, appears unlikely with known or incrementally improved technology. Therefore, some fundamentally new optical solutions that build on top of today's WDM optical networks will be urgently required in order to ensure that the global Internet is able to continue to serve as the engine for economic growth. It is estimated that without new technology developments, Internet growth will be constrained starting between 2013 and 2014.

Information Processing

The main, and growing, use of optics in information processing is to connect information within and between information switching and processing machines. There are substantial physical reasons—specifically, reducing power dissipation and increasing the density of information communications—why optics is preferable and ultimately possibly essential for such connections.[10] The idea of using optics for performing the logical switching in information processing—as in some kind of optical transistor—is one that has continuing research interest, although the criteria for success there are challenging,[11] and no such use appears imminent. The last decade has seen sustained research interest in very advanced ideas such as quantum computing,[12] although such ideas remain very much in fundamental research. Optics, however, is likely to have very important roles in any such quantum computing or quantum communications—for example, as the best means of communicating quantum states over any meaningful distance—and optics might also be important in future quantum gates. Quantum encryption, a means of sending key information with immunity to any eavesdropping, has seen first commercial optical systems. In the meantime, efforts are needed to create systems that utilize the best combination of optics and electronics to enable integrated systems for efficient information processing.

In the practical connection of information, optics has increasingly taken over the role of data communications within local networks for information processing systems such as data centers, supercomputers, and storage area networks, for

---

[10] Miller, D.A.B. 2009. Device requirements for optical interconnects to silicon chips. *Proceedings of the IEEE* 97:1166-1185.

[11] Miller, D.A.B. 2010. Are optical transistors the next logical step? *Nature Photonics* 4:3-5.

[12] Nielsen, M.A., and I.L. Chuang. 2000. *Quantum Computation and Quantum Information.* Cambridge, U.K.: Cambridge University Press.

distances from approximately 100 meters to 10s of kilometers. Commercial approaches include optics in networking architectures such as Ethernet, Fiber Channel, and Infiniband, especially as data rates and data densities rise, and in active optical cables that are internally optical but externally have electrical inputs and outputs. Low cost is particularly critical for shorter distances. Consequently, such data communications have often used technologies different from the long-haul telecommunications approaches, such as vertical cavity surface-emitting lasers and laser arrays, and multi-mode (rather than single-mode) optical fibers for the shorter distances. The use of optics in such networks is likely to steadily increase because of continuing increases in network traffic and information processing.

As can be seen in Figure 3.1, the ability to compute information, as indicated by the growth of the computational power (expressed in millions of instructions per second [MIPS] that can be executed by the general-purpose silicon chips that are sold),[13] has approximately the same growth rate as that of Internet traffic. The ability to connect information within information processing systems is, however, hitting increasingly severe limits. The ability of wires to interconnect does not scale well to higher densities, especially for information communications outside the silicon chips themselves ("off-chip" communications). This point is well understood by the semiconductor industry, as shown, for example, in its projections of an off-chip interconnect bottleneck.[14] The interconnect capability of electrical wires between chips, or processor cores, is not going to scale to keep up with the ability to compute—sometimes called the bytes per flops (floating point operations per second) gap.[15] Thus, although Internet bandwidth and raw computational power in MIPS or flops might continue to scale, the systems performance in between will not, unless a new interconnect technology is introduced that can scale in capacity.

High-performance silicon chips today, as characterized, for example, by the International Technology Roadmap for Semiconductors (ITRS) numbers,[16] already have interconnect capabilities well into the range of multiple terabits per second per chip. To scale to keep up with processing power, future chips in the 2020 time frame would need 100s of terabits per second of interconnect capability, an amount of interconnect for one chip that is comparable to the entire Internet traffic today. It is worth noting that generally there is likely much more information sent inside information processing systems than is sent between them. One order-of-magnitude

---

[13] Hilbert M., and P. Lopez. 2011. The world's technological capacity to store, communicate, and compute information. *Science* 332(6025):60-65.

[14] Miller, D.A.B. 2009. Device requirements for optical interconnects to silicon chips.

[15] Miller, D.A.B. 2009. Device requirements for optical interconnects to silicon chips.

[16] More information is available through the International Technology Roadmap for Semiconductors at http://www.itrs.net/Links/2010ITRS/Home2010.htm. Accessed June 26, 2012.

estimate[17] is that each byte traversing the Internet causes approximately 1 million bytes of data communications within data centers.

The large and increasing interconnected bandwidth inside information processing systems creates problems both for density of interconnects and for power sourcing and dissipation. Interconnecting one bit of information off a chip currently involves several to 10s of picojoules (pJ) of energy. A substantial portion (e.g., 50 to 80 percent) of all power dissipation on silicon chips is used to interconnect the information rather than to perform logic.[18] Power dissipation in general is a severe limitation on information processing; for example, the clock rate on mainstream silicon chips has not changed substantially in recent years because higher clock rates would lead to too much power dissipation. A recent NRC study[19] on supercomputing also emphasizes this problem with power dissipation and the need, in particular, to reduce interconnection power while maintaining or even increasing the byte-per-flops ratio.

Power dissipation in information processing is already environmentally significant. Electricity used in data centers in 2010 likely accounted for between 1.1 and 1.5 percent of total electricity use globally, and between 1.7 and 2.2 percent of electricity use in the United States.[20] The power dissipation in interconnects in servers alone has been estimated to exceed solar power generation capacity.[21] Of course, information processing and communication can have substantial environmental benefits in improving the efficiency of many activities and in reducing travel.[22] But, **if current growth trends in Internet traffic and in information processing continue, reducing the energy per bit processed and/or communicated will be crucial.** Even though sending a bit over the Internet might take approximately 10 nanojoules (nJ)/bit[23] compared to 10s of pJ/bit for off-chip interconnects, the vastly greater number of bits being sent inside the machine (approximately 10^6

---

[17] G. Astfalk. 2009. Why optical data communications and why now? *Applied Physics A: Materials Science and Processing* 95(4): 933-940.

[18] Miller, D.A.B. 2009. Device requirements for optical interconnects to silicon chips.

[19] National Research Council. 2010. *The Future of Computing Performance: Game Over or Next Level?* Washington, D.C.: The National Academies Press.

[20] Koomey, Jonathan. 2011. Growth in Data Center Electricity Use 2005 to 2010. Oakland, Calif.: Analytics Press. Available at http://www.analyticspress.com/datacenters.html. Accessed October 27, 2011.

[21] Miller, D.A.B. 2009. Device requirements for optical interconnects to silicon chips.

[22] The Climate Group. 2008. *SMART 2020: Enabling the Low Carbon Economy in the Information Age.* A report by the Climate Group on behalf of the Global eSustainability Initiative. Available at http://www.smart2020.org/_assets/files/02_Smart2020Report.pdf. Accessed June 26, 2012.

[23] Tucker, R.S., R. Parthiban, J. Baliga, K. Hinton, R.W.A. Ayre, and W.V. Sorin. 2009. Evolution of WDM optical IP networks: A cost and energy perspective. *Journal of Lightwave Technology* 27:243-252.

times[24]) means that the power dissipation inside information processing machines may substantially dominate over telecommunications power. Hence reducing interconnect power inside machines is crucial for controlling power dissipation in information processing.

The research understanding of the reasons for, and the requirements on, optics and optoelectronics for such shorter connections is now relatively complete and consistent.[25,26] Optics helps in two ways. First, optics can carry information at much higher density than is possible in electrical wires, as is essential for future scaling of interconnect capacity. Second, optics can fundamentally save energy in interconnects because it completely avoids the need for charging wires (the charging of electrical wires to the signal voltage is the dominant source of energy dissipation in electrical interconnects).

The key challenges in short-distance, high-density, low-energy optical interconnect are technological: optoelectronic devices must operate with very low energies—for example, 10 femtojoules (fJ)/bit—much lower than have been required for telecommunications and data communications until now. For cost and performance reasons, the optical and optoelectronic technologies have to be integrated with electronics in a mass-manufacturable process.

New technologies have been advancing in research and early product introductions over the last decade, especially in terms of devising ways of combining optics, optoelectronics, and electronics in silicon chips and platforms (silicon photonics).[27,28] These technological opportunities are discussed below.

Data Storage

The past decade has seen the maturation of technologies such as the DVD (digital versatile disk) for the distribution of consumer video, and the emergence of third-generation disk technologies such as Blu-ray for higher density. Research continues on possible fourth-generation optical disks, including multi-layer and possible holographic recording. Optics may also have crucial roles to play in other potential emerging technologies such as HAMR (heat-assisted magnetic

---

[24] Astfalk, G. 2009. Why optical data communications and why now? *Applied Physics A: Materials Science and Processing* 95:933-940.

[25] Miller, D.A.B. 2009. Device requirements for optical interconnects to silicon chips.

[26] Astfalk, G. 2009. Why optical data communications and why now? *Applied Physics A: Materials Science and Processing* 95:933-940.

[27] Lipson, M. 2005. Guiding, modulating, and emitting light on silicon—Challenges and opportunities. *Journal of Lightwave Technolology* 23(12):4222-4238.

[28] Jalali, B., and S. Fathpour. 2006. Silicon photonics. *Journal of Lightwave Technolology* 24(12): 4600-4615.

recording)[29] for next-generation magnetic hard drives, for which near-field optics using nanophotonic technologies may provide the precise heating required.

The picture for the future of optical disks like DVDs is less clear because of the competition from other forms of storage, such as flash memory (memory sticks), and from networks. Whether optical disks will have the role in the distribution of data and entertainment, such as movies and games, which they have had in the past is debatable because of the growth of broadband networks as an alternative distribution channel—as, for example, in the growing use of video on demand over the Internet.

IMPACT EXAMPLE: THE INTERNET

As is now readily self-evident, the Internet has transformed the way in which society operates. Previous conceptualizations of distance and geography have disappeared in a maze of ubiquitous cellular telephones, e-mail, web browsing, and social networking. Indeed, the productivity of people coming of age after the mid-1990s is intertwined with the easily available communication and information found through the Internet.

Initially the Internet was not reliant on optical communications. However, data transfer rates during those formative times were painfully slow. Without optical technologies, conventional methods were utilized: users depended on slow, dial-up modems, e-mails occasionally took hours before arriving at their destination, and making intercontinental phone calls involved an annoying delay between speaking and listening.

Optics changed the Internet and data transfer by increasing the capacity of the system by nearly 10,000-fold over the past two decades. The nearly instantaneous nature of current optical-based communication allows for real-time video chats. Telepresence, telemedicine, and tele-education would not be possible without optics. In short, without optics, the Internet as we know it would not exist. As it was stated in 2006 by the NRC Committee on Telecommunications Research and Development: "Telecommunications has expanded greatly over the past few decades from primarily landline telephone service to the use of fiber optic, cable, and wireless connections offering a wide range of voice, image, video, and data services."[30]

The same report continues:

> Yet it is not a mature industry, and major innovation and change—driven by research—can be expected for many years to come.

---

[29] Kryder, M.H., E.C. Gage, T.W. McDaniel, W.A. Challener, R.E. Rottmayer, G. Ju, Y.-T. Hsia, and M.R. Erden. 2008. Heat assisted magnetic recording. *Proceedings of the IEEE* 96:1810-1835.

[30] National Research Council. 2006. *Renewing U.S. Telecommunications Research*. Washington, D.C.: The National Academies Press, pp. 1-20.

> Without an expanded investment in research, however, the nation's position as a leader is at risk. Strong competition is emerging from Asian and European countries that are making substantial investments in telecommunications R&D.
>
> For many telecommunications products and services that are now commodities, the United States is at a competitive disadvantage compared with countries where the cost of doing business is lower. Continued U.S. strength in telecommunications, therefore, will require a focus on high-value innovation that is made possible only by a greater emphasis on research. Expansion of telecommunications research is also necessary to attract, train, and retain research talent.
>
> Telecommunications research has yielded major benefits such as the Internet, radio frequency wireless communications, optical networks, and voice over Internet Protocol. Promising opportunities for future research include enhanced Internet architectures, more trustworthy networks, and adaptive and cognitive wireless networks.[31]

The 2002 NRC study Atoms, Molecules, and Light: AMO Science Enabling the Future stated:

> Internet optical backbone link capacity increased a hundredfold between 1995 and 1998, to 20,000 trillion bits per second. While the communications industry has recently suffered a downturn, the potential for increasing demand for capacity in the 21st century remains, with features and services such as HDTV, broadband communications, and advanced home security systems becoming more available through the use of optical fiber, wireless, satellite, and cable connections.[32]

The use of optics, however, is not always obvious to the casual observer. Often someone can be seen using a cell phone to perform an Internet search. Where is the optics? The cell phone uses a wireless radio connection to a local cell tower, but that radio signal is converted to an optical data stream and sent along the fiber optic network across the planet. The data search itself, such as through Google, relies on data centers in which clusters of co-located computers talk to each other through high-capacity optical cables. In fact, there can be as many as a million lasers in a given data center.

During the introduction to the 2009 Nobel Prize Lecture by Dr. Charles Kao on the innovation of optical fiber communications, the Physics Committee Chair said: "The work has fundamentally transformed the way we live our daily lives."[33]

---

[31] National Research Council. 2006. *Renewing U.S. Telecommunications Research.* pp. 1-20.

[32] National Research Council. 2002. *Atoms, Molecules, and Light: AMO Science Enabling the Future.* Washington, D.C.: The National Academies Press, pp. 7-8.

[33] More information on the Nobel Prize is available at http://www.nobelprize.org/nobel_prizes/physics/laureates/2009/. Accessed June 26, 2012.

TECHNICAL ADVANCES

Communications

The past 20 years have seen an explosion in technical achievements in the field of optical communications. Whereas an undersea optical cable was deployed in 1988 and could accommodate 280 Mb/s,[34] there exist today commercial systems that can carry 5 Tb/s per fiber strand. This represents an astonishing increase of four orders of magnitude in performance in commercial systems, and R&D laboratories can transmit more than an order of magnitude higher than that.[35] This technological improvement was driven primarily by the exponential growth in capacity demand, primarily due to Internet usage. Looking to the future, it is expected, as stated above, that the network traffic demand will grow by a factor of 100 over the next decade,[36] necessitating technology advances that can keep pace. However, the research community is struggling to envision technical paths for achieving this vision.

In the mid-1990s, leading laboratories achieved the transmission of 1 Tb/s and 0.1 Tb/s respectively over hundreds and thousands of kilometers of single-mode optical fiber.[37] The key advances that enabled these results and that were captured by the NRC's 1998 *Harnessing Light*[38] study include those described below.

Wavelength Division Multiplexing (WDM)

By simultaneously transmitting different channels of data on different wavelengths by means of WDM, the system capacity can be dramatically increased. Moreover, wavelength-selective components enable highly efficient network routing, such that specific wavelength channels can be added or dropped at intermediate nodes without disrupting the other optical channels.[39]

---

[34] Runge, P.K. 1990. Undersea lightwave systems. *Optics and Photonics News* 1(11):9-12.

[35] Zhu, B., T.F. Taunay, M. Fishteyn, X. Liu, S. Chandrasekhar, M.F. Yan, J.M. Fini, E.M. Monberg, and F.V. Dimarcello. 2011. 112-Tb/s space-division multiplexed DWDM transmission with 14-b/s/hz aggregate spectral efficiency over a 76.8-km seven-core fiber. *Optics Express* 19(17):16665-16671.

[36] Winzer, P.J. 2010. Beyond 100G Ethernet. *IEEE Communications Magazine* 48(7):26-29.

[37] Kerfoot, F.W., and W.C. Marra. 1998. Undersea fiber optic networks: Past, present, and future. *Journal on Selected Areas in Communications* 16:1220-1225.

[38] National Research Council. 1998. *Harnessing Light.*

[39] Kaminow, I.P., C.R. Doerr, C. Dragone, T. Koch, U. Koren, A.A.M. Saleh, A.J. Kirby, C.M. Ozveren, B. Schofield, R.E. Thomas, R.A. Barry, D.M. Castagnozzi, V.W.S. Chan, B.R. Hemenway, D. Marquis, S.A. Parikh, M.L. Stevens, E.A. Swanson, S.G. Finn, and R.G. Gallager. 1996. A wideband all-optical WDM network. *Journal on Selected Areas in Communications* 14:780-799.

Erbium-Doped Fiber Amplifiers

Broadband (>3 THz), low-noise optical amplifiers—erbium-doped fiber amplifiers (EDFAs)—enabled the amplifying of many channels simultaneously and the cascading of numerous stages, thereby making long-distance WDM transmission practical.[40]

Dispersion Management

Chromatic dispersion and nonlinearities are fiber-based effects that significantly degrade transmitted data. However, by judiciously alternating between positive and negative dispersion element, the accumulated dispersion can be compensated, and the phase-matching needed for generating inter-channel nonlinear effects can be minimized.[41]

New Engineering Approaches Needed

Similar to the experience in other areas of technology, new engineering approaches are required in order to achieve new eras of technical advancement in data transmission. Some of the key directions over the past decade[42] that have enabled the continued achievement include approaches borrowed from the radio community, although these approaches were not possible until recent device performance increases. They are described below.

Higher-Order Modulation Formats. Until recently, deployed systems employed simple, amplitude on/off-keyed digital data encoding. However, phase-shift keying (PSK) of the optical carrier wave produces a better receiver sensitivity and is relatively tolerant to degrading effects such as nonlinearities.[43] The fiber bandwidth is precious because the available spectrum is rapidly filling up, and so spectral efficiency, measured in bits per second per hertz, becomes more critical; as another benefit, spectrally narrow channels are more robust to dispersion effects. A typical example is quadrature-phase-shift-keying (QPSK), which encodes two bits of information on each symbol, thereby doubling the spectral efficiency. This can be

[40] Li, T.Y. 1993. The impact of optical amplifiers on long-distance lightwave telecommunications. *Proceedings of the IEEE* 81:1568-1579.

[41] Kurtzke, C. 1993. Suppression of fiber nonlinearities by appropriate dispersion management. *IEEE Photonics Technology Letters* 5:1250-1253.

[42] Kaminow, I.P., Tingye Li, and A.E. Willner. 2008. *Optical Fiber Telecommunications V B: Systems and Networks.* New York, N.Y.: Academic Press.

[43] Winzer, P. J., and R.-J. Essiambre. 2007. Advanced modulation formats for high-capacity optical transport networks. *Journal of Lightwave Technology* 24(12):4711-4720.

further extended to even higher I/Q (in-phase/quadrature) data constellations—for example, quadrature-amplitude modulation (QAM) of 16 to 512,[44,45,46] as seen in Figure 3.2, in which data are encoded on both phase and amplitude of the optical carrier. Additionally, independent data can be transmitted along the two orthogonal polarization axes of the optical fiber, that is, polarization-division-multiplexing, doubling again the capacity and spectral efficiency.[47]

Coherent Systems Utilizing Digital Signal Processing. In heterodyne receivers, ubiquitous in the radio world, a weak data signal is mixed with a local oscillator that has much higher power. Following the same approach, an optical coherent system utilizes a narrow line-width laser as the local oscillator to mix with a weak incoming optical data signal. The balanced detectors can be used to recover both amplitude and the phase information, that is, the time history of the data channel's wave. Coherent systems (1) exhibit better receiver sensitivity than direct detection and (2) can utilize sophisticated electronic digital-signal-processing (DSP) techniques to equalize and compensate for many impairments, such as chromatic- and polarization-based degradations.[48,49,50]

Concepts That Might Improve Capacity

Recently, due to technology advances, two other concepts have emerged that might further improve capacity for the next several years, as discussed below.

---

[44] Zhou, X., and J. Yu. 2009. Multi-level, multi-dimensional coding for high-speed and high-spectral-efficiency optical transmission. *Journal of Lightwave Technology* 27(16):3641-3653.

[45] Schmogrow, R., D. Hillerkuss, S. Wolf, B. Bäuerle, M. Winter, P. Kleinow, B. Nebendahl, T. Dippon, P.C. Schindler, C. Koos, W. Freude, and J. Leuthold. 2012. 512QAM Nyquist sinc-pulse transmission at 54 Gbit/s in an optical bandwidth of 3 GHz. *Optics Express* 20:6439-6447.

[46] Okamoto, S., K. Toyoda, T. Omiya, K. Kasai, M. Yoshida, and M. Nakazawa. 2010. "512 QAM (54 Gbit/s) Coherent Optical Transmission over 150 km with an Optical Bandwidth of 4.1 GHz." Conference paper. 36th European Conference and Exhibition on Optical Communication, Turin, Italy.

[47] Gnauck, A.H., P.J. Winzer, A. Konczykowska, F. Jorge, J.Y. Dupuy, M. Riet, G. Charlet, B. Zhu, and D.W. Peckham. 2012. Generation and transmission of 21.4-Gbaud PDM 64-QAM using a novel high-power DAC driving a single I/Q modulator. *Journal of Lightwave Technology* 30:532-536.

[48] Taylor, M.G. 2004. Coherent detection method using DSP for demodulation of signal and subsequent equalization of propagation impairments. *IEEE Photonics Technology Letters* 16(2):674-676.

[49] Savory, S.J. 2010. Digital coherent optical receiver: Algorithms and subsystems. *Journal of Selected Topics in Quantum Electronics* 16(5):1164-1179.

[50] Salsi, M., O. Bertran Pardo, J. Renaudier, W. Idler, H. Mardoyan, P. Tran, G. Charlet, and S. Bigo. 2011. "WDM 200 Gb/s Single-Carrier PDM-QPSK Transmission Over 12,000 km." Conference paper. 37th European Conference and Exhibition on Optical Communication, Geneva, Switzerland. Available at http://ieeexplore.ieee.org/xpls/abs_all.jsp?arnumber=6065939. Accessed June 26, 2012.

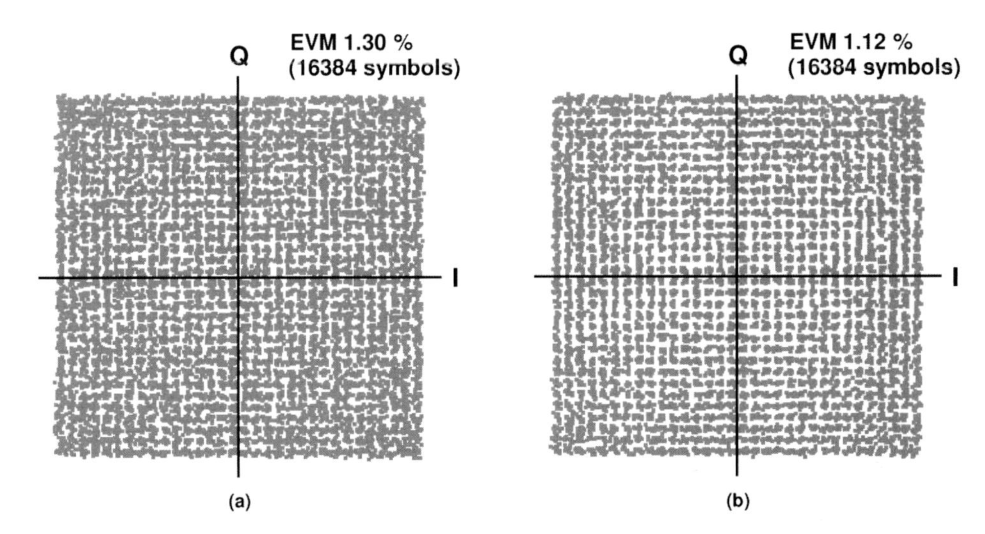

FIGURE 3.2 Recorded experimental data constellation of optical 1024-quadrature-amplitude-modulation (QAM) signal (60 Gb/s) using coherent optical transmission over 150 km. The two axes are for in-phase and quadrature directions, with the total constellation representing data points encoded in amplitude and phase without (a) and with (b) digital nonlinear compensation. SOURCE: Reprinted, with permission, from Koizumi Y., K. Toyoda, M. Yoshida, and M. Nakazawa. 2012. 1024 QAM (60 Gb/s) single-carrier coherent optical transmission over 150 km. *Optics Express* 20(11):12508-12514.

Space Division Multiplexing (SDM). After time, wavelength, and polarization multiplexing, space multiplexing is now being investigated to further increase the transmission capacity. In a space-division-multiplexing system, each independent data channel is carried by an orthogonal spatial dimension. Two emerging approaches include using the following: (1) a special multi-core fiber, each individual core of which transmits an independent data stream;[51] and (2) a few-mode fiber, with each independent data channel on one of the orthogonal spatial modes (see Figure 3.3). In both approaches, the key challenge is the crosstalk. Unique challenges for multi-core fiber systems include the following: (1) further increasing the number of cores, (2) decreasing the inter-core nonlinear effects, and (3) developing multi-mode/multi-core network elements, such as an erbium-doped fiber that has matched multi-cores, so that all cores can be amplified simultaneously in a single fiber element. For multi-mode systems, unwanted mode conversions among different spatial modes are a natural occurrence which leads to several-channel crosstalk.

[51] The Climate Group. 2008. *SMART 2020: Enabling the Low Carbon Economy in the Information Age*. A report by the Climate Group on behalf of the Global eSustainability Initiative. Available at http://www.smart2020.org/_assets/files/02_Smart2020Report.pdf. Accessed June 26, 2012.

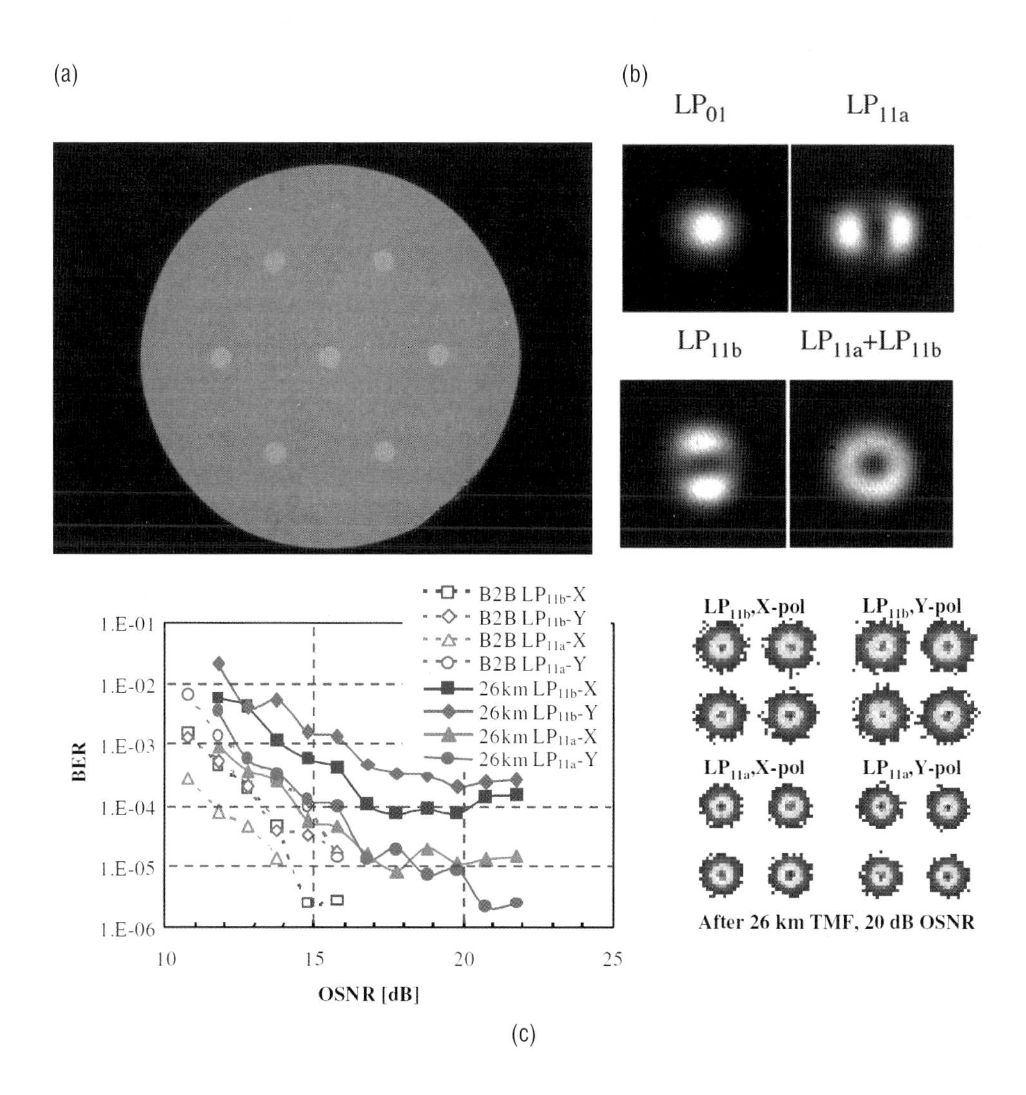

FIGURE 3.3 (a) Space division multiplexing (SDM) using multi-core fiber. SOURCE: Reprinted, with permission, from Matsuo, S., K. Takenaga, Y. Arakawa, Y. Sasaki, S. Tanigawa, K. Saitoh, and M. Koshiba. 2011. Crosstalk behavior of cores in multi-core fiber under bent condition. *IEICE Electronics Express* 8(6):385-390. (b) Multi-mode fiber. SOURCE: Reprinted, with permission, from Ryf, R., S. Randel, A. Gnauck, C. Bolle, A. Sierra, S. Mumtaz, M. Esmaeelpour, E. Burrows, R. Essiambre, P. Winzer, D. Peckham, A. McCurdy, and R. Lingle. 2012. Mode-division multiplexing over 96 km of few-mode fiber using coherent 6×6 MIMO processing. *IEEE Journal of Lightweight Technology* 30:521-531. (c) Bit error rate performance and recovered data constellations of the quadrature-phase-shift-keying (QPSK) signal tributaries after SDM transmission using four spatial modes in a single few-mode fiber. SOURCE: Reprinted, with permission, from Al Amin, Abdullah, An Li, Simin Chen, Xi Chen, Guanjun Gao, and William Shieh. 2011. Dual-LP11 mode 4x4 MIMO-OFDM transmission over a two-mode fiber. *Optics Express* 19:16672-16679.

This problem can be partially solved by utilizing a multiple-input, multiple-output (MIMO) approach. MIMO is a popular technique in radio-frequency (RF) systems and can untangle some of the crosstalk between modes using digital signal processing.[52,53] However, it is unclear as to the upper limit of performance and number of modes that can be supported.

Orthogonal Frequency Division Multiplexing (OFDM). OFDM uses a single carrier wave and multiple orthogonal subcarrier waves; it has produced some of the most exciting results in the RF world. Each subcarrier that carries independent data can be densely packed in the spectrum, since orthogonality is maintained by advanced electronic processing in the transmitter/receiver.[54] OFDM is spectrally efficient and is tolerant to many fiber-based impairments, but it does require a fair amount of high-speed electronic data signal processing, such as fast Fourier transforms and analog-digital converters.[55]

Device and Subsystem Achievements

These recent systems-level advances have been enabled by device and subsystem achievements, which include those described below.

High-Speed Local and Access Area Networks. The advances of local area networks (LANs) made it possible to deliver the high bandwidth from long-haul optical networks to desktops. Started with light-emitting diodes (LEDs) and Fabry-Perot lasers as transmitters, multi-mode optical fibers and fiber ribbon cables became major LAN standards in the mid-1990s. In the late 1990s, direct modulated gigabits

[52] Kaminow et al. 1996. A wideband all-optical WDM network.

[53] Ryf, R., A. Sierra, R. Essiambre, A. Gnauck, S. Randel, M. Esmaeelpour, S. Mumtaz, P.J. Winzer, R. Delbue, P. Pupalaikis, A. Sureka, T. Hayashi, T. Taru, and T. Sasaki. 2011. "Coherent 1200-km 6 x 6 MIMO Mode-Multiplexed Transmission over 3-CoreMicrostructured Fiber." Conference paper. European Conference and Exposition on Optical Communications. Available at http://www.opticsinfobase.org/abstract.cfm?URI=ECOC-2011-Th.13.C.1. Accessed June 26, 2012.

[54] Shieh, W., and I. Djordjevic. 2009. *Orthogonal Frequency Division Multiplexing for Optical Communications.* New York, N.Y.: Elsevier/Academic Press.

[55] Liu, X., S. Chandrasekhar, B. Zhu, P.J. Winzer, A.H. Gnauck, and D.W. Peckham. 2011. 448-Gb/s reduced-guard-interval CO-OFDM transmission over 2000 km of ultra-large-area fiber and five 80-GHz-Grid ROADMs. *IEEE Journal of Lightwave Technology* 29:483-490.

per second vertical cavity surface emitting laser (VCSEL)[56,57,58,59,60] transmitters became the dominant transmitters due to the low-drive voltage and low manufacturing cost. At present, 10 Gb/s VCSELs are being deployed and 40 Gb/s have been demonstrated in laboratories.[61,62] Research efforts toward wavelength-tunable and 100 Gb/s direct modulated lasers will be important for next-generation broadband communications.[63,64]

High-Speed Electronic Circuits. Electronic circuits are an "optical system's best friend." High-speed, high-output-power and linear electronics are clearly needed to drive the optical modulators, and high-speed logic circuits are crucial to achieving the signal processing in coherent, MIMO, and OFDM systems.[65,66] Good electronics can help mitigate many of the problems produced in the optical domain, thereby enabling better system performance.

Photonic Integration. In general, a system with better performance requires more components, which might make the system even more complex. For example, the transceiver in a system using higher-order modulation formats is much more com-

[56] Soda, H., K. Iga, C. Kitahara, and Y. Suematsu. 1979. GaInAsP/InP surface emitting injection lasers. *Japanese Journal of Applied Physics* 18:2329-2330.

[57] Watanabe, I., F. Koyama, and K. Iga. 1986. Low temperature CW operation of GaInAsP/InP surface emitting laser with circular buried heterostructure. *Electronics Letters* 22:1325-1327.

[58] Jewell, J.L., S.L. McCall, Y.H. Lee, A. Scherer, A.C. Gossard, and J.H. English. 1989. Lasing characteristics of GaAs microresonators. *Applied Physics Letters* 54:1400-1402.

[59] Chang-Hasnain, C.J., J.P. Harbison, G. Hasnain, A. Von Lehmen, L.T. Florez, and N.G. Stoffel. 1991. Dynamic, polarization, and transverse mode characteristics of vertical cavity surface emitting lasers. *IEEE Journal of Quantum Electronics* 27:1402-1409.

[60] Maeda, M.W., C.J. Chang-Hasnain, J.S. Patel, C. Lin, H.A. Johnson, and J.A. Walker. 1991. Use of a multi-wavelength surface-emitting laser array in a 4-channel wavelength-division-multiplexed system experiment. *IEEE Photonics Technology Letters* 3:268-269.

[61] Müller, M., W. Hofmann, A. Nadtochiy, A. Mutig, G. Böhm, M. Ortsiefer, D. Bimberg, and M.-C. Amann. 2010. B1.55 m high-speed VCSELs enabling error-free fiber-transmission up to 25 Gbit/s. Proceedings of International Semiconductor Laser Conference (ISLC), September 26-30, Kyoto, Japan, pp. 156-157.

[62] Westbergh, P., J.S. Gustavsson, B. Kögel, A. Haglund, and A. Larsson. 2011. Impact of photon lifetime on high-speed VCSEL performance. *IEEE Journal of Selected Topics in Quantum Electronics* 17(6):1603-1613.

[63] Chang-Hasnain, C.J. 2000. Tunable VCSEL. *IEEE Journal of Selected Topics in Quantum Electronics* 6(6):978-987.

[64] Taubenblatt, M.A. 2012. Optical interconnects for high-performance computing. *Journal of Lightwave Technology* 30(4):448-457.

[65] Li, T.Y. 1993. The impact of optical amplifiers on long-distance lightwave telecommunications. *Proceedings of the IEEE* 81:1568-1579.

[66] Kurtzke, C. 1993. Suppression of fiber nonlinearities by appropriate dispersion management. *IEEE Photonics Technology Letters* 5:1250-1253.

plicated than an on-off keying (OOK) system,[67] because it requires multiple lasers, modulators, couplers, and balanced detectors, as seen in Figure 3.4. This scenario has benefited greatly from advances in photonic integrated circuits on both III-V[68] and silicon materials.[69]

Optical Fiber Modifications. Although the attenuation of optical fiber reached its near-optimum point of 0.2 decibels per kilometer (dB/km) decades ago, modifications to the fiber itself have produced critical changes to the optical wave propagation properties that enabled systems advances. By varying the core and cladding size and composition, key changes in the amount and spectral slope of chromatic dispersion have managed the accumulation of dispersion and nonlinear effects.[70] Moreover, large-effective-area fiber has reduced the power density, thereby reducing the build-up of nonlinear effects.[71] Furthermore, creating nanostructures in the cladding of the fiber has enabled the fiber to be very tightly wound without incurring bending losses, thus enabling ubiquitous deployment of fiber without the need for highly skilled technicians.

Potential Innovation. Device and subsystem innovations that herald the next wave of communication systems could include those described below.

- *Silicon photonics.* At the time of the publication of *Harnessing Light* in 1998,[72] the community did not pay much attention to silicon as a photonics material. In terms of performance and the ability to interact with light, III-V materials were far superior in performance. However, significant advances in the past few years in creating silicon-based photonic elements have excited the industry. Modulators, waveguides, couplers, detectors, multiplexers and filters have all been demonstrated.[73] (See Figure 3.5.)

---

[67] OOK is the simplest form of amplitude-shift keying (ASK) modulation, in which digital data are represented by the presence or absence of a carrier wave.

[68] Welch, D.F., F.A. Kish, R. Nagarajan, C.H. Joyner, R.P. Schneider, V.G. Dominic, M.L. Mitchell, S.G. Grubb, T.-K. Chiang, D. Perkins, and A.C. Nilsson. 2006. The realization of large-scale photonic integrated circuits and the associated impact on fiber-optic communication systems. *Journal of Lightwave Technology* 24:4674-4683.

[69] Asghari, M., and A.V. Krishnamoorthy. 2011. Silicon photonics: Energy-efficient communication. *Nature Photonics* 5:268-270.

[70] Miller, D.A.B. 2009. Device requirements for optical interconnects to silicon chips. *Proceedings of the IEEE* 97:1166-1185.

[71] Lewotsky K. 1996. Large-effective-area fiber minimizes nonlinearities. *Laser Focus World* 32:16.

[72] National Research Council. 1998. *Harnessing Light.*

[73] Xu, Q., B. Schmidt, S. Pradhan, and M. Lipson. 2005. Micrometre-scale silicon electro-optic modulator. *Nature* 435:325-327.

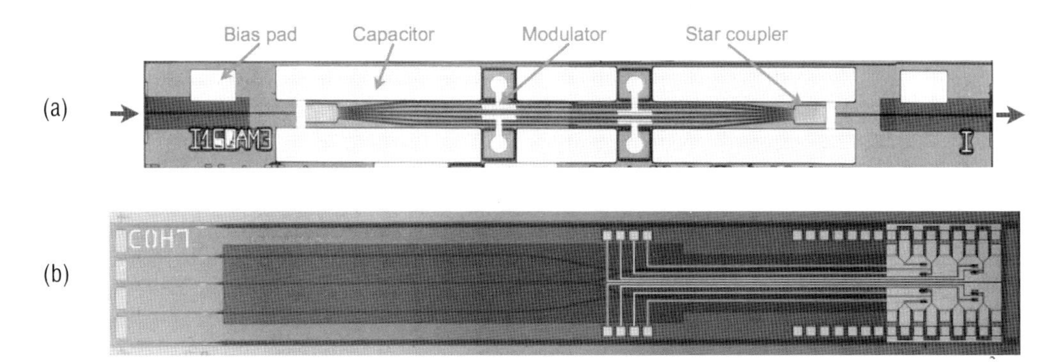

FIGURE 3.4 (a) 28-Gbaud InP square or hexagonal 16-QAM modulator. SOURCE: Reprinted, with permission, from Doerr, C.R., L. Zhang, P. Winzer, and A.H. Gnauck. 2011. "28-Gbaud InP Square or Hexagonal 16-QAM Modulator." Conference paper. Optical Fiber Communication Conference (OFC), Los Angeles, Calif., March 6, 2011. Available at http://www.opticsinfobase.org/abstract.cfm?URI=OFC-2011-OMU2. (b) Monolithically integrated InP dual-port coherent receiver for 100-Gb/s PDM-QPSK signal. SOURCE: Reprinted, with permission, from Houtsma, V.E., N. Weimann, T. Hu, R. Kopf, A. Tate, J. Frackoviak, R. Reyes, Y. Chen, C.R. Doerr, L. Zhang, and D. Neilson. 2011. "Manufacturable Monolithically Integrated InP Dual-Port Coherent Receiver for 100G PDM-QPSK Applications." Conference paper. Optical Fiber Communication Conference (OFC), Los Angeles, Calif., March 6, 2011. Available at http://ieeexplore.ieee.org/xpls/abs_all.jsp?arnumber=5875296.

Even active elements such as lasers have been reported by wafer bonding[74] or direct growth.[75,76] This makes it possible to mass-produce cost-effective, chip-scale integrated photonic circuits by using the massive silicon manufacturing infrastructure. An important application is in the mass production of active cables, in which silicon photonics and fiber take the place of coaxial cable to provide tens of gigahertz of bandwidth with extremely low loss over 10s of meters.

- *Low power consumption.* As data communications increase at an exponential rate, the power consumption by the communication infrastructure is growing rapidly. Moreover, it has been well known that although electronic power consumption scales with increased data rate, the power consumption of photonics does not. In view of both of these facts, the optical communications community is taking very seriously the mantle of using optics in novel ways to dramatically reduce the power consumption. This will require new integrated device technologies, such as photonic-crystal

[74] Liang, D., and J.E. Bowers. 2010. Recent progress in lasers on silicon. *Nature Photonics* 4:511-517.

[75] Kunert, B., S. Zinnkann, K. Volz, and W. Stolz. 2008. Monolithic integration of Ga(NAsP)/(BGa)P multi-quantum well structures on (0 0 1) silicon substrate by MOVPE. *Journal of Crystal Growth* 310(23):4776-4779.

[76] Chen et al. 2011. Nanolasers grown on silicon.

(a)

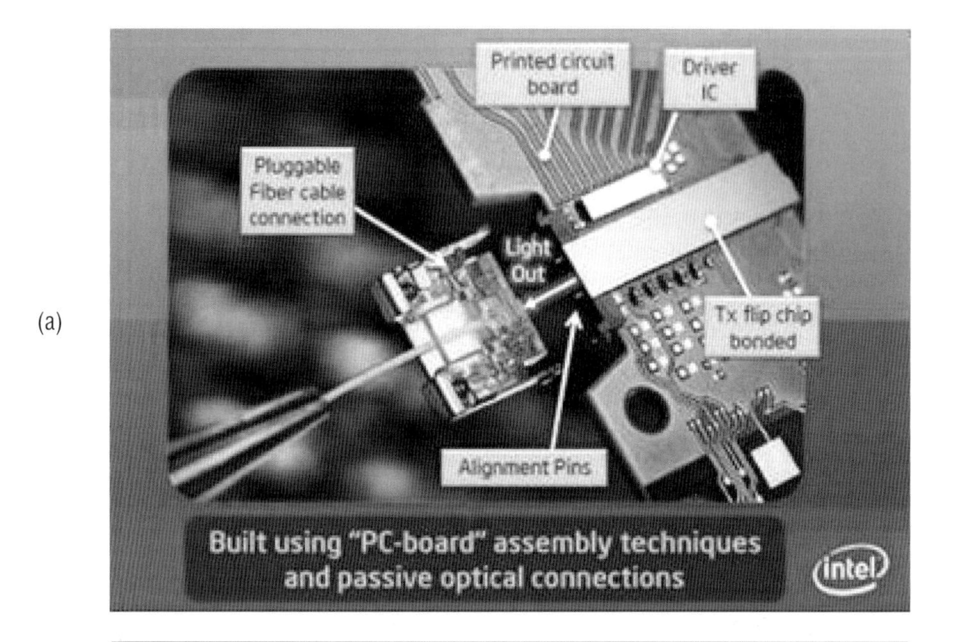

(b)

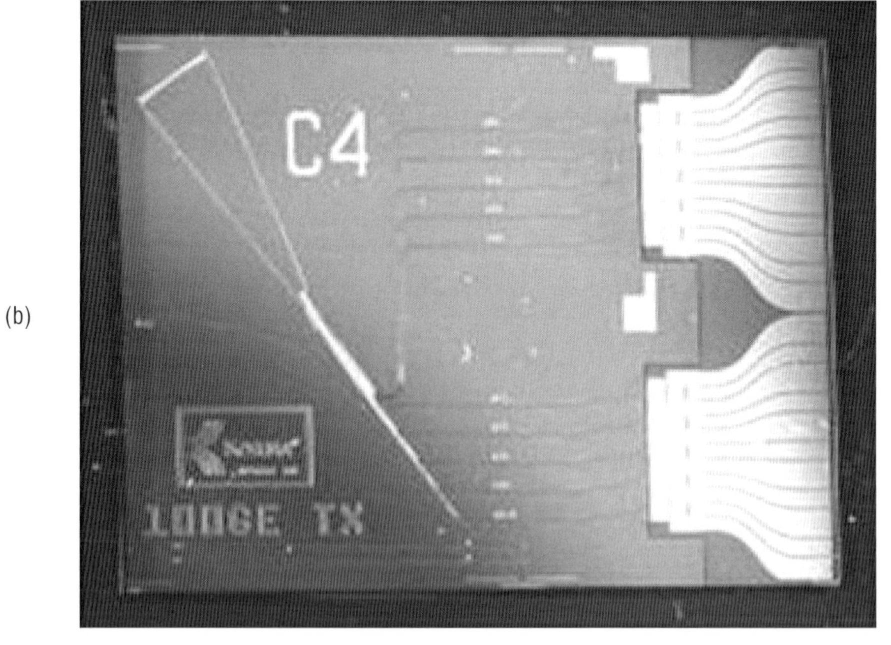

FIGURE 3.5 (a) Intel creates the world's first end-to-end silicon photonics connection with integrated lasers. SOURCE: Image available at http://www.demonstech.com/2010/07/intel-creates-worlds-first-end-to-end.html. (b) Kotura's chip-scale 10×10 Gb/s integrated silicon transmitter. SOURCE: Image available at http://www.gazettabyte.com/home/2009/10/26/photonic-integration-bent-on-disruption.html. Reprinted with permission.

structures, as well as novel system and switching architectures to create a "green" network.[77,78]

- *Reconfigurable network elements.* Electronic systems are not static, and yet optical systems tend to be deployed in a fairly static fashion. Tunable and reconfigurable network elements will provide needed flexibility to optical systems, such that the network can optimize its bandwidth allocation and quality of service in a robust fashion.[79,80] Filters, (de)multiplexers, routers, switches, dispersion compensators, and lasers all fall into this category.[81]

Networking

The topics discussed above deal primarily with the core issues of increasing the capacity and distance performance of optical communications systems—the solutions to this need has fueled the dramatic momentum of worldwide communications. However, the power of optics has also been unleashed to enable dramatic advances in multi-user networking. Not only do parallel wavelengths readily provide capacity enhancement, they also enable wavelength-dependent routing.[82] In a WDM network, the wavelength can define the network routing such that a specific wavelength channel can appear at its destination by traversing wavelength-dependent network elements. A network node can detect only the specific wavelength and traffic meant for that node and pass the other wavelengths not meant for that destination without disturbance. Key benefits include these: (1) lower network delay and latency for non-dropped traffic, and (2) lower "cost" to detect only the dropped traffic rather than all the traffic entering the node.

A key element in this network revolution is the wavelength-selective optical add/drop multiplexer (OADM). Moreover, the OADM has evolved to become dy-

---

[77] Nishimura, S., K. Shinoda, Y. Lee. G. Ono, K. Fukuda, F. Yuki, T. Takemoto, H. Toyoda, M. Yamada, S. Tsuji, and N. Ikeda. 2011. Components and interconnection technologies for photonic-assisted routers toward green networks. *IEEE Journal of Lightwave Technology* 17:347-356.

[78] Kilper, D. 2011. "Tutorial: Energy Efficient Networks." Conference paper. Optical Fiber Communication Conference (OFC), Los Angeles, Calif., March 6. Available at http://www.opticsinfobase.org/abstract.cfm?uri=OFC-2011-OWI5. Accessed June 26, 2012.

[79] Gringeri, S., B. Basch, V. Shukla, R. Egorov, and T.J. Xia. 2010. Flexible architectures for optical transport nodes and networks. *IEEE Communications Magazine* 48(7):40-50.

[80] Saleh, A.A.M., and J.M. Simmons. 2011. Technology and architecture to enable the explosive growth of the internet. *IEEE Communications Magazine* 49(1):126-132.

[81] Poole, S., S. Frisken, M. Roelens, and C. Cameron. 2011. "Bandwidth-Flexible ROADMs as Network Elements." Conference paper. Optical Fiber Communication Conference (OFC), Los Angeles, Calif., March 6, 2011. Available at http://www.opticsinfobase.org/abstract.cfm?URI=OFC-2011-OTuE1. Accessed June 26, 2012.

[82] Feuer, M.D., S.L. Woodward, P. Palacharla, X. Wang, I. Kim, and D. Bihon. 2011. Intro-node contention in dynamic photonic networks. *Journal of Lightwave Technology* 29(4):529-535.

namic, such that a reconfigurable OADM (ROADM) can accommodate changing traffic patterns.[83] A myriad of wavelength-selective and -tunable components have enabled such advances.[84,85] The speed of reconfiguration might be as long as once a day or as short as nanoseconds, depending on the application.

These applications include circuit and packet switching. Circuit switching has limited speed requirements and is considered fairly easy to implement but is somewhat inefficient in terms of capacity utilization. As bandwidth becomes more precious in the network, packet switching, which is common in today's Internet, has fast switching requirements and is much harder to implement optically but is quite capacity-efficient. In the 5- to 10-year horizon, it is likely to become clearer as to the specific technologies that might be used for optical packet switching as well as the perceived benefits of implementing it in the optical domain.

As time has passed, optical fiber has moved ever closer to the end user. Even 30 years ago, it was clear that optical fiber would dominate long-distance communications. At present, optical fiber is being deployed around the world to enable individual users to access the broadband Internet. Fiber to the x, in which x can represent many things, including curb, home, or office, will only accelerate as costs decrease.[86] In fact, access networks are quite sensitive to cost, given that only a few users will share the cost of the optical components, which are generally assumed to be more expensive at present than electronic ones. An exciting area of R&D is in passive optical networks, such that gigabits per second data rates can be deployed at low cost to many simultaneous users in a neighborhood.[87] Advances in low-cost integrated optical components and efficient optical access architectures will be crucial to achieving this vision; note that RF-over-fiber, in which a radio-frequency subcarrier is encoded and is transmitted on an optical carrier wave, represents an interesting and potentially exciting approach to bridge the gap between wireless signals and WDM networks.[88] It should be emphasized that the exponential growth in wireless traffic only accelerates the need for optics in access networks, since wireless hubs will typically connect in the ground to the larger network through optical fiber. Clearly, the optics and photonics community needs

[83] Lewotsky, K. 1996. Large-effective-area fiber minimizes nonlinearities. *Laser Focus World* 32:16.

[84] Asghari, M., and A.V. Krishnamoorthy. 2011. Silicon photonics: Energy-efficient communication. *Nature Photonics* 5:268-270.

[85] Lewotsky, K. 1996. Large-effective-area fiber minimizes nonlinearities. *Laser Focus World* 32:16.

[86] Lee, C., W.V. Sorin, and B. Kim. 2006. Fiber to the home using a PON infrastructure. *IEEE Journal of Lightwave Technology* 24:4568-4583.

[87] Skubic, A., J. Chen, J. Ahmed, L. Wosinska, and B. Mukherjee. 2009. A comparison of dynamic bandwidth allocation for EPON, GPON, and next-generation TDM PON. *IEEE Communications Magazine* 47(3):S40-S48.

[88] Chen, L., K. Preston, S. Manipatruni, and M. Lipson. 2009. Integrated GHz silicon photonic interconnect with micrometer-scale modulators and detectors. *Optics Express* 17(17):15248-15256.

to bring forward technologies for the next-order-of-magnitude capacity increases in optical networks very soon.

R&D Example Areas

In general, the field of optical communications is quite rich in terms of innovations and applications. Brief descriptions of example areas of R&D that have the potential for large impact include the following:

- *Quantum communications.* Quantum optical communications typically use the quantum state of a single photon or a pair of entangled photons—which can be widely separated while maintaining their inherent quantum connection—in order to achieve ultimately secure and highly efficient communications.[89] For example, if information is encoded on the photon's state of polarization (SOP), then any eavesdropper would necessarily disturb the photon's SOP and be detected. Quantum key distribution (QKD) might be extremely important for links requiring ultimate security. Moreover, for systems in which single photons are carrying information, the bits per photon can be quite efficient. Practical single-photon transmitters and detectors are being actively pursued, and the quantum amplifier/repeater remains an elusive research goal.[90,91]
- *Free-space communications.* Free-space communications have existed for decades as radio. However, there has been the desire, also for decades, to make use of optics in free-space communications due to its much higher frequency range and data transmission capacity.[92] Lasers are also highly directional, such that they are power-efficient in terms of divergence and are relatively secure since they are difficult to intercept. Especially for satellite communications at higher bandwidths, there is the potential for dramatic decreases in the crucial parameters of size, weight, and power (SWaP). Satellite or high-altitude free-space laser communications are attractive because of the lack of weather attenuation issues, even at optical wavelengths. The Department of Defense is especially interested in free-space laser commu-

[89] Treiber, A., A. Poppe, M. Hentschel, D. Ferrini, T. Lorünser, E. Querasser, T. Matyus, H. Hübel, and A. Zeilinger. 2009. A fully automated entanglement-based quantum cryptography system for telecom fiber networks. *New Journal of Physics* 11:045013. Available at http://iopscience.iop.org/1367-2630/11/4/045013. Accessed June 26, 2012.

[90] Kimble, H.J. 2008. The quantum Internet. *Nature* 453:1023-1030.

[91] Gisin, N., and R.T. Thew. 2010. Quantum communication technology. *Electronics Letters* 14: 965-967.

[92] Killinger, D. 2002. Free space optics for laser communication through the air. *Optics and Photonics News* 13(10):36-42.

nications, as discussed in Chapter 4 of this report, because of the need for high-bandwidth communication with moving platforms.

Information Processing

In the technology for optics for interconnection, the past decade has seen significant progress in three main areas:

- Evolutionary advances in optoelectronics made from III-V semiconductors,
- The emergence of silicon photonics as a potential new technology base, and
- Nanophotonics for very-high-performance optics and optoelectronics, some with quite novel capabilities and opportunities.

Evolutionary Progress in III-V Device Technology

Technology for optical data links continued to evolve in the last decade, with continued improvements in speed and reduced power dissipation for device technologies such as VCSELs. Integration of devices in larger functional units has continued in III-V technology.[93] Such predominantly III-V approaches can integrate multiple active optoelectronic devices, such as lasers, optical amplifiers, modulators, switches, and detectors for highly functional units for optical networking.

Silicon Photonics

The potential for the use of silicon-based optical technologies (so-called silicon photonics, as mentioned above) has advanced substantially.[94] Such silicon photonics can be manufactured in silicon complementary metal oxide semiconductor (CMOS) fabrication facilities, thereby potentially reducing manufacturing costs and allowing direct integration with low-cost silicon electronics. Such an approach offers serious possibilities for an integrated electronics/optoelectronics/optics platform that could address problems of manufacturability and cost for future high-volume applications such as local optical networks or interconnects.

Silicon itself and/or related CMOS materials such as silicon nitride or silicon dioxide can all be used as very effective optical waveguides on the surface of a

[93] Blumenthal, D.J., J. Barton, N. Beheshti, J.E. Bowers, E. Burmeister, L.A. Coldren, M. Dummer, G. Epps, A. Fang, Y. Ganjali, J. Garcia, B. Koch, V. Lal, E. Lively, J. Mack, M. Masanovic, N. McKeown, K. Nguyen, S.C. Nicholes, H. Park, B. Stamenic, A. Tauke-Pedretti, H. Poulsen, and M. Sysak. 2011. Integrated photonics for low-power packet networking. *IEEE Journal of Selected Topics in Quantum Electronics* 17:458-471.

[94] Soref, R. 2006. The past, present, and future of silicon photonics. *IEEE Journal of Selected Topics in Quantum Electronics* 12:1678-1687.

silicon chip. Additionally, very close integration of devices such as photodetectors and output devices has the potential to improve system performance, especially when power dissipation is particularly important, as it is in optical interconnect applications.

Optical modulators using silicon as the active material can be made in linear (e.g., Mach-Zehnder) forms that have been integrated into products or in more aggressive high-quality-factor ring-resonator forms that also can allow the selection and switching of different wavelength channels. Ring resonators are still largely in research, in part because of the requirements for precise tuning and thermal stabilization. Most silicon photonics work is targeted at the telecommunications wavelengths (e.g., in the so-called C band near 1550 nanometers [nm]) where silicon itself is transparent. Incorporating photodetectors requires the addition of germanium, which absorbs light efficiently in some of the C band. Germanium is already known to be compatible with silicon CMOS technology, as it is already in use for other purposes in modern CMOS systems.

Other silicon-compatible modulator concepts based on optical absorption changes in germanium structures are starting to emerge from research. These devices might offer very low operating energies without the tuning of resonators. Approaches include Franz-Keldysh bulk germanium modulators[95] and, more aggressively, so-called Quantum-Confined Stark Effect (QCSE) modulators[96] based on the use of thin germanium "quantum well" layers.

At the time of this writing, there is still no room-temperature-operated electrically-driven laser that can be grown directly on silicon, although research is continuing on various approaches, including attempts to make germanium or related materials (such as germanium-tin alloys) into the so-called direct bandgap materials that are generally required for efficient lasers. Lasers have been successfully demonstrated based on the hybridization of III-V materials onto silicon after growth.[97] Viable III-V light sources on silicon by direct growth onto the

[95] Liu, J., M. Beals, A. Pomerene, S. Bernardis, R. Sun, J. Cheng, L.C. Kimerling, and J. Michel. 2008. Waveguide-integrated, ultralow-energy GeSi electro-absorption modulators. *Nature Photonics* 2:433-437.

[96] Roth, J.E., O. Fidaner, E.H. Edwards, R.K. Schaevitz, Y.-H. Kuo, N.C. Helman, T.I. Kamins, J.S. Harris, and D.A.B. Miller. 2008. C-band side-entry Ge quantum-well electroabsorption modulator on SOI operating at 1 V swing. *Electronics Letters* 44:49-50.

[97] Fang, A.W., H. Park, O. Cohen, R. Jones, M.J. Paniccia, and J.E. Bowers. 2006. Electrically pumped hybrid AlGaInAs-silicon evanescent laser. *Optical Express* 14:9203-9210.

silicon are still a subject of research, although some nanophotonic structures offer promise.[98,99]

It is an open question whether the laser light sources need to be integrated on the silicon. Use of off-chip lasers removes some power dissipation from the chip and allows centralized stabilization of wavelengths and/or pulse timing, at the expense of having to align the laser light to the chip with some additional optics. Approaches exist for hybridizing III-V materials onto silicon substrates and/or optical structures, and such approaches offer one intermediate approach for integrating light sources on the chip. However, amplifiers on the chip may be very desirable in order to increase functionalities, given that the Si-based optical components are still fairly lossy, which limits the number of components that can be integrated. This argues for monolithic integration of active components with optical gain.

Although the various output device options (modulators and lasers) for direct modulation are still emerging from research, they do offer serious possibilities for devices that could meet the energy per bit requirements[100] to allow optics to replace wires even at relatively short distances such as chip-to-chip interconnects. It is conceivable that one technology platform could emerge here that would allow the manufacture of systems for applications ranging from chip-to-chip all the way to long-distance optical networks, but much research and technology development remain to be done.

Nanophotonics

The ability to controllably fabricate structures on a deeply sub-wavelength scale—for example, using the same lithographic techniques employed for silicon CMOS electronics—has opened up a broad range of possibilities in optics and optoelectronics. Areas include photonic crystal structures, metamaterials, nanometallics/plasmonics, nanoresonators, nanolasers, and new classes of optical components based on sub-wavelength nonperiodic design.

The basic physics of various novel nanophotonic structures has advanced substantially. Dielectric nanoresonators of various kinds achieved very high quality factors. Nanometallic structures have allowed the concentration of light to very small, sub-wavelength volumes and the guiding of light in sub-wavelength circuits.[101] Metamaterials, based on sub-wavelength patterning, allow the engineering of effective materials with optical properties unlike those of conventional

---

[98] Kunert, B., S. Zinnkann, K. Volz, and W. Stolz. 2008. Monolithic integration of Ga(NAsP)/(BGa)P multi-quantum well structures on (0 0 1) silicon substrate by MOVPE. *Journal of Crystal Growth* 310(23):4776-4779.

[99] Chen et al. 2011. Nanolasers grown on silicon.

[100] Miller, D.A.B. 2009. Device requirements for optical interconnects to silicon chips.

[101] Brongersma, M.L., and V.M. Shalaev. 2010. The case for plasmonics. *Science* 328:440-441.

materials.[102] Novel nanophotonics approaches to optical components such as very compact wavelength splitters have emerged, including the concept of nonperiodic design for function.[103] Extremely sensitive optical and optoelectronic device structures have been demonstrated, including lasers, photodetectors, modulators, and switches, with some down to single-photon sensitivity.[104] With such nanophotonics approaches, there is demonstrated potential for very low threshold lasers,[105] very low energy modulators, and very low capacitance photodetectors. Novel approaches to optoelectronic devices exploiting nanoscale growth techniques may circumvent some of the difficulties (such as crystal lattice constant matching) that often limit more conventional approaches to device fabrication.[106] As the size of many of these new optics and photonics elements shrinks, it certainly opens up for the possibility to produce systems that utilize the best of optics and electronics to enable integrated systems to seamlessly provide solutions in many of today's fields.

Data Storage

The advent of nanometallic structures to concentrate light to deeply sub-wavelength volumes has opened up new opportunities for optics in data storage, both for sub-wavelength optical reading and writing and for the use of optics to concentrate light for other storage approaches, as in heat-assisted magnetic recording, whereby the light provides the very localized heating above the Curie temperature to enable correspondingly localized changes in magnetic state. This emerging technology is apparently a serious contender for future mainstream magnetic hard-drive technologies.[107]

---

[102] Chen, H., C.T. Chan, and P. Sheng. 2010. Transformation optics and metamaterials. *Nature Materials* 9:387-396.

[103] Liu, V., Y. Jiao, D.A.B. Miller, and S. Fan. 2011. Design methodology for compact photonic-crystal-based wavelength division multiplexers. *Optics Letters* 36:591-593.

[104] Fushman, I., D. Englund, A. Faraon, N. Stolz, P. Petroff, and J. Vuckovic. 2008. Controlled phase shifts with a single quantum dot. *Science* 320(5877):769-772.

[105] Ellis, B., M.A. Mayer, G. Shambat, T. Sarmiento, J. Harris, E.E. Haller, and J. Vučković. 2011. Ultralow-threshold electrically pumped quantum-dot photonic-crystal nanocavity laser. *Nature Photonics* 5:297-300.

[106] Chen et al. 2011. Nanolasers grown on silicon.

[107] Kryder, M.H., E.C. Gage, T.W. McDaniel, W.A. Challener, R.E. Rottmayer, G. Ju, Y.-T. Hsia, and M.R. Erden. 2008. Heat assisted magnetic recording. *Proceedings of the IEEE* 96:1810-1835.

MANUFACTURING

Communications

Optical communications network equipment includes functional elements such as transmissions systems; optical networking elements, including add/drop multiplexers; crossconnects; and network-management software systems. Substantial numbers of circuit boards for electrical multiplexing, electrical cross-connects, and control functions are also required. These optical systems and network elements are assembled from optical modules (for example, optical transmitters and receivers) that in turn are built from optical components, including lasers, optical modulators, photodetectors, optical fiber amplifiers, and others.

At the systems' vendor level, which is a roughly $15 billion annual business, there have been significant changes in the industry since *Harnessing Light*[108] was written. Nortel, a major Canadian vendor, has gone out of business. Lucent Technologies has been acquired by Alcatel to form Alcatel-Lucent, with headquarters in Paris, France. Of the top 10 optical network systems vendors, two—Cisco and Tellabs—are headquartered in the United States. Alcatel-Lucent, with significant R&D operations in the United States, and Huawei, headquartered in Shensen, China, have vied for the market-leader position over the last several years. ZTE, also from China, is also in the top 5.[109] In the optical module and component business, nearly half of the market share is owned by U.S.-based companies: Finisar and JDSU are the market number one and number two leaders, respectively.[110] The growth and fabrication of the III-V semiconductor chips to build the basic laser and detector discrete devices are done in the United States, Japan, and Taiwan and are increasingly in areas like Singapore and Southeast Asia. There is little evidence that companies in China have mastered this highly specialized technology at this time, but work in the research laboratory is ongoing there.

At the circuit board and optical module level, manufacturing assembly is increasingly being done in Asia. However, some U.S. companies continue to do assembly of leading-edge boards—such as 100 Gb/s transmitters and receivers—in the United States. Infinera, a metropolitan and long-haul optical networking company based in the United States and a leading proponent of photonic integrated circuits (PICs) to provide transmitter and receiver arrays for WDM transmission

---

[108] National Research Council. 1998. *Harnessing Light.*

[109] Ovum. 2011. "Ovum: ZTE Ranked Third in Growing Global Optical Networking Market." Ovum. Available at http://www.tele.net.in/news-releases/item/7010-ovum-zte-ranked-third-in-growing-global-optical-networking-market. Accessed July 26, 2012.

[110] Ovum. 2012. "Oclaro Combines with Opnext to Challenge Finisar for No. 1." Available at http://ovum.com/2012/03/27/oclaro-combines-with-opnext-to-challenge-finisar-for-no-1/. Accessed July 26, 2012.

systems, produces the PIC chip in the United States. Because PICs provide on-chip integration of basic components that otherwise would be done by hand, this technology eliminates much of the manually intensive activity of assembling and connecting individual components that is currently done more cost-effectively in low-labor-cost regions. More highly integrated optical modules will be needed for next-generation systems from a functional point of view. By being the technology leader in this area, the United States could potentially increase the value added in the United States that, with today's modules made up of discrete devices, is moving offshore.

Optical components for communications networks still involve several technologies—indium phosphide (InP), CMOS, lithium niobate, and silica. At the same time, volumes addressed by each technology are rather limited. The company and/or nation that finds a consolidation technology (if possible) could have substantial advantage. Some might suggest that silicon photonics has the potential to displace other passive technologies as well as non-lasing electro-optic technologies such as lithium niobate.

Also worth noting is anecdotal evidence which suggests that as the manufacturing of optical systems equipment or optical modules moves to lower-cost manufacturing regions, the development function tends eventually to move to that region as well.

Information Processing

Current optoelectronic systems are dominated by III-V semiconductors in systems with generally small integration levels. To make the transition to a new technology such as silicon-based photonics with high integration levels is difficult, and it appears to need high volumes to justify it.[111] There is, therefore, a "chicken and egg" issue with this major potential technology shift. There is substantial medium-to-long-term potential for a very high volume market for optical interconnections, one that could help retain a U.S. technology and market leadership in information processing equipment, but it may need a shift to a low-cost, possibly silicon-based technology to satisfy that market. The same silicon photonics platform, if successfully adopted, also has the capability to create a broad range of nanophotonic structures.

[111] Fuchs, E., R. Kirchain, and S. Liu. 2011. The future of silicon photonics—Not so fast?: The case of 100G Ethernet LAN transceivers. *Journal of Lightwave Technology* 29(15):2319-2326.

Data Storage

At present, the substantial majority of manufacturing for optical data storage products, such as DVDs, is likely in Asia.[112] Given that the impact of a next generation of optical disk storage faces strong competition from networks as a distribution medium for video, it is not clear that a major push to manufacture such a technology in the United States is a strong choice. There is significant research in the United States in future areas such as the HAMR optically assisted hard-drive technology, which may influence some future manufacture of such magnetic hard drives in the United States.

At this time, beyond the strong possibilities in HAMR and some continuing use of optical disks such as DVDs, the path for other optical data storage approaches is less clear, although there is continuing development of holographic storage for commercial archiving, with possible capacities in the 500 Gb to 1 Tb range in a disk of a size comparable to that of current DVDs.[113] Other approaches remain in research at the present.

ECONOMIC IMPACT

Communications networks have taken on a role well beyond people-to-people voice communications; they provide the information and trade routes of the new global digital economy. They will also likely provide the sensor integration and distributed control for the power grid, transportation, and freight networks and smart enterprises and cities in the future. Optical interconnects are very likely to have a key role in future information processing technology and systems, allowing performance to continue to scale. Leadership in communications networks and interconnects, including the underpinning optical networks, both in the R&D and in commercialization, deployment, and effective use, will be absolutely essential to maintaining and enhancing economic growth and improving the quality of people's lives in the 21st century. These optical technologies will be increasingly indispensable for the technology of the information age.

This critical importance of communications networks to the future has driven action in countries around the world. Several countries, including Australia for example, have mandated broadband fiber to the home. Europe continues to pro-

---

[112] Esener S.C., M.H. Kryder, W.D. Doyle, M. Keshner, M. Mansuripur, and D.A. Thompson. 2012. International Technology Research Institute, World Technology (WTEC) Division. 1999. *WTEC Panel Report on the Future of Data Storage Technologies*. Available at http://www.wtec.org/pdf/hdmem.pdf. Accessed August 1, 2012.

[113] More information is available at General Electric's Global Research website at http://ge.geglobalresearch.com/blog/breakthrough-in-micro-holographic-data-storage/. Accessed June 26, 2012.

vide substantial research funding, as do Korea and Japan. However, the single most important threat to U.S. leadership, especially with respect to network equipment development and manufacture, is likely to come from China, where the government-sponsored Huawei ranks as number 1 or number 2 in global market share,[114] followed closely by ZTE, another systems vendor from China. Already extremely innovative in its products, China is now focused on research and innovation. Its paper submissions to international optics journals and to the premier global optical communications conferences have increased substantially in the last 5 years.[115] There is particularly strong research in Europe and recently also in Japan in the emerging silicon photonics platforms for networks and interconnects.

Given that labor-intensive industries will continue to migrate their manufacturing to low-labor-cost regions, it is imperative that the United States stay at the leading edge of optical technology at the component, platform, and system levels. This approach appears to have served the U.S. electronics industry well. To keep a substantial portion of the value chain in the United States, it will be important for U.S. enterprises to have the critically enabling intellectual property that provides a barrier of entry without financial compensation. To have that intellectual property, it is essential to be at the leading edge of enabling the fundamental and applied research. Suggested areas of focus on the component side include very high speed electronics and optical components, including modulators and detectors at operating rates of 400 Gb/s and 1 Tb/s; advanced signal processing to overcome transmission impairments for coherent systems; and on-chip integration that provides increased functionality while reducing size, power consumption, and cost. Such integrated chips also do not require the substantial manual assembly that is now being performed in low-labor-cost regions. In the systems and networks area, finding a new approach to cost-effectively achieve several-orders-of-magnitude long-haul and metropolitan distance transmission capacity and to rapidly get that technology to market will be critical in order for the United States to maintain a strong global leadership position. Such technology evolution will also underpin the increasing move to optics inside information processing that will be essential for the continued scaling of an information-driven economy. The successful development of an integrated platform technology, such as some version of silicon photonics that can service a broad range of applications and integration with electronics, could be a major enabler for U.S. economic impact.

---

[114] Huawei. 2010. Milestones. Available at http://www.huawei.com/en/about-huawei/corporate-info/milestone/index.htm. Accessed July 26, 2012.

[115] Cao, J. 2012. A new journal in optics and photonics—*Light: Science and Applications*. Editorial. *Light: Science and Applications* 1:Online. Available at http://65.199.186.23/lsa/journal/v1/n3/full/lsa20123a.html. Accessed July 26, 2012.

Comparison Between the United States and the Rest of the World

For decades the United States has been the envy of the entire world in terms of R&D in high-tech fields, including optics and photonics for communications. For decades the most prestigious international journals, conferences, and professional societies have been based in the United States. Many of the most impactful R&D advances came from the U.S. industrial laboratories (e.g., AT&T Bell Laboratories, Corning, RCA Sarnoff Labs, IBM), with several Nobel Prizes as well. The U.S. university system was unrivaled, with student researchers flocking to the United States from countries around the world. The combined R&D in optics from corporate laboratories and federally funded university labs set the bar high in terms of quality and quantity. Critical advances were made in the United States, including low-loss fibers, semiconductor lasers, optical amplifiers, and information theory.

The scenario has changed somewhat in the past 20 years. In general, corporate laboratories no longer enjoy steady, long-term funding, and the main sources of university research funding have not grown at a pace consistent with that in the rest of the world. The main sources of university funding in communications and information processing, such as the NSF and the Department of Defense (DOD), tend to fund a relatively small percentage of new proposals, particularly with a 10-year focus.

In the area of optical research for information processing, U.S. efforts are generally comparable in size to efforts in Europe and Japan in the emerging technologies for optics in information processing, but the United States has no overall lead against these research competitors. The United States lacks the larger European framework projects that help tie together a broad range of players from academia to industry. Both Europe and Japan are making substantial investments in silicon photonics technology research now.

Trends in terms of the United States in relation to the rest of the world include the following:

- *Research funding.* Nations all over the world view the United States as a benchmark in terms of research funding. In order for countries to compete, many nations focus on specific, strategic areas in which to invest in long-term photonics research. The United States tends not to make such strategic, long-term "bets." The effect is that in any given area, the United States will share leadership with, and perhaps be surpassed by, research in specific countries. High-speed communications in Germany, integrated photonics in Japan, and access technologies in China are all examples of thrusts centered in various countries.
- *Professors.* Several nations have invested in hiring distinguished researchers to prominent, well-paid professorial positions. Oftentimes, these research-

ers have made their reputations in the United States, only to be drawn away in their prime. Countries including Australia, Canada, China, and Germany have hired excellent researchers in communications with generous research funding. This trend, if it accelerates, could have a profound impact.

- *China.* When observing statistics of prestigious optical communications journals, it is undeniable that the quantity of research from China is growing rapidly. Looking at "optical communication" in the Scopus database, there are 15,003 journal publications from 1969 to 2012 from 21 common optical communications journals;[116] in total, the United States represents 3,909 of those, and China, 1,303. However, since 2000 the U.S. scientists have published 42 percent of their work, or 1,642 publications, whereas Chinese scientists have published 85 percent of their work, or 1,108 publications. Most of the Chinese publications have appeared since 2000. Moreover, the quality gap, according to rejection ratio statistics, is narrowing. Using the number of citations that a paper receives as a metric of quality also points to China's gaining on the United States. A comparison between the two countries with respect to publications related to the three keywords (Optics, Photonics, and Communication) in the Scopus database, using the highest-cited papers, gives the following results: currently there are 2.3 citations of a U.S. paper for each citation that a Chinese paper receives; however, 5 years ago this number was 2.8, or 16 percent higher.[117] Furthermore, China is aggressively funding research in optical communications and has a foothold as home to some of the world's largest communications companies. China is poised to make great strides in the coming decade.

FINDINGS AND CONCLUSIONS

In spite of great progress in optical communications over the last three decades, the optics and photonics community faces a great challenge if optical communications networks are to continue to satisfy the insatiable global demand for informa-

[116] The selected journals were these: *Optics Communications, IEEE Photonics Technology Letters, Optics Express, Journal of Lightwave Technology, Optical Engineering, Microwave and Optical Technology Letters, Journal of Optical Communications, Physical Review A Atomic Molecular and Optical Physics, Optics Letters, Photonic Network Communications, Applied Optics, Applied Physics Letters, Journal of the Optical Society of America B Optical Physics, Fiber and Integrated Optics, Optical and Quantum Electronics, Proceedings of SPIE the International Society for Optical Engineering, IEEE Transactions on Communications, IEEE Transactions on Microwave Theory and Techniques, IEEE Proceedings Optoelectronics, Journal of Optical Networking,* and *Photonics Spectra.*

[117] Data from SciVerse Scopus, www.scopus.com. Data derived using the top 25 papers and averaging over 5-year periods. Accessed March 21, 2012.

tion processing, data storage, bandwidth, and ubiquitous connectivity that, in turn, drive global economic growth.

Key Finding: The growth in bandwidth demand over the next 10 years is expected to be at least another 100-fold, possibly much more. It is important to note that the previous 100-fold gain in capacity that came very naturally with wavelength division multiplexing has been used up; growth by means of higher bit rates per wavelength comes more slowly; hence without a new breakthrough, increases in data transmission capacity will stall.

Key Finding: Cloud services not only drive capacity demand, but also make critical the role of large data centers. This will usher in a new and important era for short-distance optical links in massive numbers to provide cost- and energy-efficient high-density interconnects inside data centers and in information processing systems generally.

Key Finding: Silicon-based photonic integration technologies offer great potential for short-distance applications and could have great payoff in terms of enabling continued growth in the function and capacity of silicon chips if optics for interconnection could be seamlessly included in the silicon CMOS platform. It is also highly likely that integrated optoelectronics is a critical development area with significant growth potential for continuing the advance of defense systems.

Finding: Nanophotonic technologies promise very compact and high-performance optics and optoelectronics that could allow such platforms to continue to scale to higher density and performance. Information processing in general and data centers will also require massive and exponentially growing data storage.

Finding: Magnetic storage will continue as the primary data storage technology, and optics-assisted magnetic techniques may play an important future role.

Finding: Communications networks and information processing consume a relatively small fraction of the global energy budget (approximately 2 percent). However, with the growing network demands, this percentage will likely grow significantly unless successful action is taken to reduce systems' energy requirements.

Finding: Leveraging networks with applications like telepresence to reduce energy-hungry activity such as travel would reduce total energy consumption.

Finding: Optics has the potential to increase the energy efficiency in networks by, for example, replacing high-energy-loss electrical conductors in data centers and wireless backhaul or, perhaps, even on silicon chips.

Conclusion: The committee believes that a strong partnership among industry, universities, and government agencies will be crucial to overcoming technical challenges and to ensuring that the United States leverages that knowledge to gain market leadership.

Finding: Many of the optical communications successes over the last 10 years are built on earlier research that came from research laboratories of vertically integrated companies, as well as strongly supporting government agencies. The industry is no longer integrated, but instead is segmented by material fabrication, components, modules, systems, network providers, and content and service providers. In this fragmented environment there has been a reduction in industrial research laboratories, because the reduced scale makes it difficult for companies to capture the value that is prudently needed to justify investing in research.

Finding: Today's broadband access by individual users in the United States is neither high-capacity nor available at reasonable cost to a large fraction of the population. Bandwidths of 1 Gb/s represent the current state of the art in broadband access to the home in leading installations today.

RECOMMENDATIONS AND GRAND CHALLENGE QUESTIONS

Key Recommendation: The U.S. government and private industry, in combination with academia, need to invent technologies for the next factor-of-100 cost-effective capacity increase in long-haul, metropolitan, and local-area optical networks.

The optics and photonics community needs to inform funding agencies, and information and entertainment providers, about the looming roadblock that will interfere with meeting the growing needs for network capacity and flexibility. There is a need to champion collaborative efforts, including consortia of companies, to find new technology—transmission, amplification, and switching—to carry and route at least another factor-of-100 capacity in information over the next 10 years.

This key recommendation leads directly to the **first grand challenge question:**

1. How can the U.S. optics and photonics community invent technologies for the next factor-of-100 cost-effective capacity increases in optical networks?

The first recommendation of the chapter states a goal for increased capacity; the next recommendation offers a path to help achieve that goal, especially with respect to very short range communication, such as that required inside a data center.

Key Recommendation: The U.S. government, and specifically the Department of Defense, should strive toward harmonizing optics with silicon-based electronics to provide a new, readily accessible and usable, integrated electronics and optics platform.

They should also support and sustain U.S. technology transition toward low-cost, high-volume circuits and systems that utilize the best of optics and electronics in order to enable integrated systems to seamlessly provide solutions in communications, information processing, biomedical, sensing, defense, and security applications. Government funding agencies, the DOD, and possibly a consortium of companies requiring these technologies should work together to implement this recommendation. This technology is one approach to assist in accomplishing the first key recommendation of this chapter concerning the factor-of-100 increase in Internet capability.

The second key recommendation in this chapter leads to the **second grand challenge question:**

2. How can the U.S. optics and photonics community develop a seamless integration of photonics and electronics components as a mainstream platform for low-cost fabrication and packaging of systems on a chip for communications, sensing, medical, energy, and defense applications?

In concert with meeting the first grand challenge, achieving the second grand challenge would make it possible to stay on a Moore's law-like path of exponential performance growth. The seamless integration of optics and photonics at the chip level has the potential to significantly increase speed and capacity for many applications that currently use only electronics, or that integrate electronics and photonics at a larger component level. Chip-level integration will reduce weight and increase speed while reducing cost, thus opening up a large set of future possibilities as devices become further miniaturized.

The size and number of data centers in the United States and globally is expected to grow dramatically over the next decade to address the needs of a global digital society, especially if cloud services become more pervasive. It is clear that these data centers will be the focal point for the development and deployment of new optics and photonics communications technologies, and as such will be very important for the economy.

Key Recommendation: The U.S. government and private industry should position the United States as a leader in the optical technology for the global data center business.

Optical connections within and between data centers will be increasingly important in allowing data centers to scale in capacity. The committee believes that strong partnering between users, content providers, and network providers, as well as between businesses, government, and university researchers, is needed for ensuring that the necessary optical technology is generated, which will support continued U.S. leadership in the data center business.

Recommendation: The U.S. government and private industry, in conjunction with academia, should strive to develop technology to have optics take over the role of communicating and interconnecting information, not just at long distances, but at shorter distances as well, such as inside information processing systems, even to the silicon chip itself, thereby allowing substantial reductions in energy consumption in information processing and allowing the performance of information processing machines and systems to continue to scale to keep up with the exploding growth of the use of information in society.

Recommendation: The U.S. government and private industry, in conjunction with academia, need to encourage the exploitation of emerging nanotechnology for the next generation of optics and optoelectronics for the dramatic enhancement of performance (size, energy consumption, speed, integration with electronics) in information communications, storage, and processing.

Recommendation: The optics and photonics community needs to position the United States in broadband to the home and business space. The U.S. government should pursue policies that will enable at least gigabits per second broadband access to the substantial majority of society at a reasonable cost by 2020.

Recommendation: A multi-agency and cross-discipline effort is recommended to identify the opportunities and optical technologies to significantly increase the energy efficiency in communications networks, information processing, and storage. In addition, new ideas for the use of energy-efficient optical approaches to displace current energy-hungry practices—for example, travel—should be identified and supported. Greater focus and support in this area, especially at the fundamental level where companies are less likely to invest and where payoff could be huge, will be important.

4

Defense and National Security

INTRODUCTION

Many technological opportunities have been made possible by the advances in optics and photonics since the National Research Council's (NRC's) publication in 1998 of *Harnessing Light: Optical Science and Engineering for the 21st Century.*[1] Because optics and photonics are playing an increasingly important role in national defense, the United States is at a critical juncture in maintaining technological superiority in these areas. The gap between sophisticated and less sophisticated adversaries is not as large as it once was, and provides little or no advantage in several key technical areas, such as conventional night-vision equipment.

Sensor systems are becoming the next "battleground" for dominance in intelligence, surveillance, and reconnaissance (ISR), with optics-based sensors representing a significant fraction of ISR systems.[2] In addition, laser weapons are poised to cause a revolution in military affairs, and integrated optoelectronics is on the verge of replacing many traditional integrated circuit functions. Sophisticated platforms have reduced the need for a large set of traditional warfighters, but there is an increased need for a high-tech workforce to support those platforms. This workforce

---

[1] National Research Council. 1998. *Harnessing Light: Optical Science and Engineering for the 21st Century.* Washington, D.C.: National Academy Press.

[2] Details of additional defense and national security technologies may be found in Appendix C in this report. Topics covered include surveillance; night vision; laser rangefinders, designators, jammers, and communicators; laser weapons; fiber-optic systems; and special techniques focusing on chemical and biological species detection, laser gyros for navigation, and optical signal processing.

relies on advanced training in technical areas with a basis in science, technology, engineering, and mathematics (STEM), which are precisely the areas in which it is becoming more difficult to find continued optics and photonics education in the United States. The ability of U.S. defense forces to leverage technology for dominance while using a small force is also threatened by an ongoing migration of optics and photonics capabilities to offshore manufacturing sites. This means that the United States may lose both first access and assured access to new optics and photonics defense capabilities.

Although conventional night-vision imagers have become commodities available to anyone with money, more sophisticated optical-based surveillance systems have made major progress in the past decade and provide a great opportunity. A number of very-wide-field-of-view passive sensor systems have been developed and are discussed in this chapter. It is now possible by using such systems to view large areas with moderate to high resolution, especially during the day. Large portions of a city can thus be continuously monitored and the data from the system stored. If something of interest occurs, it is possible to re-examine that event to determine exactly what happened. Once areas of interest have been detected, it would be useful to have exquisite detail in certain critical areas, highlighted by the wide-area detection sensor. There have recently been long-range identification demonstrations using active electrooptical (EO) systems called laser radar, or ladar. Although synthetic aperture radar (SAR) has been around for decades, it is only recently that synthetic aperture ladar systems have been flown. These are briefly discussed below. Multiple sub-aperture-based, potentially conformal, active sensor developments are also discussed. This is a developing technology that will allow lighter-weight, long-range imaging systems that can also be applied to laser weapons. After an object has been detected and identified, it may be recognized as a threat that has to be dealt with. "Speed-of-light" weapons are ideal choices for certain applications, such as for a boosting missile. These laser weapons can destroy a boosting ballistic missile, causing whatever warhead is on the missile to fall back on the nation that fired the missile. Recently the Airborne Laser Test Bed (ALTB)[3] shot down a boosting ballistic missile with an onboard laser for the first time. Although this was a highly successful test, it was done with a chemical laser, using a mixture of oxygen iodine as the gain medium. There is strong interest in and great potential for laser weapons that run on electricity. If sufficient electricity can be generated from onboard fuel, one could use the same fuel, already in use. Multiple all-electric laser options are briefly discussed below.

The three areas just referred to have made major progress over the last decade,

[3] More information on the Airborne Laser Laboratory is available at http://www.fas.org/spp/starwars/program/all.htm (accessed November 22, 2011) and http://boeing.mediaroom.com/index.php?s=43&item=1075 (accessed November 7, 2012).

but they have been pursued as stovepiped activities. A laser weapon needs to detect a potential target, the target must be identified, and an aim point must be selected and maintained. Additionally, as communication between sensors producing "input" about a situation and systems taking action (output) needs to be faster, there is a technological need to put these sensors and systems as close together as possible. The committee believes that there is significant synergy between these activities and that photonics technologies will be an integral part of this new integrated system capability.

Over the last decade, significant work has evolved on silicon photonics, to closely integrate optics and electronics in a cost-effective manner, as was discussed in the previous chapter, on communications. Most of this work has been driven by communications needs, but it will be an enabler for the defense arena as well. Optics is becoming integrated in defense systems other than optical systems, such as into microwave radars, using radio-frequency (RF) photonics. It is anticipated that more and more areas of defense "electronics" will become defense optoelectronics.

OPTICS AND PHOTONICS: IMPACT ON DEFENSE SYSTEMS

There is virtually no part of a modern defense system that is not impacted in some way by optics and photonics, even when the system is not optically based. Modern defense systems are migrating toward optically based imaging, remote sensing, communications, and weapons. This trend makes maintaining leadership in optics and photonics vital to maintaining the U.S. position in defense applications. Additional areas of impact include the following: precision laser machining, optical lithography for electronics, optical signal interconnects, solar power for remote energy needs, and generation of a stable timebase for the Global Positioning System (GPS). Even when the actual sensor is not optics-based, in many cases optics plays an important role, such as the migrating of RF photonics into microwave radar systems mentioned above.

TECHNOLOGY OVERVIEW

There have been significant advances in optics and photonics for national defense both in components and in systems since the publication of *Harnessing Light* in 1998.[4] Some of the key areas include surveillance, night vision, laser systems, fiber-optics systems, chemical and biological detection, and optical processing. One example of a significant advance in component technologies is laser diode efficiency, which has directly impacted the efficiency of laser systems. In addition, there has been significant progress in both laser power and available wavelengths

[4] National Research Council. 1998. *Harnessing Light.*

for applications important to national defense. These advances in laser technology have also enabled several significant system advances. One example is the area of optical aperture synthesis, which rapidly went from the laboratory to flight system demonstrations within several years during the period since *Harnessing Light* appeared.

While the advances described above have enabled new capabilities for the United States, they have also narrowed the technology gap for adversaries. Importantly, the proliferation of low-cost, high-powered lasers has provided inexpensive countermeasures for adversaries. One example is the use of high-powered handheld laser pointers as laser dazzlers against helicopter pilots, causing a bedazzled pilot to become temporarily blinded or disoriented. The low cost and abundance of these devices put them in anyone's reach.

Changes Since the *Harnessing Light* Study

This section briefly discusses the changes in each of the areas that were addressed in Chapter 4, "Optics in National Defense" in the NRC's 1998 report *Harnessing Light*. For a more detailed discussion of these topics, see Appendix C in this report. The most significant changes have been due to the advances made in optical components that have enabled new sensors to be developed and demonstrated (see the section below entitled "Identification of Technological Opportunities from Recent Advances"). The following subsections provide an update for the areas of surveillance, night vision, laser systems operating in the atmosphere and in space, fiber-optic systems, and special techniques (e.g., chemical and biological species detection, laser gyros, and optical signal processing).

Surveillance

Surveillance still plays a critical role in detecting and assessing hostile threats to the United States. The progress in optical sensors over the past decade has created an exponential growth in ISR data from both passive and active sensors, including an increase in area coverage rate and an increase in sensor capabilities and performance. As the Defense Advanced Research Projects Agency (DARPA) chief Regina Dugan puts it, "We are swimming in sensors and drowning in data."[5] Materials advances have made collection at new wavelengths feasible, and improved components provide new data signatures, including vibrometry, polarimetry, hyper-spectral signatures, and three-dimensional data that mitigate camouflage for targets of interest.

[5] Comment can be found in Norris, P. 2010. *Watching Earth from Space: How Surveillance Helps Us—and Harms Us.* Chichester, U.K.: Praxis Publishing.

Night Vision

The proliferation of night-vision equipment over the past few decades has led to a significant amount of surplus equipment available at very low cost. This equipment has eroded the tactical advantage that the United States previously had in this area of warfare during the night.

Laser Rangefinders, Designators, Jammers, and Communicators

The significant increase in laser diode efficiency coupled with the decrease in cost has enabled recent advances in the area of laser designators. One of the key motivators for moving to, for example, optical communications, is that they minimize the probability of interception, jamming, and detection, while dramatically minimizing the power needed. The improved efficiency and availability of high-powered lasers at a broader range of wavelengths has also enabled the development of countermeasure systems for several applications, including defense against the now-prolific man portable air defense systems (MANPADS) capabilities that threaten commercial and military aircraft. The specific developments are not covered in detail in this report.

Laser Weapons

The Missile Defense Agency demonstrated the potential use of directed energy to defend against ballistic missiles when the Airborne Laser Test Bed successfully destroyed a boosting ballistic missile on February 11, 2010. As discussed in a Missile Defense Agency news release,[6] this revolutionary use of directed energy is very attractive for missile defense, with the potential to attack multiple targets at the speed of light, at a range of hundreds of kilometers, and at a low cost per interception attempt compared to current technologies (see Figure 4.1). Since publication of the 1998 NRC report, there have also been other successful demonstrations, including the Tactical High Energy Laser (THEL),[7] the Mobile Tactical High Energy Laser (MTHEL), and the Maritime Laser Demonstrator (MLD).[8]

---

[6] Missile Defense Agency, U.S. Department of Defense. 2010. "Airborne Laser Test Bed Successful in Lethal Intercept Experiment." MDA news release. Available at http://www.mda.mil/news/10news0002.html. Accessed August 2, 2012.

[7] Shwartz, J., J. Nugent, D. Card, G. Wilson, J. Avidor, and E. Behar. 2003. Tactical high energy laser. *Journal of Directed Energy* 1(1):34-47.

[8] More information is available through the Office of Naval Research, at http://www.onr.navy.mil/Media-Center/Press-Releases/2011/Maritime-Laser-MLD-Test.aspx. Accessed June 4, 2012.

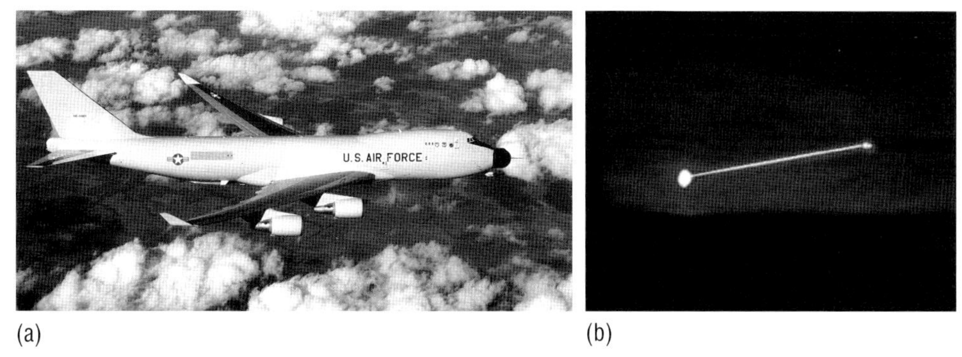

(a) (b)

FIGURE 4.1 Airborne Laser Test Bed (ALTB). (a) The ALTB is a platform for the Department of Defense's directed-energy research program. Two solid-state lasers and a megawatt-class chemical oxygen iodine laser (COIL) are housed aboard a modified Boeing 747-400 Freighter. (b) An infrared image of the Missile Defense Agency's Airborne Laser Test Bed (at right in the image) destroying a threat representative short-range ballistic missile (at left in the image). SOURCE: Images available from the Missile Defense Agency, at http://www.mda.mil/news/gallery_altb.html.

Fiber-Optic Systems

Fiber-optic systems have continued to evolve to achieve higher performance with lower power in a smaller volume. In addition, fiber-based supercontinuum sources have significantly advanced since the advent of photonic-crystal fibers (PCFs) in 1996.[9] PCFs simultaneously provide high nonlinearity and a variable zero dispersion wavelength for a broadband continuum that can span more than an octave. Since the NRC's 1998 report *Harnessing Light*, these sources have gone from concept, to demonstration, and finally to commercial products.

Special Techniques

The special techniques (i.e., chemical and biological species detection and optical signal processing) evaluated in the 1998 *Harnessing Light* report have evolved in different ways. Optical signal processing has also advanced, but not at the pace forecasted at that time. Importantly, recent advances in optical integrated circuits should enable significant advances in optical signal processing over the next decade.

Chemical and Biological Species Detection. Weapons of mass destruction, including nuclear, biological, and chemical weapons, continue to be a high-priority threat. Long-range chemical and biological detection has advanced considerably since the

[9] Knight, J., T. Birks, P. Russell, and D. Atkin. 1996. All-silica single-mode optical fiber with photonic crystal cladding. *Optics Letters* 21:1547.

NRC's 1998 *Harnessing Light* report. One example is the Joint Biological Stand-off Detection System (JBSDS), a light detection and ranging (lidar)-based system that is designed to detect aerosol clouds out to 5 kilometers (km) in a 180-degree arc and to discriminate clouds with biological content from clouds without biological material at distances of 1 to 3 km or more.

Optical Signal Processing. Optical signal processing has not changed very much since the NRC's 1998 report was issued. Optical processing continues to be very promising, since some mathematical functions can be performed very rapidly using optical analog techniques. One example is optical correlations that rely on Fourier transforms. Optical correlators compare two-dimensional image data at very high speeds. The most promising advances are discussed below, in the section entitled "Integrated Optoelectronics."

Identification of Technological Opportunities from Recent Advances

As the military capabilities of other countries have been expanding quickly, sensor systems are becoming the next battleground for dominance in ISR, as noted above. Advanced systems have reduced the reliance on the traditional warfighter so that now there is a need for a more technologically focused personnel. The data generated by deployed sensor systems have grown significantly due to the advances in sensor capabilities. This change has allowed new intelligence data products, but it has also driven the need for more sophisticated data processing and transmission in order to handle these data rates. The following subsections provide an overview of opportunities in synthetic aperture laser radar, multi-mode laser sensing, sparse aperture laser sensing, wide-area surveillance sensors, Geiger-mode imaging, and hyper-spectral sensing.

Synthetic Aperture Laser Radar

The diffraction limit presents a significant limitation on cross-range resolution for long-range remote sensing applications. Synthetic aperture sensing and analysis techniques provide a method of overcoming this limitation in some applications requiring high-resolution coherent images at great distances. These techniques have been employed in the RF domain for many years in synthetic aperture radar systems. Only in the past several years have advances in simultaneously stable and widely tunable coherent optical systems enabled the application of SAR techniques to the optical domain, allowing a potential for greatly improved illumination efficiency and image acquisition time. As pointed out by Beck et al. in their paper "Synthetic-Aperture Imaging Laser Radar: Laboratory Demonstration and Signal Processing," the first Synthetic Aperture Imaging Ladar (SAIL) image of a fixed, diffusely scattering target with a moving aperture[10] demonstrated the use of a chirped optical source to provide a demonstration with 60-micron-range resolution and 50-micron cross-range resolution. The earliest synthetic aperture experiments[11] in the optical domain were performed at the United Aircraft Research Laboratories in the late 1960s, using inverse techniques to focus a moving point target. Another experiment[12] used a technique to perform synthetic aperture imaging in two dimensions. A later effort[13] used a continuous wave Nd:YAG microchip laser to demonstrate inverse-SAIL imaging in one dimension together with diffraction limited conventional imaging in the other dimension (with an asymmetric high-aspect-ratio aperture) to produce two-dimensional images. Other demonstrations[14,15] achieved one-dimensional SAIL imaging of a point target, using a continuous wave CO_2 system, and a two-dimensional inverse-SAIL image of a translated target.[16] More recent efforts include the DARPA Synthetic Aperture Ladar for Tactical Imaging (SALTI) Program.[17]

The progress from laboratory to flight demonstration within several years shows that optical synthetic aperture imaging is becoming a valuable technology, with several recent additional advancements in combining synthetic aperture imag-

[10] Beck, S.M., J.R. Buck, W.F. Buell, R.P. Dickinson, D.A. Kozlowski, N.J. Marechal, and T.J. Wright. 2005. Synthetic-aperture imaging laser radar: Laboratory demonstration and signal processing. *Applied Optics* 44:7621-7629.

[11] Lewis, T.S., and H.S. Hutchins. 1970. A synthetic aperture at 10.6 microns. *Proceedings of the IEEE* 58:1781-1782.

[12] Aleksoff, C.C., J.S. Accetta, L.M. Peterson, A.M. Tai, A. Klossler, K.S. Schroeder, R.M. Majewski, J.O. Abshier, and M. Fee. 1987. Synthetic aperture imaging with a pulsed CO_2 TEA laser. In *Laser Radar II*. Becherer R.J., and R.C. Harney, eds. *Proceedings of the SPIE* 783:29-40.

[13] Green, T.J., S. Marcus, and B.D. Colella. 1995. Synthetic-aperture-radar imaging with a solid-state laser. *Applied Optics* 34(30):6941-6949.

[14] Yoshikado, S., and T. Aruga. 2000. Short-range verification experiment of a trial one-dimensional synthetic aperture infrared laser radar operated in the 10-μm band. *Applied Optics* 39(9):1421-1425.

[15] Bashkansky, M., R.L. Lucke, E. Funk, L.J. Rickard, and J. Reintjes. 2002. Two-dimensional synthetic aperture imaging in the optical domain. *Optics Letters* 27:1983-1985.

[16] Using 10 nm of near-linear optical chirp at 1.5 microns with an analog reference channel (to mitigate waveform uncertainties) with path length exactly matched to the target channel's path length.

[17] Dierking, M., B. Schumm, J.C. Ricklin, P.G. Tomlinson, and S.D. Fuhrer. 2007. Synthetic aperture LADAR for tactical imaging overview. *Proceedings of the 14th Coherent Laser Radar Conference* 191-194. Available at http://toc.proceedings.com/05549webtoc.pdf. Accessed June 27, 2012.

ing and digital holography.[18,19,20] Another flight demonstration[21] has shown that the techniques are well understood and can be implemented with near-real-time processing.

Applying aperture synthesis techniques to the optical domain provides two significant advantages: (1) improved image acquisition time and (2) illumination efficiency.[22] Improvement in illumination efficiency can be understood by considering that illuminating only a 10-m-diameter area of interest with a 10 gigahertz (GHz) signal at a range of 100 km would require a transmit aperture of approximately 300 m for the RF case, compared to approximately 1.5 centimeters (cm) for an optical system with a wavelength of 1.5 microns. It is clear that this technology has the potential to be further developed and that it can provide additional benefits if it is combined directly with defensive systems.

Multi-Mode Laser Sensing

Multi-function sensors seek to exploit the maximum information that a laser-based sensor can obtain by incorporating several functions into a single sensor (e.g., three-dimensional imaging, vibrometry, polarimetry, aperture synthesis, agile apertures, etc.). Ideally these sensors utilize waveforms matched to the requirements of both the hardware (e.g., optical amplifiers, modulators) and the targets being imaged. Recent demonstrations[23] have achieved 7-millimeter (mm)-range resolution (0.1-mm-range precision) along with simultaneous vibrometry.[24] The inherent multi-functionality of these systems allows maximal use of available aperture, volume, and power. Therefore, a multi-function system will enable practical, high-performance ladar remote sensing systems with scalable, reconfigurable

---

[18] Stafford, J.W., B.D. Duncan, and M.P. Dierking. 2010. Experimental demonstration of a stripmap holographic aperture ladar system. *Applied Optics* 49:2262.

[19] Duncan, B.D., and M.P. Dierking. 2009. Holographic aperture ladar. *Applied Optics* 48:1168.

[20] Rabb, D.J., D.F. Jameson, J.W. Stafford, and A.J. Stokes. 2010. Multi-transmitter aperture synthesis. *Optics Express* 18:24937.

[21] Krause, B., J. Buck, C. Ryan, D. Hwang, P. Kondratko, A. Malm, A. Gleason, and S. Ashby. 2011. "Synthetic Aperture Ladar Flight Demonstration." Conference paper. CLEO: Applications and Technology (CLEO: A and T), Baltimore, Md., May 1, 2011. Available at http://www.opticsinfobase.org/abstract.cfm?uri=CLEO:%20A%20and%20T-2011-PDPB7. Accessed June 27, 2012.

[22] For a given cross-range resolution, the image acquisition time scales with the carrier wavelength. For an RF system with $\lambda = 3$ cm, $v_{platform} = 100$ m/s, $R = 100$ km, and $\delta x = 1$ cm, the image collection time would be 1500 sec for the required baseline of 150 km. For an optical system with $\lambda = 1550$ nm, the same specifications require a baseline of 7.75 m with an image collection time of 78 ms.

[23] Buck, J., A. Malm, A. Zakel, B. Krause, and B. Tiemann. 2007. High-resolution 3D coherent laser radar imaging. *Proceedings of the SPIE* 6550:655002.

[24] Buck, J., A. Malm, A. Zakel, B. Krause, and B. Tiemann. 2007. Multi-function coherent ladar 3D imaging with S3. *Proceedings of the SPIE* 6739:67390F.

operating modes to obtain spectral, spatial, and temporal information about a target along with information about the target's depolarization properties. This combined information set can provide an unprecedented ability to characterize targets with a single sensor and shows a possible path for the future development of more complex systems.

Sparse Aperture Laser Sensing

The use of small-aperture modules can lead to revolutionary optical sensing and communications approaches, eliminating large, complex, and expensive apertures.[25,26,27,28] Many sub-aperture modules have a much shorter focal length than one large EO aperture with the same F number. As a result, the overall aperture array will be much shallower and will weigh much less than the monolithic system. Phased-array approaches will enable using optical apertures along the surface of a vehicle because of the shallow aperture depth. Small, standard modules can enable responsive space, with an array of modules stored and ready to be configured and launched. Conformal and structural optical sensing and communications approaches (i.e., those that can be implemented on a platform without modifying its skin and thereby avoiding the impacting of platform aerodynamics) can be developed. RF systems already have conformal and structural weight-bearing RF apertures. A conformal array system is robust to element failure, which is important for system operation in hazardous environments. In conformal systems, beam focusing and retargeting can be performed using fast control of wave-front phase tip and tilt at each conformal system sub-aperture.[29] This would allow orders-of-magnitude faster retargeting of the outgoing or received optical waves. With conformal optical systems, atmospheric turbulence-induced phase distortions can be pre-compensated using adaptive optics (AO) elements that are directly integrated

[25] McManamon, P.F. 2008. "Long Range ID Using Sub-Aperture Array Based Imaging." Conference paper. Coherent Optical Technologies and Applications (COTA), Boston, Mass., July 13, 2008.

[26] McManamon, P.F., and W. Thompson. 2003. Phased array of phased arrays (PAPA) laser systems architecture. *Fiber and Integrated Optics* 22(2):79-88.

[27] McManamon, P.F. 2004. "The Vision of Optical Phased Array and Phased Array of Phased Arrays." Conference paper. SPIE Great Lakes Photonics Symposium, Cleveland, Ohio, June 8, 2004.

[28] McManamon, P.F., and W. Thompson. 2002. Phased array of phased arrays (PAPA) laser systems architecture. *IEEE Aerospace Conference Proceedings* 3:1465-1472.

[29] Vorontsov, M.A., T. Weyrauch, L. Beresnev, G. Carhart, L. Liu, S. Lachinova, and K. Aschenbach. 2009. Adaptive array of phase-locked fiber collimators: Analysis and experimental demonstration. *IEEE Journal of Selected Topics in Quantum Electronics* 15:269-280.

into individual sub-apertures. This enables the cost-effective integration of AO capabilities into conformal optical systems.[30,31]

Spatial heterodyne, a form of digital holography, is also being developed as a method of active imaging with high-resolution multiple sub-apertures and framing cameras.[32,33,34,35] This is a new area with potential significant advantages, and it is anticipated that multiple-sub-aperture-based imaging will grow over time.

Wide-Area-Surveillance Sensors

One of the developments that will make a difference in U.S. military capability is wide-area surveillance, especially for cities. The U.S. military has a great surveillance capability in open spaces, but cities, in which there are many noncombatants, present a new problem. From a security point of view, in cities there is a large amount of "clutter" in terms of buildings and people. The ability to watch everything all the time in a city improves the ability to do surveillance, especially when one can store the imagery and replay it at any time. A number of wide-area-surveillance systems are being developed. One example is an airborne system developed in concert with the Philadelphia Police Department to show the value of such a system.[36] The system was flying over a troubled neighborhood for one day as a test. When a woman got home from work she called the police to report that her house had been broken into during the day. The police reviewed the imagery collected that day and could see someone enter and leave the house around 2:00 p.m. They could trace the person who left the house to another house 8 blocks away. This imagery provided sufficient proof for a warrant to search the house to which the person was traced, and there the stolen goods were found and an arrest

---

[30] Vorontsov, M.A., and S.L. Lachinova. 2008. Laser beam projection with adaptive array of fiber collimators. I. Basic considerations for analysis. *Journal of the Optical Society of America A* 25:1949-1959.

[31] Lachinova, S.L., and M.A. Vorontsov. 2008. Laser beam projection with adaptive array of fiber collimators. II. Analysis of atmospheric compensation efficiency. *Journal of the Optical Society of America A* 25:1960-1973.

[32] Marron, J.C., and R.L. Kendrick. 2007. Distributed aperture active Imaging. *Proceedings of the SPIE* 6550:65500A.

[33] Rabb, D.J., D.F. Jameson, A.J. Stokes, and J.W. Stafford. 2010. Distributed aperture synthesis. *Optics Express* 18:10334-10342.

[34] Marron, J.C., R.L. Kendrick, N. Seldomridge, T.D. Grow, and T.A. Höft. 2009. Atmospheric turbulence correction using digital holographic detection: Experimental results. *Optics Express* 17:11638-11651.

[35] Miller, N.J., J.W. Haus, P. McManamon, and D. Shemano. 2011. Multi-aperture coherent imaging. *Proceedings of the SPIE* 8052:8052-8056.

[36] Written communication to the committee, October 25, 2011, from the President of Persistent Surveillance Systems.

was made. This example illustrates the power of surveillance that is conducted 100 percent of the time. Whether it is employed in a military or security application, wide-area high-resolution surveillance can have a major impact.

There have been a number of efforts to develop wide-area-surveillance systems in the visible region. The largest system so far by pixel count is the DARPA-funded ARGUS (Autonomous Real-time Ground Ubiquitous Surveillance) system, with 1.8 billion pixels (Figure 4.2).

One possible next step could be to build a large-format mid-wavelength infrared (MWIR) system in order to image during moonless nights (see Figure 4.3 and

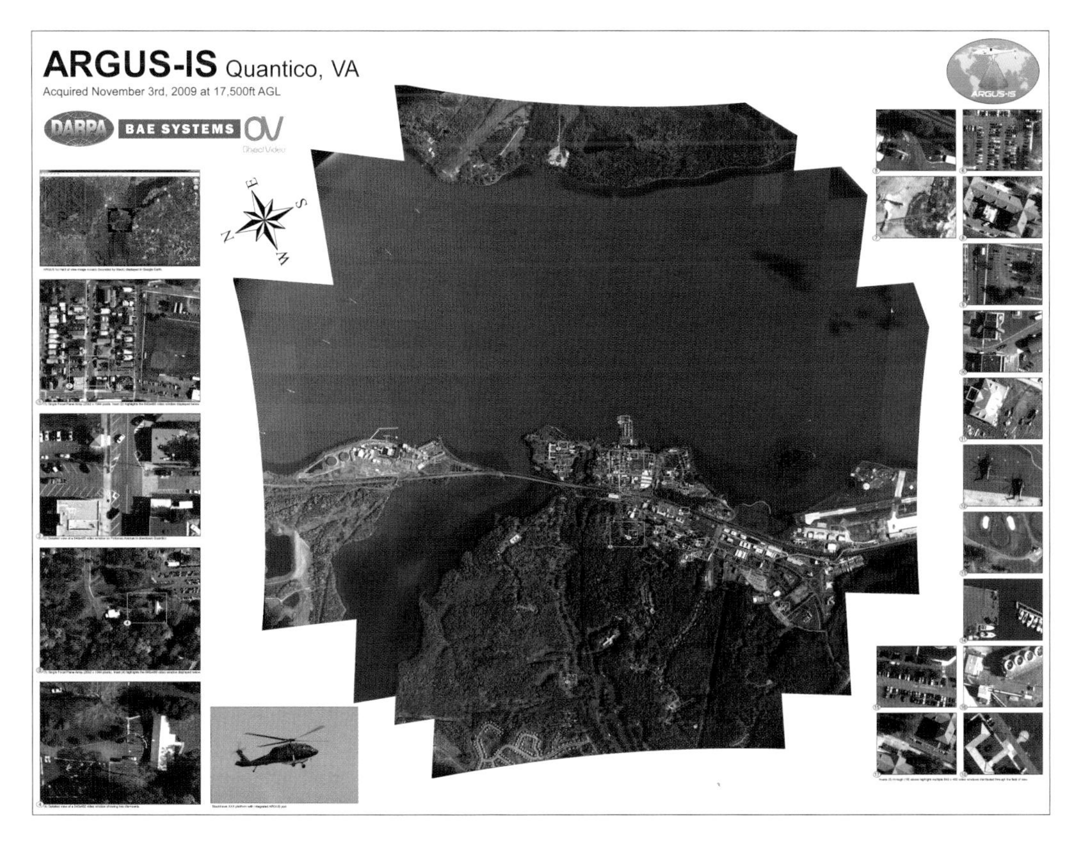

FIGURE 4.2 A sample of the ARGUS-IS imagery. The system was mounted under a YEH-60B helicopter flying at 17,500 feet over Quantico, Va. The Argus-IS images an area more than 4 km wide. The ARGUS system uses a large number of inexpensive cell phone cameras to create a whole new imaging modality with a huge format. SOURCE: Image available from DARPA Information Innovation Office, Autonomous Real-time Ground Ubiquitous Surveillance-Imaging System (ARGUS-IS), at http://www.darpa.mil/Our_Work/I2O/Programs/Autonomous_Real-time_Ground_Ubiquitous_Surveillance-Imaging_System_%28ARGUS-IS%29.aspx.

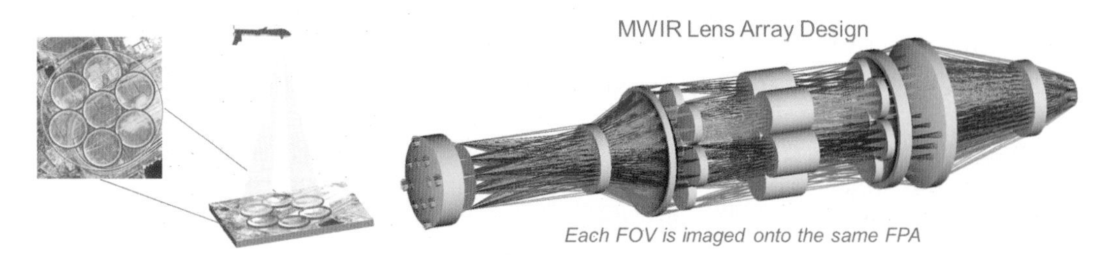

FIGURE 4.3 *(Right)* Wide-area mid-wavelength infrared (MWIR) step and stare imager. *(Left)* Each element of the Hex-7 lens array has a 6° × 6° field of view (FOV). Each FOV is imaged onto the same focal plane array (FPA) to provide seven distinct viewing angles. The effective FOV of the system is 17.5° × 17.5°. SOURCE: Reprinted, with permission, from Masterson, H., R. Serati, S. Serati, and J. Buck. 2011. MWIR wide-area step and stare imager. *Proceedings of the SPIE 8052*, Acquisition, Tracking, Pointing, and Laser Systems Technologies XXV.

BOX 4.1
MWIR Step and Stare Wide-Angle Image

A step and stare mid-wavelength infrared (MWIR) imager, recently developed by Boulder Nonlinear Systems (BNS) for the Air Force Research Laboratory, switches between fields of view in a Hex-7 pattern to achieve 0.1 milli-radian resolution within a 17.5° × 17.5° field of view. Historically, step and stare techniques required a nominal 5 milliseconds (msec) to stare, and then 150-200 msec to move from one angular location to another. By using switchable shutters to move from one angular location to another, the BNS system dramatically changed the ratio of staring time to stepping time, reducing the step time to approximately 1 msec. The demonstration was accomplished using a 1 km × 1 km MWIR focal plane array, which covers an area of 3 km × 3 km using a Hex-7 scan pattern; 4 km × 4 km MWIR cameras have been developed, and with a Hex-19 pattern would cover a 20 km × 20 km circular pixel area. For a 4 msec staring time and 1 msec step time, 5 msec per hex element yields 95 msec per frame, for a rate of >10 Hz over a 20 km × 20 km region. An 8 km × 8 km array would provide a 40 km × 40 km picture with the same frame rate, which could be scaled to larger frames with a Hex-37 step pattern at a slower frame rate.

Box 4.1). One approach to wide-area MWIR surveillance is to use the largest MWIR camera available, and then to use a step and stare technique to cover a larger area.[37]

Another example of a wide-area MWIR imager is the DARPA Large Area Coverage Optical Search-while-Track and Engage (LACOSTE) program. As discussed on DARPA's webpage,[38] this is a wide field-of-view (FOV) coded aperture imaging

[37] Masterson, H., R. Serati, S. Serati, and J. Buck. 2011. "MWIR wide-area step and stare imager." Proceedings of SPIE 8052, Acquisition, Tracking, Pointing, and Laser Systems Technologies XXV, 80520N. doi 10.1117/12.884290:

[38] More information is available at web-ext2.darpa.mil. Accessed August 6, 2012.

technology approach for the single-sensor day/night persistent tactical surveillance of all moving vehicles in a large urban battlefield. LACOSTE coded aperture imaging technology focused on achieving a very wide instantaneous FOV using multiple simultaneous wide-FOV images.[39,40,41,42,43,44] Discussions within the references cited in footnotes 39 through 44 make it clear that in coded apertures, a structured mask of pinhole cameras is created such that the image from each individual pinhole falls across a common focal plane array (FPA). With a large-area mask centered above a small FPA, the pinhole camera structure opens as a lens in the desired look direction, while the remainder of the mask remains opaque. With a known mask structure, the multiple images on the same focal plane are digitally deconvolved to form an image, which provides several unique and enabling features.

Geiger-Mode Imaging

Geiger-mode detectors based on avalanche photodiodes (APDs) are highly sensitive semiconductor electronic devices that exploit the photoelectric effect to convert light to electricity. They can have high quantum efficiencies and can be thought of as photodetectors that provide a built-in first stage of gain through avalanche multiplication. Since the publication of the NRC's 1998 report, Geiger-mode arrays have been developed and used for several demonstrations, including the DARPA Jigsaw program,[45] which provided a three-dimensional flash imaging system.

---

[39] Mahalanobis, A., C. Reyner, H. Patel, T. Haberfelde, D. Brady, M. Neifeld, B.V.K. Vijaya Kumar, and S. Rogers. 2007. IR performance study of an adaptive coded aperture diffractive imaging system employing MEMS eyelid shutter technologies. *Proceedings of the SPIE* 6714:67140D.

[40] Mahalanobis, A., C. Reyner, T. Haberfelde, M. Neifeld, and B.V.K. Vijaya Kumar. 2008. Recent developments in coded aperture multiplexed imaging systems. *Proceedings of the SPIE* 6978:6978G-69780G-8.

[41] Mahalanobis, A., M. Neifeld, B.V.K. Vijaya Kumar, and T. Haberfelde. 2008. Design and analysis of a coded aperture imaging system with engineered PSFs for wide field of view imaging. *Proceedings of the SPIE* 7096:7096C-70960C-11.

[42] Mahalanobis, A., M. Neifeld, B.V.K. Vijaya Kumar, T. Haberfelde, and D. Brady. 2009. Off-axis sparse aperture imaging using phase optimization techniques for application in wide-area imaging systems. *Applied Optics* 48(28):5212-5224.

[43] Slinger, C., M. Eismann, N. Gordon, K. Lewis, G. McDonald, M. McNie, D. Payne, K. Ridley, M. Strens, G. DeVilliers, and R. Wilson. 2007. An investigation of the potential for the use of a high-resolution adaptive coded aperture system in the mid-wave infrared. *Proceedings of the SPIE* 6714:671408.

[44] McNie, M.E., D.J. Combes, G.W. Smith, N. Price, K.D. Ridley, K.L. Lewis, C.W. Slinger, and S. Rogers. 2007. Reconfigurable mask for adaptive coded aperture imaging (ACAI) based on an addressable MOEMS microshutter array. *Proceedings of the SPIE* 6714:67140B.

[45] More information on the Foliage-Penetrating 3D Imaging Laser Radar System is available at http://www.ll.mit.edu/publications/journal/pdf/vol15_no1/15_1jigsaw.pdf. Accessed June 27, 2012.

Hyper-Spectral Sensing

Hyper-spectral imaging is an extreme form of color imaging. People are very familiar with color imaging. We all know that spotting a bright red object lying on a green lawn is much easier than seeing a green object on a green lawn. Color in an image can be divided into many wavebands for more resolution. One of the significant issues associated with multi- or hyper-spectral imaging is whether or not one needs to see at night. Daytime viewing uses the visible and near-infrared regions of the spectrum, where as nighttime viewing requires detectors for longer wavelengths. Although there is a phenomenon called night glow[46] in the near-infrared, and often in man-made lighting or light from the Moon, reliable viewing requires moving to the mid- or long-wave infrared (IR). Most of the current commercial applications of spectral, or hyper-spectral, imaging use the visible and near-IR regions.

Multi- or hyper-spectral imaging can be used for telling the status of crops, for finding minerals, and for surveillance. Spectral information is always valuable for looking at surface material properties. In addition, hyper-spectral imaging technology is very useful for search-and-rescue applications.[47] One of the disadvantages of hyper-spectral imaging is signal availability, because the narrow bands provide limited signal for a passively illuminated scene. For hyper-spectral imaging, the resolution of the sensor must be traded with the available signal levels.

Defense Systems

Defense systems (laser weapons) have made great progress since the 1998 NRC report was issued. The Airborne Laser Laboratory (ABL) intercepted two ballistic missiles in February 2010 (see Figure 4.1)[48] with the megawatt-class oxygen iodine laser emitted from the nose of the aircraft. After this successful test, ABL was converted to the Airborne Laser Test Bed to explore issues associated with potential follow-on activities. As of this writing, no follow-on activity has been identified. Until April 2009, ABL was on a path to deployment in small numbers. However,

---

[46] Barber, D.R. 1957. A very early photographic observation of the spectrum of the night glow. *Nature* 179(4556):435.

[47] Eismann, M.T., A.D. Stocker, and N.M. Nasrabadi. 2009. Automated hyperspectral cueing for civilian search and rescue. *Proceedings of the IEEE* 97(6):1031-1055.

[48] Wolf, Jim, and David Alexander. 2010. "U.S. Successfully Tests Airborne Laser on Missile." Available at http://www.reuters.com/article/2010/02/12/usa-arms-laser-idUSN1111660620100212?type=marketsNews. Accessed October 26, 2011.

at that time the second ABL aircraft was recommended for cancellation, with the program to return to a research and development effort.[49]

Another major laser weapons effort was the advanced tactical laser (ATL), a short-range weapon for use on a gunship-like aircraft, with the laser replacing a gun. In August 2008, the first test-firing of the "high-energy chemical laser" mounted in a Hercules transport plane was announced. In August 2009, a ground target was "defeated" from the air with the ATL aircraft.[50] This laser weapon is also based on an oxygen iodine laser, requiring hauling hazardous chemicals to the field. At the time of this writing, there is no planned follow-on effort.

Because of the interest in electric-powered lasers that do not require a specialized logistics tail, the High Energy Laser-Joint Technology Office (HEL-JTO) initiated a program to demonstrate a 100-kilowatt (kW)-output, electric-powered laser capable of being used in a laser weapon system, called the Joint High Power Solid State Laser (JHPSSL). As discussed in the December 2010 news release from DARPA entitled "Compact High-Power Laser Program Completes Key Milestone,"[51] JHPSSL operated above the rated 100-kW power level for 6 hours.[52,53] The goal of the High Energy Liquid Laser Area Defense System (HELLADS) is to demonstrate 150 kW of power in a lightweight package. In June 2011, DARPA completed the laboratory testing of a fundamental building block for HELLADS, a single laser module that successfully demonstrated the ability to achieve high power and beam quality from a significantly lighter and smaller laser.[54] Another DARPA program that is developing an approach to laser weapons is the Adaptive Photonic Phase Locked Elements (APPLE) program.[55] APPLE uses a modular system (Figure 4.4) to scale the available power, which requires high-powered lasers with sufficiently

---

[49] Gates, Dominic. 2009. "Boeing Hit Harder than Rivals by Defense Budget Cuts. Available at http://seattletimes.nwsource.com/html/localnews/2008997361_defensecuts07.html. Accessed October 26, 2011.

[50] Boeing. 2009. "Boeing Advanced Tactical Laser Defeats Ground Target in Flight Test." Available at http://boeing.mediaroom.com/index.php?s=43&item=817. Accessed June 27, 2012.

[51] DARPA. 2011. "Compact High-Power Laser Program Completes Key Milestone." Available at http://www.darpa.mil/NewsEvents/Releases/2011/2011/06/30_COMPACT_HIGH-POWER_LASER_PROGRAM_COMPLETES_KEY_MILESTONE.aspx. Accessed October 26, 2011.

[52] JHPSSL first achieved the 100-kW power levels in March 2009. More information is available at http://www.irconnect.com/noc/press/pages/news_releases.html?d=161575. Accessed June 4, 2012.

[53] Optics. 2010. "Northrop's 100 kW Laser Weapon Runs For Six Hours." Available at http://optics.org/news/1/7/13. Accessed October 26, 2011.

[54] DARPA. 2011. "Compact High-Power Laser Program Completes Key Milestone."

[55] Dorschner, Terry A. 2007. "Adaptive Photonic Phase Locked Elements: An Overview." Raytheon Network Centric Systems presentation. Available at http://www.dtic.mil/cgi-bin/GetTRDoc?Location=U2&doc=GetTRDoc.pdf&AD=ADA503733. Accessed June 27, 2012.

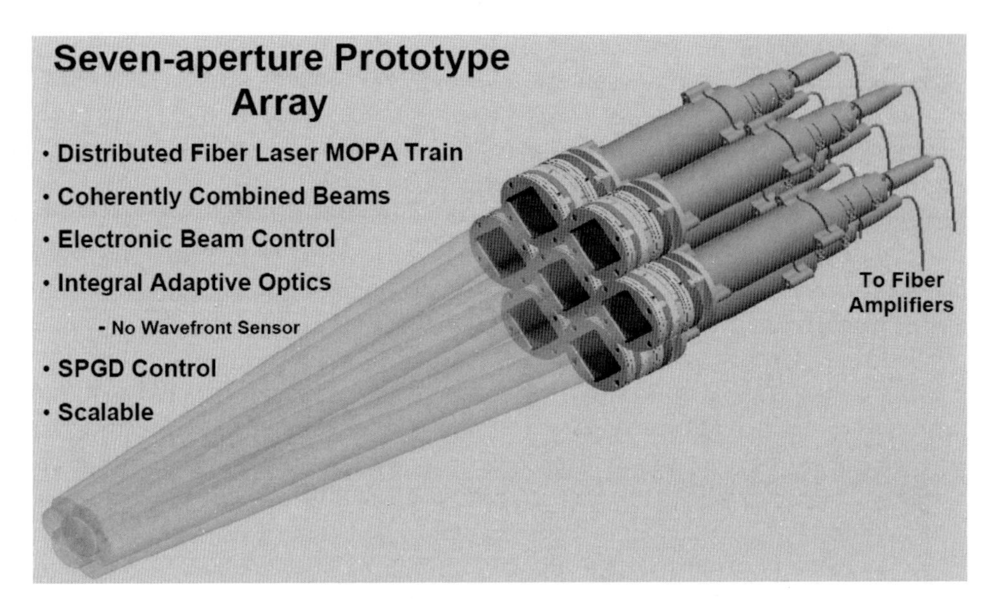

FIGURE 4.4 Overview of the Adaptive Photonic Phase Locked Elements (APPLE) system, which uses a distributed fiber laser with a master oscillator power amplifier (MOPA) train and coherently combined beams to scale the optical power while correcting for wavefront errors. NOTE: Stochastic parallel gradient descent (SPGD). SOURCE: Raytheon Co. Reprinted with permission.

narrow linewidth for phasing. There are several other efforts to investigate electric laser-based defense systems.[56,57,58,59]

Free-Space Laser Communications

The Air Force 405B program started the thrust toward free-space laser communications in 1971 with a goal of 1 gigabit per second (Gb/s) free-space laser communications, with a flight demonstration in 1979. The Department of Defense (DOD) is interested in free-space laser communications for high-speed communications with mobile platforms (e.g., aircraft, satellites, ground vehicles, dismounted solders). Although the DOD can and does make use of the Internet, there is a strong need to extend high-bandwidth communications to mobile platforms, with RF communications used as the baseline for mobile DOD communications.

[56] Optics. 2010. "Northrop's 100 kW Laser Weapon Runs for Six Hours."

[57] DARPA. 2011. "Compact High-Power Laser Program Completes Key Milestone."

[58] Dorschner, Terry A. 2007. "Adaptive Photonic Phase Locked Elements: An Overview."

[59] Page, Lewis. 2007. "DARPA Looking to Kickstart Raygun Tech." Available at http://www.theregister.co.uk/2007/08/23/darpa_laser_blast_cannon_plan. Accessed October 26, 2011.

At the 2011 Defense Security and Sensing Symposium Fellows luncheon, Larry Stotts, from DARPA, pointed out that the DOD owns 300 megahertz (MHz) of RF bandwidth for communications, which represents the total extent of RF bandwidth legally available to the DOD for communications. The primary disadvantage of free-space optical communications is the limited penetration of significant cloud depths, resulting in DOD programs combining RF and optical communications to maintain continuous link management. Because the technology for laser communications is very similar to that of laser radar sensors, both areas have jointly benefited from the advances in each.[60] However, it is also believed that there is possible synergy by fully merging optical surveillance technology, laser weapon technology, and free-space laser technology based on the reduced communication paths, close integration of sensors, and increased reliability. Traditionally these areas have been addressed as separate technologies, but since progress has been made in many of these fields, integration can clearly result in additional synergies.

Solar Power for Military Applications

The military uses a substantial amount of energy in various forms, with significant logistics complications due to the remote deployment of forces. Reducing the overall cost of energy along with simplifying remote energy-supply methods represents a promising application of solar technologies. The dismounted soldier going into remote areas typically carries a heavy backpack, with batteries representing a significant fraction of the weight. Therefore, the remote charging of batteries is the simplest application of solar power. Because solar power represents one of the primary energy sources for space-based platforms, solar power for space has been one of the sources of funding for very high efficiency solar cells.

There is a strong military interest in developing long-dwell platforms, such as the DARPA Vulture program.[61] Vulture is supposed to fly continuously for 5 years at an altitude above 60,000 ft and is expected to be solar-powered. Another, similar effort is the NASA Helios work, which has performed demonstration flights of a prototype (see Figure 4.5). Integrated Sensor Is Structure (ISIS) is another long-dwell, high-altitude DARPA program; it uses a blimp with solar cells for power.[62]

---

[60] Stotts, L.B., L.C. Andrews, P.C. Cherry, J.J. Foshee, P.J. Kolodzy, W.K. McIntire, M. Northcutt, R.L. Phillips, H.A. Pike, B. Stadler, and D.W. Young. 2009. Hybrid optical RF airborne communications. *Proceedings of the IEEE* 97(6):1109-1127.

[61] *Defense Industry Daily.* 2010. "DARPA's Vulture: What Goes Up, Needn't Come Down." Available at http://www.defenseindustrydaily.com/DARPAs-Vulture-What-Goes-Up-Neednt-Come-Down-04852/. Accessed October 26, 2011.

[62] *Defense Industry Daily.* 2011. "USA's HAA & ISIS Projects Seek Slow, Soaring Surveillance Superiority." Available at http://www.defenseindustrydaily.com/darpas-isis-project-seeks-slow-soaring-surveillance-superiority-updated-02189/. Accessed October 26, 2011.

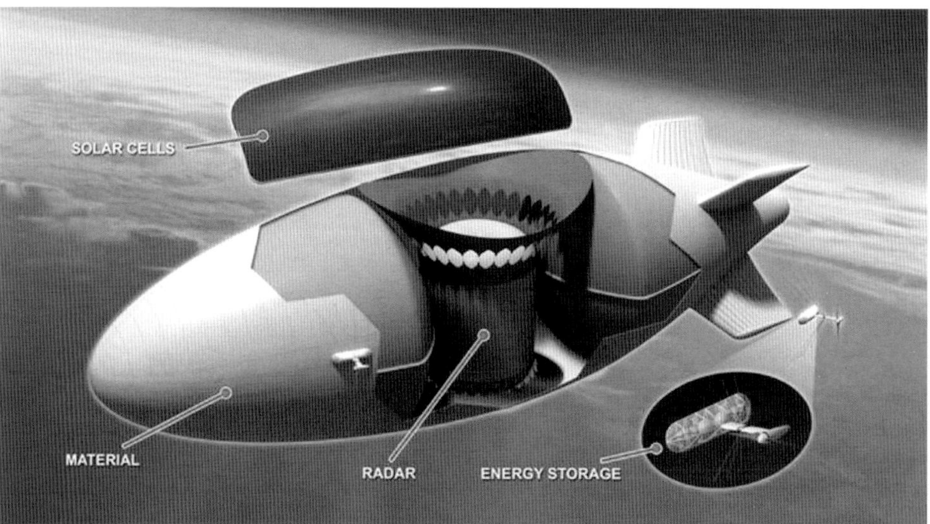

FIGURE 4.5 Overview of long-dwell platforms. (*Top*) The NASA Helios prototype during its test-flight over the Pacific Ocean. SOURCE: *Defense Industry Daily.* 2010. "DARPA's Vulture: What Goes Up, Needn't Come Down."Available at http://www.defenseindustrydaily.com/DARPAs-Vulture-What-Goes-Up-Neednt-Come-Down-04852/. (*Bottom*) A depiction of the DARPA Integrated Sensor Is the Structure (ISIS) concept. SOURCE: *Defense Industry Daily.* 2011. "USA's HAA & ISIS Projects Seek Slow, Soaring Surveillance Superiority." Available at http://www.defenseindustrydaily.com/darpas-isis-project-seeks-slow-soaring-surveillance-superiority-updated-02189/.

Integrated Optoelectronics

For more than 40 years, fulfillment of the promise of a truly integrated optoelectronic circuit—that is, a single-crystal monolithically integrated circuit combining lasers, waveguides, modulators, detectors, and amplifiers—has been awaited. Such an advance would enable unprecedented capability for optical systems, much in the same way that electronics have evolved using integrated circuits.[63] It is also known that electronic components are susceptible to being influenced by stray electromagnetic radiation, whereas optical components are not affected by most microwave radiation. As more and more of the "circuit" components are converted from electronics to optics, the vulnerability of the U.S. military's electronic systems to electromagnetic pulse and other electronic vulnerabilities is being reduced.

In order for computing power to continue to adhere to Moore's law, it is likely that the integration of optics and electronics in a single chip will be required (see the discussion in Chapter 3 in this report). However, as speed increases, it is important to integrate optics and electronic functions seamlessly in very close proximity, reducing communication time between functions best done in optics and functions best done in electronics. The most promising advances for defense applications have been in the development of indium phosphide (InP)-based subsystems, which, although more costly than silicon, allow the needed subsystems to be created with a single material. There is currently very little InP work being done in the United States, and limited trusted foundries are suitable for these applications. The potential for scaling processing power beyond that possible with electronics, along with the reduction in electromagnetic interference and power requirements, makes this a very promising technology for dealing with the large data rates being generated with new optical sensors.

MANUFACTURING

In order for the United States to maintain leadership in advanced defense systems, it is critical for the nation to be at the forefront of both research and manufacturing. Defense systems have unique needs that require both first access and assured access to important technology components, and both types of access are compromised if the manufacturing capabilities do not exist within the United States. There has been a steady migration of photonics manufacturing overseas[64] at precisely the same time that these technologies are becoming critical in defense applications. Some of this migration has been driven by the need to cut costs for high-volume consumer products, but there is an alarming shift of manufacturing

[63] Yariv, A. 1981. Integrated optoelectronics. *Engineering and Science* 44(3):17-20.

[64] NAS-NAE-IOM. 2007. *Rising Above the Gathering Storm: Energizing and Employing America for a Brighter Economic Future.* Washington, D.C.: The National Academies Press.

for critical items due to the International Traffic in Arms Regulations (ITAR). Although these regulations were originally designed to keep critical technologies out of the hands of adversaries, their current implementation has created an incentive for companies to move manufacturing overseas in order to be well positioned for commercial applications of their technologies. This has resulted in some companies moving the related research and development efforts to be near the manufacturing for a more rapid development cycle. A previous National Research Council study has reported on the impacts of ITAR controls on U.S. technology.[65] With insufficient funding from the DOD to maintain first and assured access to many critical photonics components, companies are unable to maintain a manufacturing capability when potentially larger commercial markets are restricted. Manufacturing for cutting-edge photonics has become increasingly globalized over the past several years, and ITAR controls have not been changed to reflect these shifts. The committee understands that the ITAR issue is very complex in scope and cannot even begin to be fully addressed in this report, but the committee also recognizes that the issue directly affects the defense optics and photonics community in a negative fashion.

Another important consideration for manufacturing related to defense technologies is the spiral threading of innovations in optics and photonics that feeds itself. For example, improvements in lasers (i.e., stability, agility, and efficiency) and detectors (i.e., arrays, expanded wavelengths, improved efficiency, bandwidth) have enabled new remote sensing capabilities over the past decade. The developments in lasers and detectors have also led to the improved manufacturing of devices, which further improves the devices for sensors, and also improves manufacturing, in a continuous loop. Thus, there are secondary impacts as a consequence of the progressive loss of photonics manufacturing in the United States.

U.S. GLOBAL POSITION

For many years, the United States has taken for granted its position as one of the leaders in defense technologies. However, several trends have been developing over the past few decades that seriously threaten that position. The military capabilities of other countries have been expanding quickly, as sensor systems become the next battleground for dominance in ISR, with optics- and photonics-based sensors representing an increasing fraction of ISR systems. During difficult economic times, long-term R&D is an easy target for cuts in favor of shorter-term applications. However, it is the long-term developments that provide the most significant

[65] National Research Council. 2008. *Space Science and the International Traffic in Arms Regulations: Summary of a Workshop*. Washington, D.C.: The National Academies Press.

advantages for defense applications. The current U.S. position relies on leadership in research, which is also being negatively impacted by the manufacturing trends.

The U.S. defense STEM workforce in photonics and other areas will be significantly diminished owing to retirements over the next decade, whereas the technical workforce of potential adversaries is expanding rapidly.[66] For example, the College of Optics and Photonics at the University of Central Florida has only approximately 40 percent U.S. nationals in its graduate optics program,[67] and according to the 2009 Program for International Student Assessment (PISA), the United States ranks 17th in science and 25th in mathematics education. Figure 4.6 shows the percent age of U.S. national PhDs in physics in U.S.-based institutions of higher learning between 1969 and 2008. Although optics accounts for only one part of the physics student population, this trend should give a good idea of the percentage of foreign optics graduate students in U.S. graduate schools.

These trends in the STEM workforce are creating a tipping point for photonics defense work, with fewer individuals who are capable of obtaining security clearances being trained in the United States. If the United States continues to shrink its STEM workforce and market share in photonics, innovations in research will bolster the economy and the defense technology of countries poised to take advantage of those advances.

The trends in manufacturing are further straining the U.S. position in photonics. It is critical for the United States to be at the forefront of both research and manufacturing in order to maintain a leadership position in photonics for defense applications. The need for first and assured access combined with ITAR controls imposes an additional need on the United States for a U.S.-based manufacturing capability for these technologies. ITAR controls have also hastened the steady migration of photonics manufacturing for advanced technologies overseas, where companies want to be positioned for commercial applications of those technologies. These companies have also begun moving research groups overseas to facilitate a rapid development cycle for such capabilities. Although the ITAR controls were meant to keep critical technologies out of the hands of adversaries, they are reducing the effectiveness of technology development for defense applications. When coupled with the workforce trends, the U.S. position in photonics for defense applications is potentially reaching a tipping point, which must be reversed in order to maintain leadership in critical areas for defense.

---

[66] NAS-NAE-IOM. 2007. *Rising Above the Gathering Storm: Energizing and Employing America for a Brighter Economic Future.* Washington, D.C.: The National Academies Press.

[67] Private communication with Dr. M.J. Soileau, Vice President for Research and Professor of Optics, Electrical and Computer Engineering, and Physics, at the University of Central Florida.

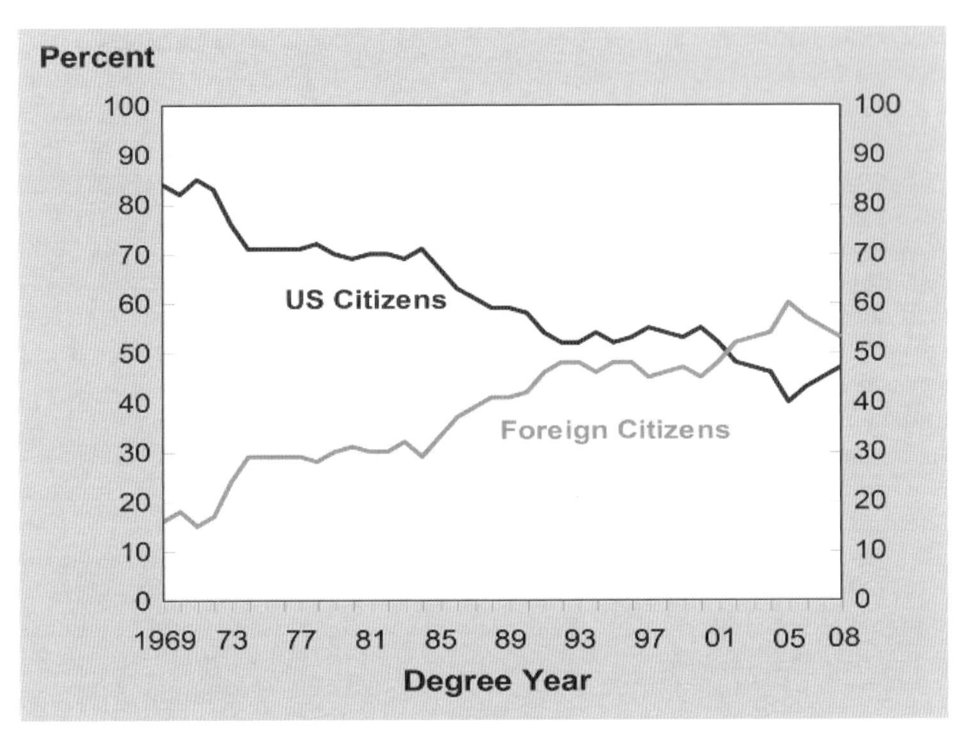

FIGURE 4.6 Citizenship of physics PhDs in U.S. institutions of higher learning from 1969 through 2008. SOURCE: Reprinted, with permission, from Mulvey, Patrick J., and Starr Nicholson. 2011. *Physics Graduate Degrees: Results from the Enrollments and Degrees and the Degree Recipient Follow-up Surveys*, available at http://www.aip.org/statistics/trends/reports/physgrad2008.pdf.

FINDINGS AND CONCLUSIONS

Finding: The committee notes that there have been several areas of optics and photonics with significant advancement for defense and security since the NRC's report *Harnessing Light: Optical Science and Engineering for the 21st Century* was published in 1998. These areas include the following:

- Long-range, laser-based identification capabilities, including multiple aperture and synthetic aperture demonstrations, wide-area passive surveillance capabilities in the visible and infrared regions, and signal processing capabilities to handle some of the new sensor data;
- Long-range, high-powered laser demonstrations from flight platforms for

intercepting ballistic missiles. Although these programs have had successful demonstrations, the DOD has not set a roadmap for this technology; and
- High-speed free-space laser communication.

Key Conclusion: There is possible synergy between optical surveillance technology, laser weapon technology, and free-space laser technology. Because of organizational and funding issues within the Department of Defense, these technical areas have been pursued mostly as separate technologies. Great progress has been made, as highlighted above, but it is likely that a higher level of cooperation can result in additional synergies.

Conclusion: The findings of previous National Research Council studies reporting the potential workforce shortages for the United States in the areas of science, technology, engineering, and mathematics are consistent for the areas of optics and photonics in relation to defense and security. There are additional constraints for the defense workforce, which requires either a sufficient number of qualified U.S. nationals or a new way of leveraging uncleared individuals in the U.S. defense workforce, and will be significantly impacted by a decrease of senior personnel due to the retirement of a disproportionately older workforce over the next 15 years.

Conclusion: It is possible that the United States is losing both first access and assured access to critical optics and photonics technologies at precisely the same time that these capabilities are becoming a crucial defense technological advantage. This problem, which is not unique to photonics within defense-related technologies and systems, is believed to be primarily due to these factors:

- The ongoing migration of optics and photonics capabilities offshore as the manufacture and assembly of these components and systems becomes increasingly globalized; and
- The inability of companies to maintain a U.S.-based manufacturing capability for critical technologies when the larger commercial markets are restricted due to ITAR controls, which have not been changed to reflect the globalization of manufacturing for cutting-edge photonics systems and components.

Key Finding: Silicon-based photonic integration technologies offer great potential for short-distance applications and could have a great payoff in terms of enabling continued growth in the function and capacity of silicon chips if optics for interconnection could be seamlessly included in the silicon complementary metal oxide semiconductor (CMOS) platform. It is also highly likely that integrated optoelectronics in InP is a critical development area with significant growth potential for continuing the advance of defense systems.

RECOMMENDATION AND GRAND CHALLENGE QUESTIONS

On the basis of the conclusions presented above, the committee makes the following recommendation in order to enable the United States to maintain a competitive position in optics and photonics for security and defense:

Key Recommendation: The U.S. defense and intelligence agencies should fund the development of optical technologies to support future optical systems capable of wide-area surveillance, exquisite long-range object identification, high-bandwidth free-space laser communication, "speed-of-light" laser strike, and defense against both missile seekers and ballistic missiles. Practical application for these purposes would require the deployment of low-cost platforms supporting long dwell times.

These combined functions will leverage the advances that have been made in high-powered lasers, multi-function sensors, optical aperture scaling, and algorithms that exploit new sensor capabilities, by bringing the developments together synergistically. These areas have been pursued primarily as separate technical fields, but it is recommended that they be pursued together to gain synergy. One method of maintaining this coordination could include reviewing the coordination efforts among agencies on a regular basis.

This key recommendation leads directly to **the third grand challenge question:**

3. How can the U.S. military develop the required optical technologies to support platforms capable of wide-area surveillance, object identification and improved image resolution, high-bandwidth free-space communication, laser strike, and defense against missiles?

Optics and photonics technologies used synergistically for a laser strike fighter or a high-altitude platform can provide comprehensive knowledge over an area, the communications links to download that information, an ability to strike targets at the speed of light, and the ability to robustly defend against missile attack. Clearly this technological opportunity could act as a focal point for several of the areas in optics and photonics (such as camera development, high-powered lasers, free-space communication, and many more) in which the United States must be a leader in order to maintain national security.

5

Energy

INTRODUCTION

This chapter discusses the use of optics and photonics in the generation of energy and the use of optics in lighting. Energy use in data centers is discussed in Chapter 3, "Communications, Information Processing, and Data Storage." Optics and photonics are very important in energy generation, in energy conservation, and in monitoring the effects of both on the environment. The main topic related to energy generation discussed in this chapter is solar energy. As noted in a previous report from the National Research Council (NRC): "Solar energy could potentially produce many times the current and projected future U.S. electricity consumption."[1] Inertial confinement fusion is a potential optical technique that could generate abundant energy in the future[2,3] A recent article discusses the National Ignition Facility, which has a near-term goal of achieving ignition (defined as 1-MJ [megajoule] output energy from a fusion burn for 1 MJ of laser energy

---

[1] NAS-NAE-NRC. 2010. *America's Energy Future: Technology and Transformation.* Washington, D.C.: The National Academies Press, p. 20.

[2] Lindl, J.D., and E.I. Moses. 2011. Special topic: Plans for the National Ignition Campaign (NIC) on the National Ignition Facility (NIF): On the threshold of initiating ignition experiments. *Physics of Plasmas* (18)5:050901-050902. DOI: http://dx.doi.org/10.1063/1.3591001.

[3] Dunne, M., E.I. Moses, P. Amendt, T. Anklam, A. Bayramian, E. Bliss, B. Debs, R. Deri, T. Diaz de la Rubia, B. El-Dasher, J.C. Farmer, D. Flowers, K. J. Kramer, L. Lagin, J.F. Latkowski, J. Lindl, W. Meier, R. Miles, G.A. Moses, S. Reyes, V. Roberts, R. Sawicki, M. Spaeth, and E. Storm. 2011. Timely delivery of laser inertial fusion energy (LIFE). *Fusion Science and Technology* 60:19-27.

input into the target system).[4] This paper mentions that plans are already underway to define the laser driver for inertial fusion energy applications. However, inertial confinement fusion is not discussed in greater detail in this chapter partially because a current NRC study is specifically addressing this topic; the interim report of that study was recently released.[5] Optics can also be used in isotope separation,[6] but very little information is available on that subject at the unclassified level, and so it was not considered in this unclassified study. Also, optical sensors can be used to assist in oil drilling and recovery, as is briefly discussed in Box 5.1.

Solar energy is discussed here with respect to its primary use by electric utilities competing against other forms of electricity generation such as new natural gas- or coal-fired electricity generation plants. For an electric utility, sunlight can be concentrated to generate solar power in a manner potentially more cost-effective, depending on the cost of the concentrating technology, than would be possible with competing forms of electricity generation in new electric power plants. Concentrating solar power (CSP) uses a heated liquid and a turbine. The heated liquid stores energy until it is converted to electricity. This is advantageous because one of the issues with solar energy generation is the storage of energy for periods of time when the Sun is not out, such as at night and in overcast situations. Concentrated photovoltaic (CPV) power generation involves the use of solar cells after the incoming light is concentrated. For CPV, the price of the actual solar cell is not as critical in that the area covered by solar cells can be reduced up to 2,000 times as compared

[4] From Willner, A.E., R.L. Byer, C.J. Chang-Hasnain, S.R Forrest, H. Kressel, H. Kogelnik, G.J. Tearney, C.H. Townes, and M.N. Zervas. 2012. Optics and photonics: Key enabling technologies. *Proceedings of the IEEE* 100(Special Centennial Issue):1604-1643: "The year 2009 saw the completion of the National Ignition Facility (NIF), a 2-MJ, single shot laser facility at Lawrence Livermore National Laboratory (Livermore, CA) designed to compress targets to generate fusion burns and ignition of a target for energy generation. The NIF laser has now been operating for three years at close to 200 shots per year, with a greater than 95% availability rate for requested shots on targets. The goal in the near term is to achieve ignition defined as 1-MJ output energy from a fusion burn for 1 MJ of laser energy input into the target system. Plans are already underway to define the laser driver for inertial fusion energy applications." Ibid. p. 1609: "We know that a fusion reaction works at even larger power scales. What we have yet to demonstrate is a nuclear burn in a laboratory under controlled conditions. When this is accomplished, it will be a "man on the moon" moment. Laser inertial fusion will open the possibility of amplifying the laser drive power by 30-100 times and in turn allow the operation of an electrical power plant with GWe output for 35-MWe laser power input. Based on our knowledge of the rate at which new infrastructure is adopted, we can predict that fusion energy will take 25-50 years to make a significant impact on our energy supply. By that time it will probably be known simply as laser energy" (p. 1608).

[5] National Research Council. 2012. *Interim Report—Status of the Study "An Assessment of the Prospects for Inertial Fusion Energy."* Washington, D.C.: The National Academies Press.

[6] Broad, W.J. 2011. "Laser Advances in Nuclear Fuel Stir Terror Fear." *New York Times.* August 20. Available at http://www.nytimes.com/2011/08/21/science/earth/21laser.html?pagewanted=all. Accessed July 24, 2012.

BOX 5.1
Oil and Gas Production

Optical systems are increasingly being used by the oil and gas industry as a means for monitoring wells, allowing increased production, and mitigating risks. Industry adoption of optics has been relatively recent, as the high pressure and temperature conditions in a well reduce the lifetime of conventional fiber optics to be substantially shorter than the 20 years over which most wells are expected to produce. Since 2000, fiber-based distributed temperature sensors have become common tools for monitoring the performance of wells, and have proven to be a robust source of information about the well performance.

Distributed temperature sensors rely on the fiber locally changing temperature and scattering light back up the fiber owing to the temperature change. Thus, when combined with a pulsed laser source, the backscattered light allows a temperature profile of the well to be determined. The temperature of the fiber is changed locally owing to fluids flowing into the well bore (Joule-Thomson effect). This information is combined with geothermal models to accurately locate and quantify fluid flows in the well.

More recently, fiber systems have been deployed for distributed acoustic sensing (DAS), to monitor well activity during several phases of well completion. DAS uses a principle similar to that for distributed temperature sensors, but uses acoustic waves generated from within the well to alter the refractive index of the fiber probe. The backscattered radiation from this index variation can be collected and processed to discriminate between various sources and depths. These systems offer reliable discrimination between perforation clusters that are active during the acid injection stage and those that are taking most of the proppant throughout the job. This technology also shows promise for many other functions, such as sand detection and gas breakthrough detection.

SOURCES: Algeroy, J., J. Lovell, G. Tirado, R. Meyyappan, G. Brown, R. Greenaway, M. Carney, J. Meyer, J. Davies, and I. Pinzon. 2010. Permanent monitoring: Taking it to the reservoir. *Oilfield Review* 22(1):34-41; Molenaar, M., D. Hill, P. Webster, E. Fidan, and B. Birch. 2011. "First Downhole Application of Distributed Acoustic Sensing (DAS) for Hydraulic Fracturing Monitoring and Diagnostics." Society of Petroleum Engineers Hydraulic Fracturing Technology Conference, January 24-26, 2011, The Woodlands, Tex.

to non-concentrating systems. With CPV, higher-cost, and more efficient, solar cells can be used. The cost of the concentrators will be the main issue, as stated in the NAS-NAE-NRC report *Overview and Summary of America's Energy Future: Technology and Transformation* published in 2010: "In general, nearly all of the costs involved in using renewable energy for power generation are associated with the manufacturing and installation of the equipment."[7] For high-concentration CPV, more concentrating options are available for larger electric plants, and so CPV may favor utility-scale plants over the smaller plants used by an individual. Photosynthesis is also briefly discussed below as another method of capturing light from the Sun and turning it into energy. However, it is clear that a successful deployment of

[7] NAS-NAE-NRC. 2010. *Overview and Summary of America's Energy Future*, p. 22.

solar power on a large scale in the United States will need the technology to have reached cost parity with current energy sources. The energy delivered to the grid as electricity will, for solar, need to reach "grid parity"; the faster solar power can do that, the better for the U.S. economy as a whole.

Solid-state (SS) lighting is clearly the next step in lighting. It has already become entrenched in many niche applications and is moving quickly toward adoption as the new standard for general lighting. The record efficiency for SS lighting today is 231 lumens per watt (lm/W),[8] compared to 4-15 lm/W for a conventional incandescent bulb, and approximately 55 lm/W for a compact fluorescent bulb. Cost is the main issue preventing widespread adoptions of SS lighting, but substantial progress is being made in lowering the cost of light-emitting diodes (LEDs).

In the 1998 NRC study *Harnessing Light: Optical Science and Engineering for the 21st Century*,[9] solar cells are discussed for space application, with a short section on terrestrial application. At the time of that publication, the cost to purchase solar cell panels for terrestrial application was $4.50 per watt (W), compared to the current cost, which is as low as $0.75/W. LEDs, also called SS lighting, were barely discussed in the 1998 report, in which Figure 3.12 shows efficiencies from 20-80 lm/W for LEDs, compared to the current record of 231 lm/W.

SOLAR POWER

Solar power has received great interest recently as a renewable energy solution capable of providing energy independence and environmental stability while beginning to deliver power at a competitive price. Because of high costs relative to other energy generation technologies, solar power currently satisfies only a small fraction of the world's energy need. Government support and private investments have led to a boom of technical advances over the past years that have continued to drive solar power toward eventually being a mainstream source of power.[10]

Solar power has great potential for home, off-grid, and utility-scale generation. Solar cells have become common in applications for which grid power is not accessible (see Figure 5.1) or convenient, such as powering remote devices or small handheld electronics. As will be seen in this chapter, utility-scale solar plants are not yet competitive with alternative sources of energy, yet the cost of solar electricity

[8] More information about Cree is available at http://www.cree.com/press/press_detail.asp?i=1304945651119. Accessed December 17, 2011.

[9] National Research Council. 1998. *Harnessing Light: Optical Science and Engineering for the 21st Century.* Washington, D.C.: National Academy Press.

[10] Tyner, C. E., G.J. Kolb, M. Geyer, and M. Romero. 2001. "Concentrating Solar Power in 2001—An IEA/SolarPACES Summary of Present Status and Future Prospects." International Energy Agency—SolarPACES. Available at http://www.solarpaces.org/Library/docs/CSP_Brochure_2001.pdf. Accessed August 1, 2012.

FIGURE 5.1 Small solar cell installation monitoring the water supply in Shenandoah National Park in Virginia.

is dropping, and it is reasonable to see a future in which solar power will be cost-competitive for new electric power plants.

It typically requires a large capital investment to build solar plants for a life cycle of a few decades, but once built they do not require period payments for fuel, like fossil-fuel-powered electric plants do. Comparisons are therefore somewhat complicated. A common metric for the economic viability of a solar installation is dollars per watt produced ($/W), which divides the module cost by the peak output power.

A more extensive measure of the economic feasibility of a solar power system is provided by the levelized cost of energy (LCOE), which is commonly measured in cents per kilowatt-hour (¢/kWh). This is a more realistic cost for an investor to use when comparing a solar plant to other potential electricity generation alternatives. The LCOE can be expanded to include known factors such as the financing structure of the plant, incentives, and system degradation. Projecting the LCOE accurately requires detailed information both on the system performance over the lifetime of the plant and on the financial structure of the installation.[11] The complexity and variability of the assumptions used in this metric highlight a need for

[11] Cambell, M. 2008. "The Drivers of the Levelized Cost of Electricity for Utility-Scale Photovoltaics." San Jose, Calif.: Sunpower Corporation. Available at http://large.stanford.edu/courses/2010/ph240/vasudev1/docs/sunpower.pdf. Accessed July 24, 2012.

better economic modeling. A solar plant is different from a fossil-fuel electricity generation plant in that most of the cost is in building the solar plant, but no fuel must be purchased on a regular basis.

In 2011, the United States launched the SunShot Initiative (see Box 5.2 and Figure 5.2.1), with the goal of advancing solar technology to be competitive with conventional sources of electricity. This cost comparability, or grid parity, is very important to the widespread adoption of solar power. Traditionally, grid parity is

BOX 5.2
The SunShot Initiative

"The DOE [Department of Energy] SunShot Initiative is a collaborative national initiative to make solar energy cost competitive with other forms of energy by the end of the decade. Reducing the installed cost of solar energy systems by about 75 percent will drive widespread, large-scale adoption of this renewable energy technology and restore U.S. leadership in the global clean energy race." (See Figure 5.2.1.)

FIGURE 5.2.1 In this photograph of a concentrating solar power (CSP) technology, stretched membrane heliostats with silvered polymer reflectors will be used as demonstration units at the Solar Two central receiver. The Solar Two project will refurbish this 10-megawatt central receiver power tower, known as Solar One. SOURCE: Image available at https://www.eeremultimedia.energy.gov/solar/.

SOURCE: The quotation above from the Department of Energy is available at http://www1.eere.energy.gov/solar/sunshot/. Accessed November 23, 2011.

defined as cost per watt (energy/time) but occasionally as cost per kilowatt-hour (energy). If watts are used, grid parity is reached when the power plants are equal in price to build; if kilowatt-hours are used, grid parity is reached when "selling" the different energy sources costs the same. In the latter case, solar includes the free energy from the Sun, whereas coal-fired electricity generation includes the cost of coal. The goal of competitive solar energy requires an LCOE of approximately \$0.06/kWh, which corresponds to approximately \$1.00/W installed for a module.[12] Installed cost will of course be more than production costs. There are already solar modules less the \$1.00/W for module cost.

Solar technologies fall primarily into two broad categories: (1) photovoltaics (PV), which convert solar radiation directly into electricity, and (2) concentrating solar power. CSP uses optical elements to focus the Sun's energy, and a heated liquid, along with a heat engine, to generate electricity. Once the Sun's light is focused, it can heat an intermediate material, which can be used to drive a turbine to generate electricity. Concentrated photovoltaics also concentrate the Sun's radiation, but not to the same extent that CSP does. With CPV, the light is concentrated so that high-efficiency solar cells can be used without there being too much concern about cell cost. The cost of the concentrating optics must of course now be considered. With current solar cells, the temperature of the cells must be kept near room temperature in CPV in order to maintain high efficiency, and so an efficient heat-removal process must be used. CPV solar cells have been measured at 1,000 Suns concentration and 43 percent efficiency,[13] and it is anticipated that at least 2,000 Suns concentration can be used, although it may cost some conversion efficiency. Below, solar conversion optics that can work at higher temperatures are discussed. These options also provide the potential to allow higher Sun concentration.

Photovoltaic Systems

As of 2010, approximately 39,611 megawatts (MW) of PV capacity had been installed around the world, with most of that in Europe.[14] Between 2010 and 2017, the installed capacity of PV is expected to increase to more than 185,000 MW, with Europe continuing to have the largest portion, but with the rest of the world gaining market share as costs continue to decrease and newer technologies come to market. The growth projections over this forecast period are shown in Figure 5.2.

---

[12] More information on the U.S. Department of Energy SunShot Initiative is available at http://www1.eere.energy.gov/solar/sunshot/. Accessed October 25, 2011.

[13] Wiemer, M., V. Sabnis, and H. Yuen. 2011. 43.5% efficient lattice matched solar cells. *Proceedings of the SPIE* 8108:810804-810804-5.

[14] In comparison, the total net electricity generation for the United States alone was 602,076 MW in 2010. More information is available at www.eia.gov/electricity/annual/pdf/table1.1.a.pdf. Accessed June 4, 2012.

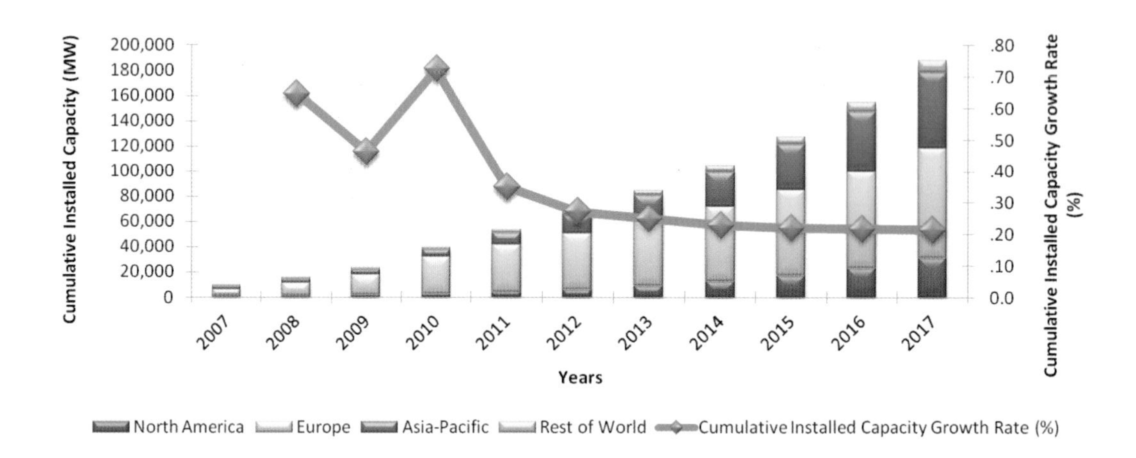

FIGURE 5.2 Forecasted installed capacity and growth rate of photovoltaic devices divided regionally. The base year is 2010. SOURCE: "Global Solar Power Market." Frost and Sullivan Analytics. July 2011. Reprinted with permission.

The compound annual growth rate of these devices is projected to be 25 percent from 2010 to 2017.

Despite the rapid growth in installed capacity over the next several years, photovoltaics are still expected to provide for only a small part of the energy demand in the regions highlighted in Figure 5.2. As a point of comparison, the total electricity generated in the United States over the last few years has ranged from about 300,000 MW to 400,000 MW.[15] The primary barrier preventing wider acceptance of utility-scale PV generation is that it is still expected to be more expensive than alternative energy sources over this 2010-2017 time frame. The PV industry is still largely reliant on government subsidies and incentives to provide power at an LCOE competitive with alternative energy sources. It is expected that a competitive solar technology must be below $1.00/W peak installed cost. Most current PV systems are substantially above this mark, even for the production cost alone, as shown in Figure 5.3. There will be installation costs and possible cost of land use for electric plants. As with all methods of generating electricity, there will be environmental considerations, such as the impact on a desert when large areas are covered by reflectors or solar cells. Current cost figures represent a drastic decrease from the average production cost of $4.50/W in 1998, when the NRC's *Harnessing Light*[16] was published.

[15] The U.S. Energy Information Administration's *Electricity Monthly Update* is available at http://www.eia.gov/electricity/monthly/update/. Accessed July 24, 2012.

[16] National Research Council. 1998. *Harnessing Light.*

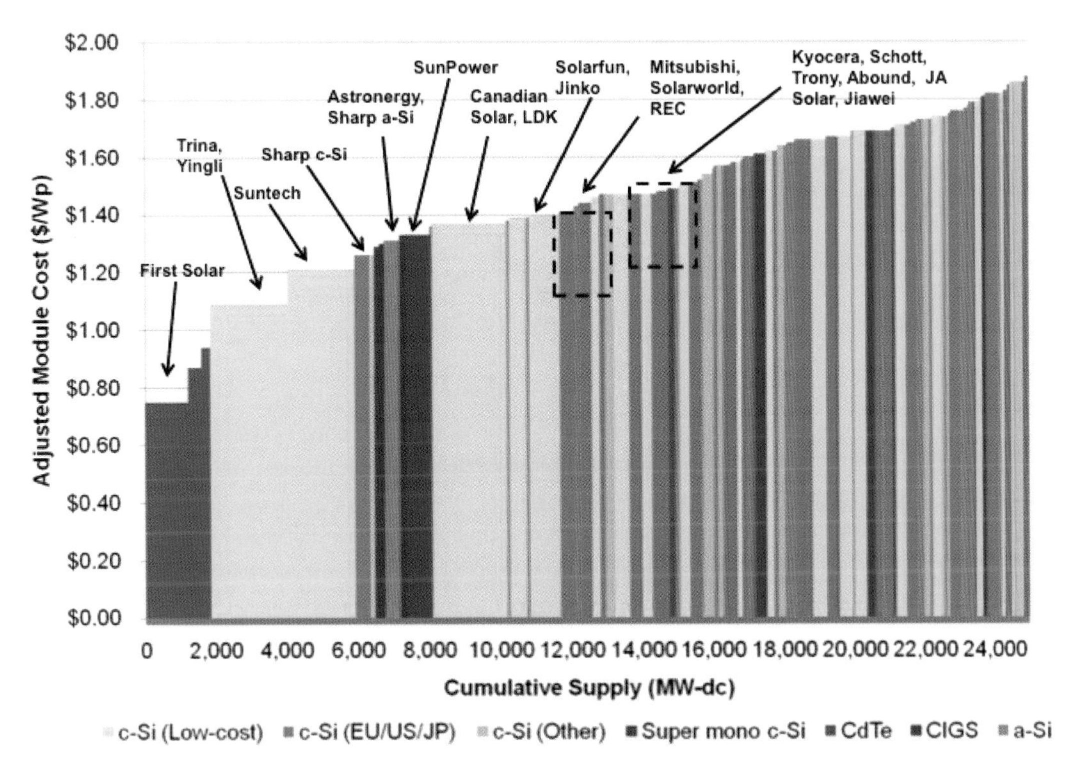

FIGURE 5.3 Average cost per watt of current widely deployed photovoltaic technologies in 2011. SOURCE: "PV Technology, Production and Cost Outlook: 2010-2015" (GTM Research). Reprinted with permission.

Investor literature for major solar cell manufacturers does promise reaching the cost parity level of $1.00/W for installed cost around 2015.[17,18] Obviously investor literature may be optimistic, but major PV solar cell companies at least have plans on how to meet the dollar per watt cost goals. The LCOE of photovoltaics continues to drop due to manufacturing infrastructure improvements and more advanced technologies becoming available.[19] Figure 5.4 shows installed cost LCOEs for a number of installed projects based on a very recent National Renewable Energy Laboratory article.[20]

[17] More information on the First Solar corporate overview, third quarter, 2011, is available at http://www.firstsolar.com/. First Solar corporate overview accessed December 7, 2011.

[18] More information on the SunPower third quarter 2011 earnings is available at http://us.sunpowercorp.com/. SunPower, November 3, 2011, Supplementary Slides accessed December 7, 2011.

[19] For more about information about installed cost time lines for solar technology, see Appendix C in this report.

[20] National Renewable Energy Laboratory. 2011. "Cost of Utility-Scale Solar: One Quick Way to Compare Projects." Available at https://financere.nrel.gov/finance/content/cost-utility-scale-solar-one-quick-way-compare-projects. Accessed July 24, 2012.

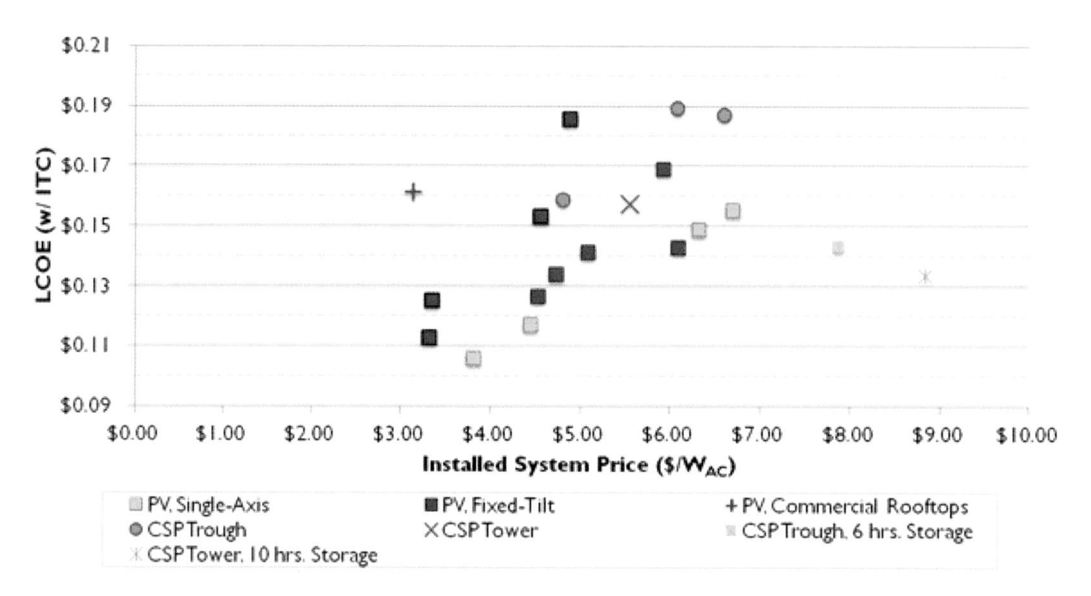

FIGURE 5.4 Levelized cost of energy (LCOE) for installed systems. SOURCE: Publicly available data from the websites of the Department of Energy Loan Program Office, Xcel Energy, *Green Fire Times*, Tri-State Generation, Getsolar, California PUC, *Business Wire*, U.S. Air Force Academy, Juwi Solar, *Winston-Salem Journal*, *iStockAnalyst*, *EarthTechling*, *Environmental Leader*, and *pv-magazine*. Figure from Renewable Energy Project Finance, and available at https://financere.nrel.gov/finance/content/cost-utility-scale-solar-one-quick-way-compare-projects. Accessed July 24, 2012. Reprinted with permission.

Even though Figure 5.4 shows the LCOE for installed systems, there is uncertainty because of assumptions such as expected lifetime.

First-Generation Silicon Cells

Photovoltaic technologies that rely on a silicon (Si) p-n junction are referred to as first-generation technologies, which encompass polycrystalline silicon, monocrystalline silicon, and silicon ribbon technologies. The typical conversion efficiency of these technologies ranges from 14 to 22 percent for a module. A laboratory record of 25 percent was achieved at the University of New South Wales.[21] The theoretical limit for cells of this family is approximately 33 percent conversion efficiency.[22]

---

[21] Green, M. 2009. The path to 25% silicon solar cell efficiency: History of silicon cell evolution. *Progress in Photovoltaics: Research and Applications* 17(3):183-189.

[22] Meillaud, F., A. Shah, C. Droz, E. Vallat-Sauvain, and C. Miazza. 2006. Efficiency limits for single-junction and tandem solar cells. *Solar Energy Materials and Solar Cells* 90(18-19):2952-2959.

In 2010, over 75 percent of the capacity of PV systems installed was first-generation solar cells.[23] Si photovoltaics are the most mature and widely accepted form of PV, and they still exhibit some of the highest conversion efficiencies of available technologies. First-generation PVs are expected to continue to dominate the market through 2017, although they are expected to lose market share to emerging technologies. Production of first-generation solar cells is highly competitive, with no single company holding a market share substantially larger than 10 percent, and different companies leading in each geographic region.[24] The focus for advancing first-generation solar cells is on reducing manufacturing costs. It is not expected that this technology will be capable of being produced for less than $1.00/W installed cost. Figure 5.5 shows first-generation solar cells for power company use.

Second-Generation Photovoltaic Technologies

The second-generation solar cells consist of thin films of silicon, or other semiconductors, deposited on glass or flexible substrates. The most prominent thin-film technologies that have emerged are cadmium telluride (CdTe) (see Figure 5.6), thin-film silicon, and copper indium gallium selenide (CIGS). These films do not convert sunlight into electricity as efficiently as the first-generation silicon devices do, but the potential for reduced manufacturing costs makes them attractive. Second-generation photovoltaics accounted for approximately 21 percent of the global installed capacity in 2010 and are projected to gain market share over the next several years, providing 38.0 percent of installed capacity in 2017.[25]

The second-generation cell accounting for most of this installed capacity is a thin-film CdTe cell. First Solar has commercialized this cell and gained a substantial market share, as the greatly reduced manufacturing cost compensates for the reduced conversion efficiency (approximately 17 percent)[26] compared to standard Si panels. Several large-scale solar projects based on CdTe technology have been built, such as the 80-MW facility at Sarnia in Canada shown in Figure 5.6.

CdTe cells have already reached the $1.00/W goal for production cost, but it is not yet clear how this translates into LCOE. For example, CdTe thin films are relatively new, so there has not been enough time to establish the lifetime and

[23] Frost and Sullivan Research Service. 2011. *Global Solar Power Market.* July. Available at http://www.frost.com/prod/servlet/report-brochure.pag?id=N927-01-00-00-00. Accessed July 24, 2012.

[24] Kann, S. 2010. *The U.S. Utility PV Market: Demand, Players, Strategy, and Project Economics Through 2015.* GTM Research. Available at https://www.greentechmedia.com/research/report/us-utility-pv-market-2015/. Accessed July 24, 2012.

[25] Frost and Sullivan Research Service. 2011. *Global Solar Power Market.*

[26] Kanellos, M. 2011. "First Solar Sets Efficiency Record: 17.3% Percent." Greentech Media. Available at http://www.greentechmedia.com/articles/read/first-solar-sets-efficiency-record-17.3 percent/. Accessed June 6, 2012.

FIGURE 5.5 Silicon solar cells deployed for utility power production. SOURCE: Courtesy of Suntech Power Holdings, Inc.

FIGURE 5.6 Eighty-megawatt thin-film cadmium telluride (CdTe)-based solar plant in Canada. SOURCE: Courtesy of First Solar.

degradation rate accurately under actual outdoor conditions. Another concern is that the materials used are rare and extremely toxic. Although initial indicators are that the production and deployment of these cells do not release cadmium into the environment, further adoption will require more extensive environmental impact testing. CdTe technology shows promise of gaining grid parity in a relatively short time frame if these obstacles can be overcome.

An accurate cost benchmark exists for CdTe solar cells. Warren Buffett recently contracted for a 550-MW facility from First Solar for $2 billion,[27] which works out to be $3.63/W. First Solar has started building this facility. If $1.00/W is equivalent to 6 ¢/kWh, then the installed facility would be 21.8 ¢/kWh. Although this is more expensive than a fossil fuel plant, California utilities are required to obtain at least 20 percent of their power from renewable sources and are on a path to a 33 percent requirement by 2020.[28,29]

An alternative embodiment of thin-film technology uses a much more efficient gallium arsenide (GaAs)-based film. A thin-film GaAs solar cell was reported in 2011 by Alta Devices to have an efficiency of 28.2 percent[30] for unconcentrated solar illumination, a record for single-junction devices. Using a process known as epitaxial liftoff in order to reduce manufacturing costs, it is projected to be competitive with CdTe cells currently in production on a dollars per watt basis while delivering almost three times the area efficiency when in production. At present this is just a projection, however.

Several immature technologies in development show promise, but they are not ready for large-scale commercialization. These exotic schemes are frequently grouped into the third generation of photovoltaics. Active research is being carried out on dye-sensitized cells, organic cells, nanocrystal cells, photoelectrochemical cells, and other technologies that have potential to dramatically reduce cost and/or increase efficiency compared to current systems.

[27] RTTNews. 2011. "Buffet's MidAmerican Energy to Buy First Solar's $2 Bln Power Plant." Available at http://www.rttnews.com/1776646/buffet-s-midamerican-energy-to-buy-first-solar-s-2-bln-power-plant.aspx. Accessed December 18, 2011.

[28] More information on the California Renewables Portfolio Standard is available at http://www.cpuc.ca.gov/PUC/energy/Renewables/index.htm. Accessed December 18, 2011.

[29] Sullivan, Colin. 2010. "Calif. Raises Renewable Portfolio Standard to 33%." *New York Times.* July 24. Available at http://www.nytimes.com/gwire/2010/09/24/24greenwire-calif-raises-renewable-portfolio-standard-to-3-24989.html. Accessed December 18, 2011.

[30] Kayes, B., H. Nie, R. Twist, S.G. Spruytte, F. Reinhardt, I.C. Kizilyalli, and G.S. Higashi. 2011. "27.6% Conversion Efficiency, a New Record for Single-Junction Solar Cells Under 1 Sun Illumination." *37th IEEE Photovoltaic Specialists Conference* 000004-000008.

Concentrated Photovoltaics

Concentrated photovoltaics offer an alternative approach, using inexpensive optics to concentrate light onto a small area of a highly efficient photovoltaic cell. By concentrating light from the Sun (see Figure 5.7), less PV material is required. A limitation for high concentration is that the module must be placed on a solar tracker in order to maintain accurate alignment with the Sun, and the accuracy of the tracker must increase as the concentration of the system increases. This limited acceptance angle also dictates that the system is only able to capture direct sunlight, meaning that it is ineffective on cloudy days. The amount of concentration provided by the optics will produce differing systems, which generally fall into the low-concentration (2-50 times) or high concentration (100-2,000 times) categories. This has some similarities to concentrated solar power, discussed below.

The low-concentration systems typically use conventional single-junction solar cells and rely on the optics being inexpensive enough that the reduced PV material required justifies the added cost of concentrating. These systems can frequently be mounted on a one-axis tracking system, and the cells can usually be kept within their normal operating temperature using passive cooling systems.

The high-concentration systems allow for the use of multi-junction cells in the system.[31] Multi-junction cells can reach higher conversion efficiencies than are possible with single-junction PV devices, with the record currently being set by the company Solar Junction at 43.5 percent,[32] eclipsing the recent record by Spire Semiconductor, LLC, of 42.3 percent[33] conversion efficiency. These cells are expensive, but by reducing the area required for a module by two or three orders of magnitude, the cost can be justified. These cells showed a peak efficiency of 43.5 percent between 400 and 600 Suns concentration and maintained efficiencies above 43 percent at 1,000 Suns concentration.[34]

A key driving cost of making these cells is the substrate on which the photovoltaic layers are grown. This material is very expensive and currently must be disposed of after a cell is grown on it. Substantial research is being done on the epitaxial liftoff technique, which allows the substrate to be reused for growing successive cells. The disadvantages of CPV systems are primarily the extra cost associated with

[31] Multi-junction solar cells contain several p-n junctions, each of which can be tuned to a different wavelength of light, thus capturing more of the incoming light.

[32] Wesoff, E. 2011. "Update: Solar Junction Breaking CPV Efficiency Records, Raising $30M." Greentech Media. Available at http://www.greentechmedia.com/articles/read/solar-junction-setting-new-cpv-efficiency-records/. Accessed December 25, 2011.

[33] Wojtczuk, S., P. Chiu, X. Zhang, D. Derkacs, C. Harris, D. Pulver, and M. Timmons. 2010. InGaP/GaAs/InGaAs 41% concentrator cells using bi-facial epigrowth. *35th IEEE Photovoltaic Specialists Conference* 001259-001264.

[34] Wiemer, M., V. Sabnis, and H. Yuen. 2011. 43.5% efficient lattice matched solar cells.

FIGURE 5.7 A high-concentration photovoltaic panel using Fresnel lens optics and III-V solar cells. SOURCE: Courtesy of Amonix.

having a precise two-axis tracker and an active cooling system required to maintain a reasonable chip temperature at high concentration, as well as the limitation to the system's use on sunny days.

Concentrated photovoltaics of both concentration regimes require performance and cost advances before they can become competitive on a utility scale. Moreover, the concentration system, including the tracker, also constitutes a large percentage of the cost. For the high-concentration systems, increasing the performance and reliability of high-efficiency cells will improve the conversion efficiency, and improved thermal management systems would allow the cells to operate at lower temperatures, at which they are more efficient.

Expected Market Share for Various PV Technologies

Over the next few years, it is expected that disruptive new technologies will gradually gain market share from the currently dominant first-generation photovoltaics. It is expected that second-generation photovoltaics will begin to rival the capacity of the first-generation PVs, and concentrating photovoltaics will begin to gain a greater market share.[35] Third-generation systems are much more difficult to predict. The estimate of relative capacities of the various technologies over the short term is shown in Figure 5.8.

Concentrating Solar Power Systems

Concentrating solar power provides an alternative to photovoltaics for producing electricity. A key deficiency of photovoltaics is energy storage, as batteries or other means of storage must be employed to store the power generated when the Sun is out.[36] The need for storage is mitigated if solar power is used to fill a peak-load requirement to meet air conditioning needs. Peak pricing in some cases has made energy storage less desirable. CSP installations have a much more efficient means of storing energy, as the heated intermediate media can be stored and used on demand. CSP systems can have energy-conversion efficiencies comparable to those of photovoltaics.

In 2010, the installed CSP capacity around the world was 1,327 MW, with Europe and North America dominating the market. The installed capacity is expected to increase more than 10-fold by 2017, with more than 15,000 MW of expected installed capacity. North America and Europe are expected to maintain dominance in this market, with the rest of the world gaining some market share. The best

[35] Frost and Sullivan Research Service. 2011. *Global Solar Power Market.*

[36] For more about information about supporting technologies associated with solar power, see Appendix C in this report.

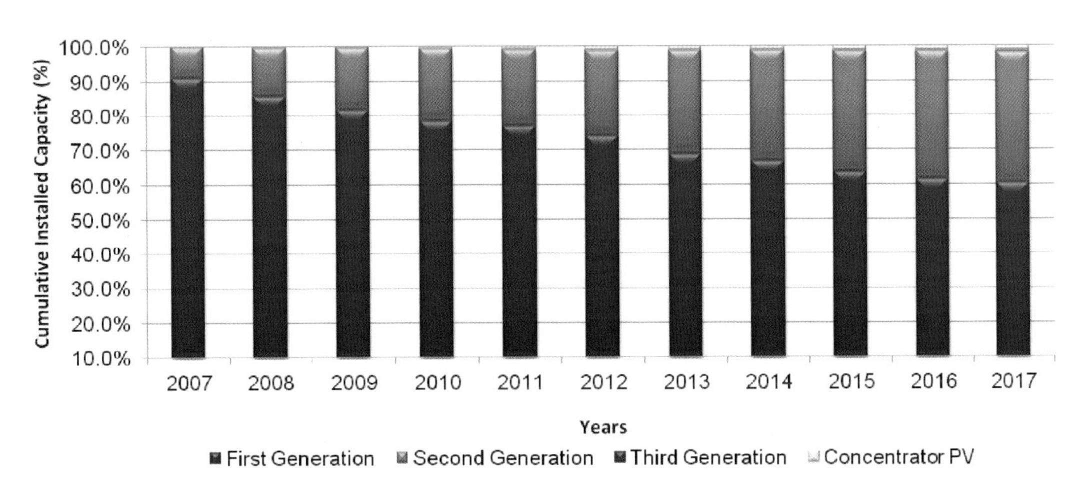

FIGURE 5.8 Market share for various photovoltaic (PV) technologies. SOURCE: "Global Solar Power Market." Frost and Sullivan Analytics. July 2011. Reprinted with permission.

currently implemented CSP installations are producing energy at an LCOE of approximately 12¢/kWh to 15¢/kWh, which is higher than for alternative energy sources. This is expected to drop to below 12¢/kWh by 2017 as manufacturing improves and new technologies are brought to market. CSP functions best on large scales, and thus building an efficient plant requires a large capital investment. The expected growth of installations broken down regionally is provided in Figure 5.9.

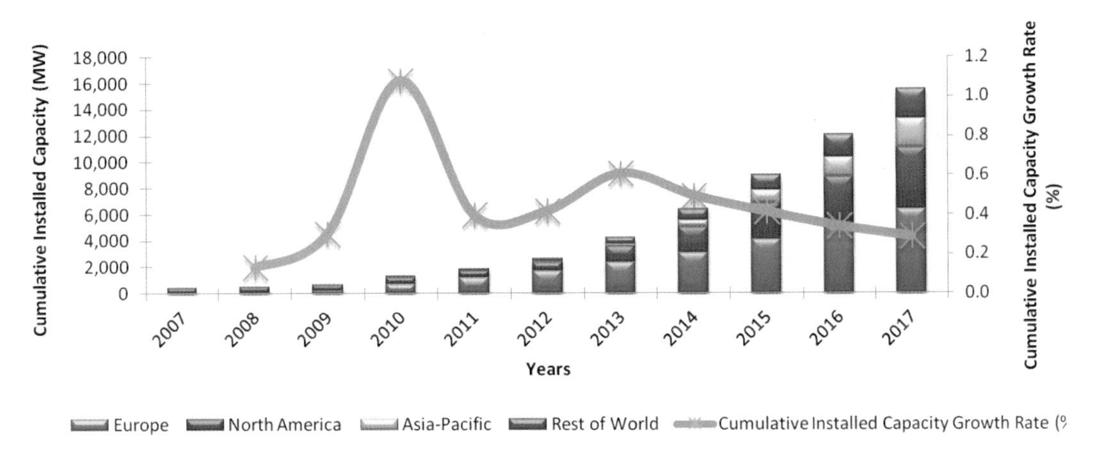

FIGURE 5.9 Expected installed capacity of concentrated solar power market divided regionally, and compound annual growth rate; 2010 is the base year. SOURCE: "Global Solar Power Market." Frost and Sullivan analysis. July 2011. Reprinted with permission.

The compound annual growth rate of the CSP market is expected to be 42.1 percent from 2010 to 2017.[37]

The most common available form of CSP is the parabolic trough concentrator, as seen in Figure 5.10. These systems are usually set up on one-axis tracking systems, which track the Sun over the course of the day. This type of plant consists of a parabolic mirror focusing sunlight onto a tube of heat-transfer medium, usually oil, which is then heated and used to create steam, which can drive an electric generator. The maximum temperature is limited by the oil-based heat-transfer medium and the lower concentration required with single-axis tracking, limiting the efficiency of these systems to approximately 15 percent.[38]

As of 2009, parabolic trough concentrators accounted for more than 83 percent of installed CSP capacity.[39] The oil intermediate is difficult to store efficiently for use during off-peak hours. Hence it is possible that trough technology with oil intermediates might lose market share and will have a harder time reaching grid parity.[40]

The solar tower is a relatively recent incarnation of CSP, using a large field of flat mirrors called heliostats to direct incident sunlight to a focal point in a central tower, as shown in Figure 5.11. A heat-exchange medium, usually a molten salt, absorbs the energy and is heated to temperatures up to 1000°C, which can then be used to generate electricity. This larger temperature difference allows more efficient electricity generation than the common trough technology, with conversion efficiencies in the range of 20 to 35 percent.[41]

Power tower systems offer a much more efficient alternative to the parabolic troughs that currently dominate the CSP landscape. In addition to the greater conversion efficiency, the power towers make more efficient use of the land, and the heliostats are less expensive to manufacture and maintain than the parabolic

[37] Frost and Sullivan Research Service. 2011. *Global Solar Power Market.*

[38] Acciona Energy. *Technology and Experience in Concentrating Solar Power.* Available at http://www.acciona-energia.com/media/315798/Technology%20and%20experience%20in%20concentrating%20solar%20power.pdf. Accessed October 25, 2011.

[39] Greenpeace International. 2009. *Global Concentrating Solar Power Outlook 09: Why Renewable Energy Is Hot.* Amsterdam, The Netherlands: Greenpeace International. Available at http://www.greenpeace.org/international/en/publications/reports/concentrating-solar-power-2009/. Accessed July 24, 2012.

[40] Feldhoff, J.F., D. Benitez, M. Eck, and K.-J. Riffelmann. 2010. Economic potential of solar thermal power plants with direct steam generation compared with HTF plants. *Journal of Solar Energy Engineering* 32:1001-1009.

[41] Hinkley, J., B. Curtin, J. Hayward, A. Wonhas, R. Boyd, C.C. Grima, A. Tadros, R. Hall, K. Naicker, and A. Mikhail. 2011. *Concentrating Solar Power—Drivers and Opportunities for Cost-Competitive Electricity.* Melbourne, Victoria, Australia: CSIRO Energy Transformed. Available at http://www.garnautreview.org.au/update-2011/commissioned-work/concentrating-solar-power-drivers-opportunities-cost-competitive-electricity.pdf. Accessed July 24, 2012.

FIGURE 5.10 A parabolic trough concentrator solar thermal plant in Nevada. SOURCE: Courtesy of Acciona Solar Power.

FIGURE 5.11 Eleven-megawatt Solar Power Tower near Seville, Spain. SOURCE: Courtesy of Abengoa Solar.

mirrors used in trough systems. A primary drawback of tower technology is the large investment required, with only large-scale plants making economic sense.

Dish concentrators (see Figure 5.12) are an emerging technology that functions on a principle similar to that of a solar tower but is scalable to smaller units. A reflective parabolic dish tracks the Sun and focuses onto a small power-conversion unit. A heat-transfer fluid is heated several hundred degrees Celsius and is subse-

FIGURE 5.12 A dish solar thermal power generation module test unit in Phoenix, Arizona. The dish is approximately 75 feet in diameter. SOURCE: Courtesy of Southwest Solar Technologies.

quently used to generate electricity. These systems are expected to have conversion efficiencies similar to those of solar tower concentrators, although dish concentrator systems have not yet been successfully commercialized.

Dish concentrators with a combined capacity of more than 1,700 MW are being developed; over 35 percent of them are in the Asia Pacific region.[42] These systems require that an efficient and inexpensive generator be developed on a much smaller scale than that needed for tower systems. These systems lack the centralized storage capability of the power tower approach, and thus alternate storage methods must be employed to store power.

Hybrid Systems

Some research is being conducted on the possibility of integrating CPV and CSP systems. In a photovoltaic system, the energy that is not converted into electrical power is dissipated as heat, which has the potential to be recycled, especially in a concentrated photovoltaic system. This waste heat can either be recycled to generate more electricity or can serve a variety of other uses, such as the heating of water.[43] The primary difficulty is that most PV chips become less efficient as they increase in temperature. Wide-bandgap materials will work better for solar cells at higher temperatures, but there will still be a loss of efficiency.[44] Researchers at Stanford University are conducting research on a "photovoltaic-like" device that becomes more efficient at higher temperatures, which would be ideal for integration into a hybrid system.[45] The Stanford approach, also discussed in a paper by Andrews et al.,[46] has calculated efficiencies for an idealized device, which can exceed the theoretical limits of single-junction photovoltaic cells. The device is based on thermionic emission of photoexcited electrons from a semiconductor cathode at high temperature. Temperature-dependent photoemission-yield measurements from gallium nitride (GaN) show strong evidence for photon-enhanced thermionic emission. As mentioned by CleanEnergyAuthority,[47] the proposed solar converter

[42] Frost and Sullivan Research Service. 2011. *Global Solar Power Market.*

[43] For information about hybrid solar/wind systems, see Appendix C in this report.

[44] Landis, G.A., D. Merritt, R.P. Raffaelle, and D. Scheiman. 2005. *High-Temperature Solar Cell Development.* Washington, D.C.: National Aeronautics and Space Administration, pp. 241-247. Available at http://ntrs.nasa.gov/archive/nasa/casi.ntrs.nasa.gov/20050206368_2005207966.pdf. Accessed July 24, 2012.

[45] Schwede, J.W., I. Bargatin, D.C. Riley, B.E. Hardin, S.J. Rosenthal, Y. Sun, F. Schmitt, P. Pianetta, R.T. Howe, Z.-X. Shen, and N.A. Melosh. 2010. Photon enhanced thermionic emission for solar concentrator systems. *Nature Materials* 9:762-767.

[46] Andrews, J.C., F. Meirer, Y. Liu, Z. Mester, and P. Pianetta. 2011. Transmission x-ray microscopy for full-field nano imaging of biomaterials. *Microscopy Research and Technique* 74(7):671-681.

[47] More information is available at www.cleanenergyauthority.com. Accessed August 6, 2012.

would operate at temperatures exceeding 200°C, enabling its waste heat to be used to power a secondary thermal engine, boosting theoretical combined conversion efficiencies above 50 percent. There will be very interesting trade-offs if one decides to optimize a hybrid system. It would be possible to make a system optimized for CSP, operating at high temperatures, and then to add high-temperature solar cells to it to boost efficiency. Alternately it would be possible to design an optimum CPV system with high PV efficiency, and then to add whatever power could be generated using a lower-temperature, and therefore less efficient, heat cycle approach to generate electricity. In general, PV makers and CSP companies have not mixed much, and so this is a somewhat unexplored design space with potential to achieve high efficiency.

LCOE Outlook for Solar Power Compared to Other Current and Possible Future Fuel Sources

The most likely initial geographical locations for commercial utility-scale solar power will have high solar irradiance and high energy prices from alternative sources. For areas with high direct solar radiation, such as the southwestern United States, high-concentration photovoltaics are expected to reach the lowest LCOE of any solar technology. These systems are only viable in areas with little cloud cover, where the direct component of solar radiation is much greater than the diffuse component, and thus can be focused. In areas such as the Southwest, energy storage capability may not be a significant factor owing to peak-load pricing considerations. Additionally, the grid needs to be considered in any final estimates of success for solar power. The NRC's *Overview and Summary of America's Energy Future: Technology and Transformation* states:

> Greatly expanding the fraction of electricity generated from renewable sources will require changes in the present electric system because of variability over space and time in the availability of renewables such as wind and the difficulty of scaling up renewable resources. Integrating an additional 20 percent of renewable electricity, whether from wind, solar, or some combination of sources, will require an expansion of the transmission system as well as large increases in manufacturing, employment in the wind power industry, and investment.[48]

The projected LCOE for various technologies deployed in optimal locations for solar energy is shown in Figure 5.13.[49] This forecast cannot predict the effect

[48] NAS-NAE-NRC. 2010. *Overview and Summary of America's Energy Future*, p. 21.

[49] More information is available on the GTM Research webinar "Concentrating Photovoltaics (CPV): Ready for Take-Off?" September 27, 2011. Available at http://www.greentechmedia.com/events/webinar/concentrated-photovoltaics-cpv-now-an-indispensable-part-of-the-energy-equa/. Accessed July 24, 2012.

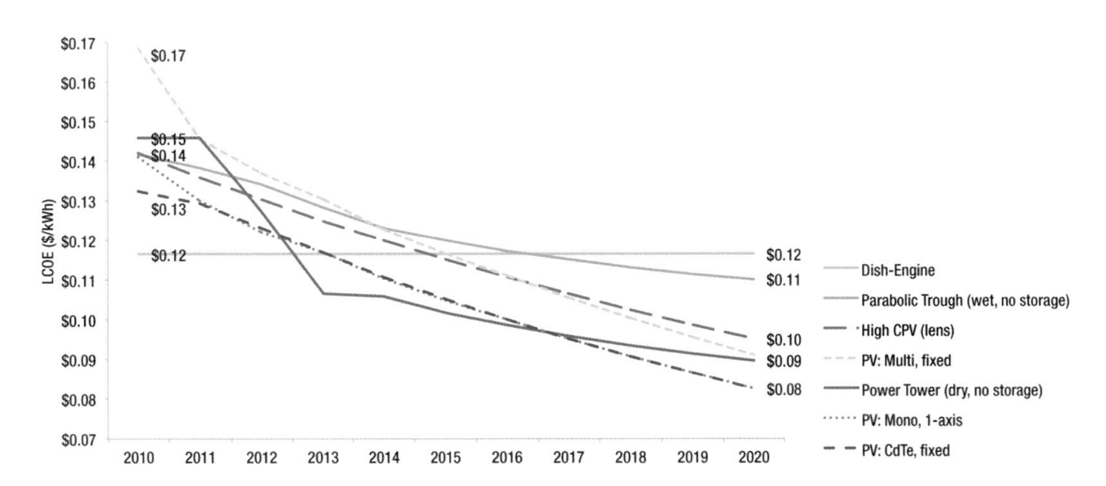

FIGURE 5.13 Levelized cost of energy (LCOE) projections for various solar technologies, 2010-2020. SOURCE: "Concentrating Solar Power 2011: Technology, Costs and Markets" (GTM Research). Reprinted with permission.

of disruptive research and development, but it shows that many solar technologies have the potential to compete with conventional energy sources. The same graph also depicts the LCOE for natural gas, which was at $0.10/kWh in 2010 and was predicted to steadily increase to about $0.14/kWh in 2020, a point at which it is possible that many of the solar technologies will be at a lower LCOE. The recent advances in the production of natural gas owing to fracking will change the cost comparison by making natural gas power plants cheaper than previously envisioned. Production of shale gas from fracking has soared 12-fold between the years 2000 and 2011, driving down the price of natural gas.[50] As can be seen in Figure 5.13, the cost of solar power is projected to keep dropping, and so even with the lower cost comparison, it is anticipated that at some point, and in certain circumstances, solar power will be cost-competitive.

Other alternative methods might become possible in the near future as well. One alternative method of harvesting solar power is through the use of a chemical process that mimics photosynthesis. The creation of fuels from abundant inputs such as water and carbon dioxide (CO_2) using sunlight has the potential to offer conventional liquid fuels while still being carbon-neutral (see Figure 5.14). This technology resembles photosynthesis, in that it involves a chemical reaction on a thin membrane, which then splits the products to be fuel and oxygen. In the case of natural photosynthesis these products are sugars as the fuel and oxygen that is

[50] *The Economist*. 2011. "Fracking Here, Fracking There." Available at http://www.economist.com/node/21540256. Accessed July 24, 2012.

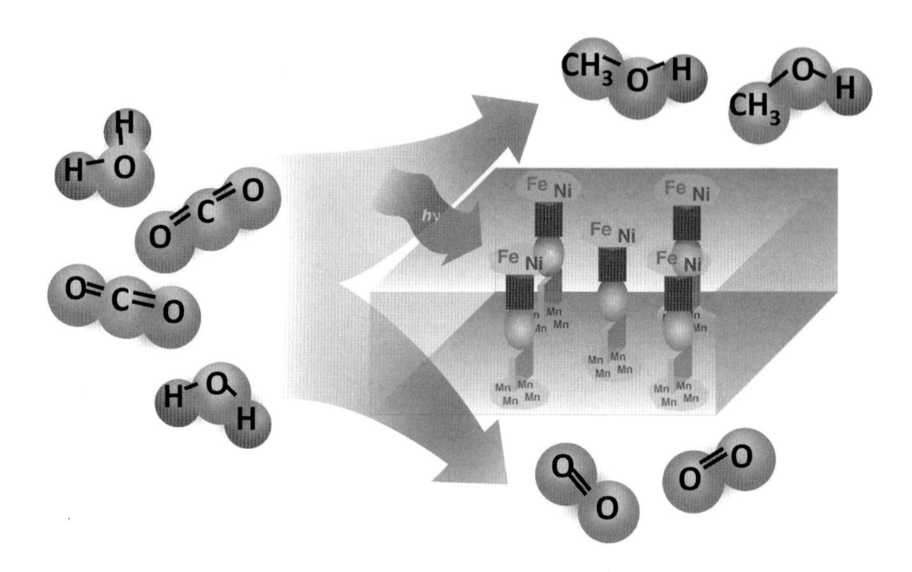

FIGURE 5.14 Water and CO_2 are converted catalytically into fossil fuels and oxygen in the presence of sunlight. The membrane separates the fuel from the oxygen. SOURCE: Joint Center for Artificial Photosynthesis, available at http://solarfuelshub.org/home. Reprinted with permission.

released into the atmosphere. Artificial photosynthesis has many different possible inputs and outputs, and research is being pursued to tune the process in order to maximize catalytic conversion into the standard n-hydrocarbons that can easily be used as fuels.

Research to advance artificial photosynthesis is being carried out at the Joint Center for Artificial Photosynthesis in California.[51]

SOLID-STATE LIGHTING

Incandescent lightbulbs have been in use for well over 100 years. English chemist Humphry Davy demonstrated the first electrically driven incandescent light in 1802. In the ensuing 80 years many people worked on improving the incandescent bulb. Thomas Edison developed the first commercially viable approach.[52] These bulbs are not efficient but are still in widespread use, and as a result, lighting

[51] The website of the Joint Center for Artificial Photosynthesis is available at www.solarfuelshub.org. Accessed October 25, 2011.

[52] Friedel, R., and P. Israel. 1986. *Edison's Electric Light: Biography of an Invention.* New Brunswick, N.J.: Rutgers University Press, pp. 115-117.

consumes more than 20 percent of the electricity used in the United States and Europe and a higher percentage in developing nations.[53] Incandescent bulbs radiate most of the input energy as infrared radiation or heat, not as visible light. Halogens have been added to some incandescent bulbs, allowing a higher operating temperature and slightly more efficiency. These are referred to as halogen bulbs, although neither the efficiency nor the lifetime has been changed substantially. In the 1930s, fluorescent bulbs became available. They are significantly more efficient then incandescent bulbs, but they contain mercury, a hazardous material.[54] Compact fluorescent bulbs fitting most lamps started to be commercially available in 1985.[55] Fluorescent lights take minutes to come to full brightness, which makes some people avoid these lights despite their higher efficiency as compared with incandescent bulbs.[56] Color balance can also be an issue for fluorescents, depending on the phosphors used.

Mercury vapor and sodium vapor lamps are another available lighting option. Mercury vapor and sodium vapor lamps can be very efficient, ranging from 100 lm/W up to 150 or even 200 lm/W.[57] Vapor lamps are used extensively in street lighting, where their long start-up time is not a significant disadvantage. Although people like the color spectrum of mercury vapor lamps, the United States banned them in 2008 and the European Union is banning them in 2015 because of the mercury usage.[58,59] Sodium vapor streetlights have a color spectrum that is less desirable. Figure 5.15 shows the lighting efficiency for various lighting types as of 2008.

A critical factor for consumers purchasing lightbulbs is the light efficiency of the bulb, in terms of both lumens[60] per watt and lumens per dollar. At a wavelength of 555 nanometers (nm), a 100 percent efficient light source would produce

---

[53] Azevedo, I.L., M.G. Morgan, and F. Morgan. 2009. The transition to solid-state lighting. *Proceedings of the IEEE* 97(3):481-510.

[54] It is unlawful to dispose of fluorescent bulbs as universal waste in the states of California, Illinois, Indiana, Michigan, Minnesota, Ohio, and Wisconsin. More information is available at http://www.epa.gov/waste/hazard/wastetypes/universal/lamps/index.htm. Accessed October 25, 2011.

[55] Kane, R, and H. Sell, eds. 2001. *Revolution in Lamps: A Chronicle of 50 Years of Progress* (2nd ed.). Lilburn, Ga.: Fairmont Press.

[56] More information is available through General Electric (GE) Lighting, "Why does my compact fluorescent light bulb flicker or appear dim when I first turn it on?" Compact Fluorescent Light Bulb (CFL) FAQs. GE Lighting. Available at http://www.gelighting.com/na/consumer/education/faqs/. Accessed July 24, 2012.

[57] de Groot, J.J., and J.A.J.M. van Vliet. 1986. *The High-Pressure Sodium Lamp.* Cambridge, Mass.: Macmillan Education.

[58] Mercury vapor banned in 2008; Energy Policy Act of 2005 (Public Law 109-58).

[59] More information on the European Union directive on street, office, and industry lighting is available at http://ec.europa.eu/energy/lumen/professional/legislation/index_en.htm. Accessed July 24, 2012.

[60] Lumen (lm) is the International System of Units (SI) unit of luminous flux, which is a measure of the total amount of visible light emitted.

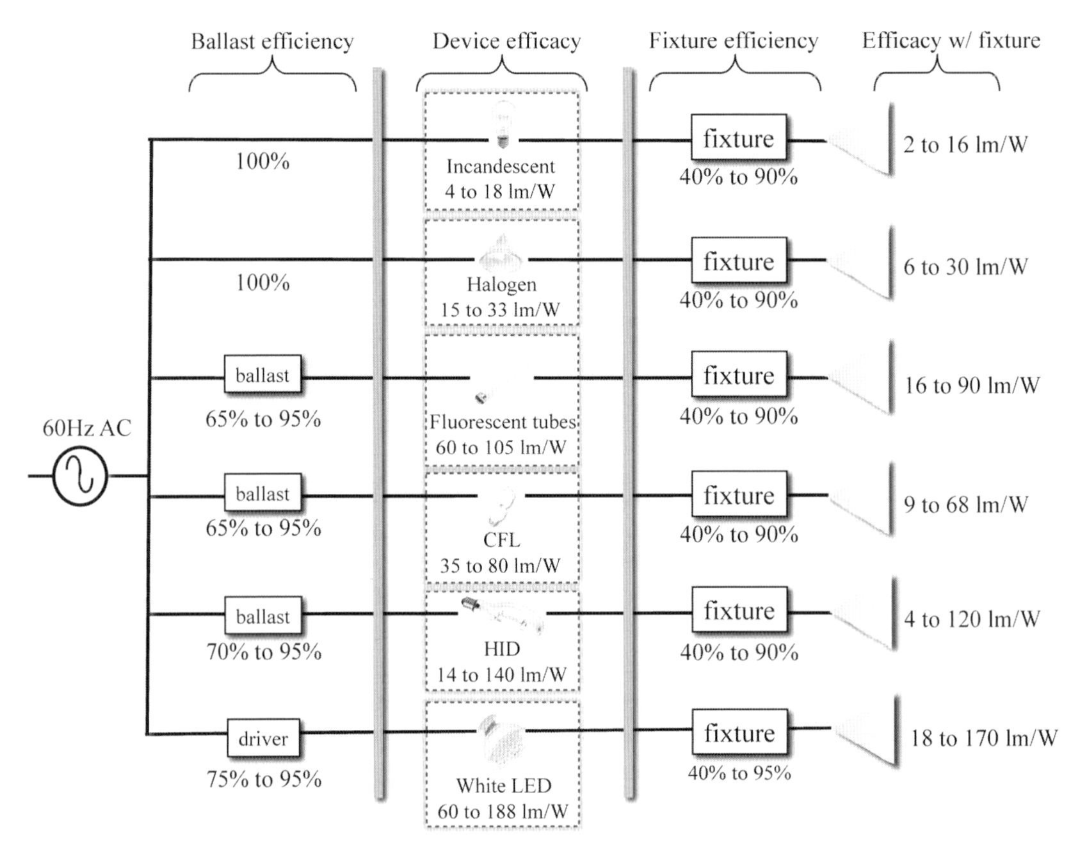

FIGURE 5.15 Efficiency of various types of lighting in 2008. SOURCE: Reprinted, with permission, from Azevedo, I.L., M.G. Morgan, and F. Morgan. 2009. The transition to solid-state lighting. *Proceedings of the IEEE* 97(3):481-510.

683 lm/W by definition, but for a white light spectrum desired by consumers, maximum light-producing efficiency will be lower, approximately 400 lm/W. A typical 100-W incandescent bulb puts out about 1,500 lm, or 15 lm/W per watt of electricity. Lower-wattage incandescent bulbs are even less efficient. Another important characteristic is the lifetime of the bulb. A longer lifetime will allow for fewer replacements over a given period of time, which is beneficial for cost, labor, and convenience. Lifetime considerations are especially important for bulbs in inconvenient places, or for those that involve a high labor cost to replace. A second benefit of much longer lifetimes will be the innovative use of lighting in areas that are only accessible when a structure is being built. It will be possible to plan lighting use as a one-time installation in areas that are difficult to reach. A pleasing color spectrum can influence the efficiency of a lightbulb; sometimes undesirable color

content can be more efficient than pleasing color content. The format of the bulb and environmental impacts are additional factors. Some bulbs contain mercury, a hazardous material, increasing their environmental impact.

Light-emitting diodes first started to be applied in niche markets like flashlights, where battery usage and ruggedness are important. Traffic lights were another early use, as the long lifetime helped reduce the labor cost of replacing bulbs. LEDs are widely used in cars and are making an entry to headlights as well, although attaining the required brightness for headlights is difficult with LEDs.[61] LEDs are frequently used in brake lights owing to the faster turn-on time of LEDs; replacing incandescent bulbs with LEDs in brake lights results in a faster turn-on time by almost 0.2 second. At 60 miles per hour, that translates to approximately 18 feet of additional stopping distance, thus helping prevent collisions. For dashboard lighting, LEDs generally last longer than the life of the vehicle and thus do not require replacement for the life of the vehicle. LED lights are also changing how lights are designed in an automobile, such as with turning-signal lights being deployed on side mirrors for better visibility.

Large flat-panel televisions are becoming reliant on LEDs for lighting, as discussed in this report in Chapter 10, "Displays." The light modulation in these large displays is still done using a liquid crystal modulator, with LEDs providing the light. Large-display manufacturers such as Samsung have made major investments in LED technology and are gaining increasing market share in the LED industry.

LEDs for lighting purposes are becoming more feasible. Walmart recently replaced the overhead lighting in many of its stores with LEDs.[62] A number of cities have replaced the street lighting with LEDs.[63,64,65,66] When the cost of labor to replace bulbs is taken into account, this can be economically reasonable even now,

---

[61] Car LED Lights. 2010. "Car LED Tail Lights: Where to Get Them." Available at http://www.car-ledlights.com/. Accessed October 25, 2011.

[62] Treehugger. 2011. "Lighting the Future: Walmart Converting Hundreds of Stores' Lot Lighting to LEDs." Available at http://www.treehugger.com/sustainable-product-design/lighting-the-future-walmart-converting-hundreds-of-stores-lot-lighting-to-leds.html. Accessed November 23, 2011.

[63] *Green Sheet, City of Ann Arbor Environmental News.* 2008. "LEDS—Ann Arbor 'Lights' the Way to Energy Savings." Available at http://www.a2gov.org/government/publicservices/systems_planning/Documents/publicservices_systems_envtlcoord_greennews_ledlights_2008_09_08.pdf. Accessed December 23, 2011.

[64] "Town of Chapel Hill." Available at http://www.townofchapelhill.org/index.aspx?NID=1986. Accessed December 23, 2011.

[65] *Los Angeles Times.* 2009. "L.A. Mayor Meets with Clinton on City Light Plan." Available at http://latimesblogs.latimes.com/lanow/2009/02/la-mayor-meets.html. Accessed December 23, 2011.

[66] City of Seattle, Washington. 2010. "Seattle City Light Begins LED Streetlight Rollout." Available at http://www.seattle.gov/light/news/newsreleases/detail.asp?ID=10888. Accessed December 23, 2011.

according to a recent article in the *Wall Street Journal*.[67] As the prices come down and efficiency increases, LEDs are likely to comprise a larger portion of lighting. The same *Wall Street Journal* article says that when Home Depot dropped the price of an LED bulb equivalent to a 40-W incandescent bulb from $21.00 to $9.97, the sales doubled. Solid-state lighting is used now in locations with inconvenient access due to the long lifetime, often from 30,000 to 80,000 hours. This is an advantage over the nominal 1,000-hour lifetime of incandescent bulbs. Halogen bulbs can have up to 5,000 or 6,000 hours of lifetime, while fluorescent bulbs can run 6,000 to 10,000 hours. Table 5.1 shows some LEDs available in August 2012 from a common supplier.

The Department of Energy issued the "L Prize" to promote the development of efficient lighting (see Box 5.3). In addition to the webpage shown in Box 5.3, a second website (see Table 5.2)[68] describes the requirements for the 60-W replacement bulb. There were two submissions for the 60-W replacement "L Prize." Philips Lumileds Lighting won this prize.

There is an upcoming "21st century" prize, which has not yet been officially published. In spite of that, one LED maker has already announced a bulb that appears capable of meeting the stringent requirements for the 21st century bulb.[69] This concept bulb emits 1,331 lumens at 8.7 W with a Color Rendering Index (CRI) of 91 and a warm color of 2800 kelvin (K). A standard 75-W incandescent bulb puts out about 1,170 lumens. Figure 5.16 shows a picture of the bulb.

The United States has two companies in the top-10 list of high-brightness LED manufacturers. These are Philips Lumileds Lighting and Cree.[70] Philips also has a strong European presence. Table 5.3, from *LED Magazine*, gives the top-10 high-brightness LED manufacturers by revenue in 2010.[71]

There are several key technical challenges to be met before LEDs can be widely deployed in homes for everyday use.

The CRI, a measure of the quality of light, has maximum value of 100. When illuminated by a low-CRI light source, objects appear to be different colors than they would if illuminated by the Sun. To obtain a high CRI, the light source must contain the right colors. This can be accomplished either using fluorescent phos-

[67] Linebaugh, Kate. 2011. "The Math Changes on Bulbs: Modern LEDS, While Expensive, Save Companies on Labor." *Wall Street Journal*. Available at http://online.wsj.com/article/SB10001424052970203537304577031912827196558.html. Accessed July 24, 2012.

[68] More information on the Department of Energy's "L Prize" is available at http://www.lightingprize.org/requirements.stm. Accessed October 25, 2011.

[69] *Jetson Green*. 2011. "Cree Demos 152 LPW LED Bulb Concept." Available at http://www.jetsongreen.com/2011/08/cree-demos-21st-century-led-bulb.html. Accessed July 24, 2012.

[70] *LED Magazine*. 2011. "Strategies Unlimited Unveils Top-Ten List of LED Manufacturers." Available at http://www.ledsmagazine.com/news/8/2/29. Accessed July 24, 2012.

[71] *LED Magazine*. 2011. "Strategies Unlimited Unveils Top-Ten List of LED Manufacturers."

TABLE 5.1 Selected Light-Emitting Diode (LED) Lightbulbs Available on the Web in August 2012

Bulb	Power (watts)	Lumens	Lumens per Watt	Lifetime (hours)	Price ($)
MR-16	4.5	348	77	30,000	19.95
A19	8.5	610	71	40,000	28.95
PAR 30	9	673	75	30,000	44.00

SOURCE: LED Waves. Available at www.ledwaves.com.

phors, or by 3 or 4 narrowband lines at the right wavelengths. In the phosphor converted (PC)-LED approach, an LED emits blue light, generally around 450 to 460 nm. Some of this light is emitted directly, and some of it is down-converted by a phosphor from the 450-to-460 nm wavelength (blue) to longer wavelengths (e.g., green, yellow, red) to produce white light.[72] The 3 or 4 line approach is often called RGB, for red, green, blue. If the efficiency of all blue, green, and red LEDs is as high as that of blue LEDs, theoretically about 20 percent higher efficiency can be gained using RGB color rendering compared to blue + phosphor LED due to the narrowband spectra, and another 20 percent can be gained in efficiency because there is no Stokes loss in the phosphor. RGB color rendering is theoretically the most efficient. Unfortunately there are no green LEDs, and also, color shifts during the lifetime of products (as R, G, and B will degrade at different rates). Currently, white LEDs are manufactured with violet/blue LEDs covered with yellow phosphors to convert part of the violet/blue light to yellow. Since the emission wavelength of an LED changes with its drive current, fixture, temperature, and age, the CRI is highly variable. Research on solving the droop problem, efficient phosphors to convert to green and red wavelengths, and efficient yellow/green/red LEDs is important. Many recent promising works are based on nanostructured materials such as InGaN nanowire LEDs and quantum-dot phosphors. For the RGB approach to color rendering, there is a lack of a green LED at the right wavelength, about 540 nm.

The cost of white LEDs is still very high due to the lack of scalability in the manufacturing processes. Often the (In, Al) GaN material is grown by metalorganic chemical vapor deposition (MOCVD) on a sapphire substrate. The sapphire substrate is relatively expensive and does not scale down with increasing wafer size. The processing steps, typically involving a flip-chip or liftoff process to remove the substrate, are complex and difficult to mass-produce. Recently, there has been increased interest in using Si as the substrate for GaN LED manufacturing. This can greatly simplify the substrate removal process, as the substrate cost would

[72] Bardsley Consulting, Navigant Consulting, Inc., Radcliffe Advisors, Inc., SB Consulting, and Solid State Lighting Services, Inc. 2012. *Solid-State Lighting Research and Development: Multi Year Program Plan March 2011 (Updated May 2011)*. Report prepared for Lighting Research and Development Building Technologies Program, DOE. Available at http://apps1.eere.energy.gov/buildings/publications/pdfs/ssl/ssl_mypp2011_web.pdf. Accessed July 24, 2012.

BOX 5.3
Department of Energy Establishes the L Prize

"The Energy Independence and Security Act of 2007 directs the U.S. Department of Energy (DOE) to establish the Bright Tomorrow Lighting Prize (L Prize) competition. The L Prize competition is the first government-sponsored technology competition designed to spur development of ultra-efficient solid-state lighting products to replace the common light bulb." (See Figure 5.3.1.)

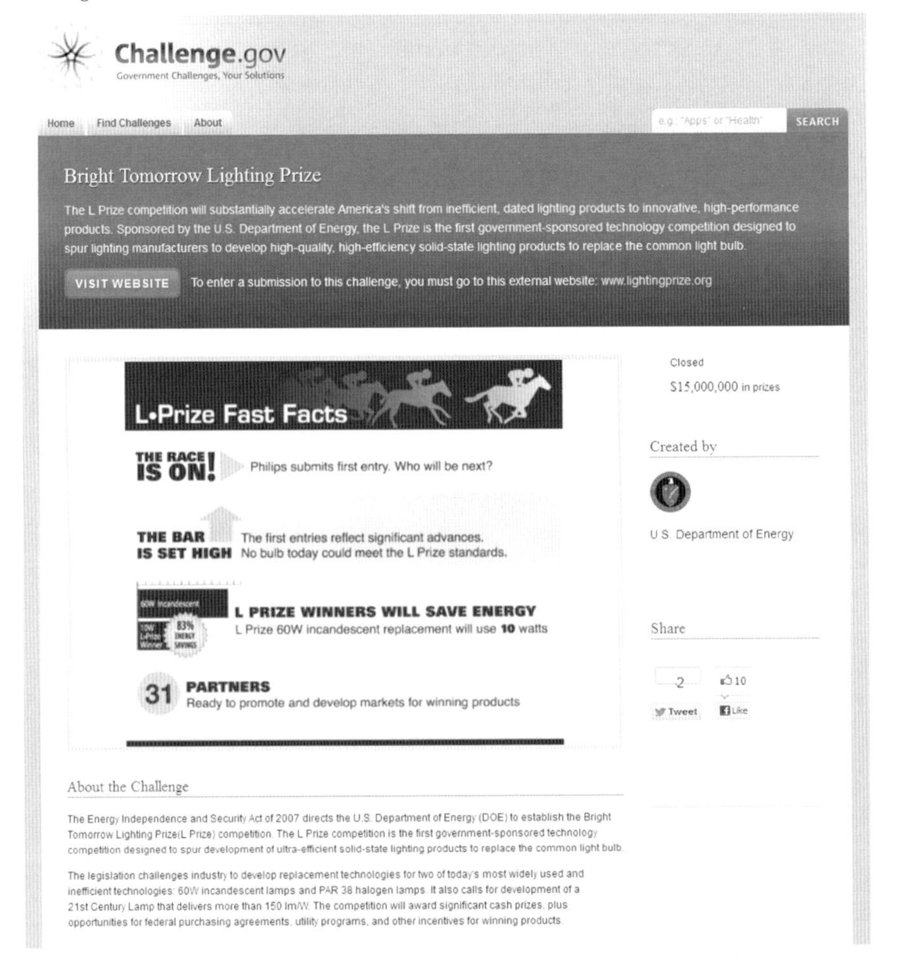

FIGURE 5.3.1 Webpage showing the "L Prize" World Wide Web page (July 2, 2011). SOURCE: Figure available at http://challenge.gov/DOE/43-bright-tomorrow-lighting-prize.

SOURCE: The quotation above from the Department of Energy is available at http://challenge.gov/DOE/43-bright-tomorrow-lighting-prize.

TABLE 5.2 Requirements for a 60-W Incandescent Replacement Lamp

More than 90 lumens per watt
Less than 10 watts
More than 900 lumens
More than 25,000 hour life
More than 90 in Color Rendering Index
Between 2700 and 3000 K in Correlated Color Temperature

SOURCE: Information available at http://www.lightingprize.org/requirements.stm.

FIGURE 5.16 Bulb that exceeds the 21st-century "L Prize" standards. SOURCE: Courtesy of Cree, Inc.

TABLE 5.3 Top-10 High-Brightness LED Manufacturers, by Revenue in 2010

2010 Ranking	Company Name
1	Nichia
2	Samsung LED
3	Osram Opto Semiconductors
4	Philips Lumileds Lighting
4	Seoul Semiconductor
6	Cree
6	LG Innotek
8	Sharp
9	Everlight
9	Toyoda Gosei

NOTE: Companies have the same ranking when the difference in revenue is within the margin of error. Revenue includes packaged LED sales only.

SOURCE: *LED Magazine*. 2011. "Strategies Unlimited Unveils Top-Ten List of LED Manufacturers." Available at http://www.ledsmagazine.com/news/8/2/29.

decrease with increasing size. Finally, the production volume can be easily scaled up, leveraging manufacturing equipment developed for silicon integrated circuits. This makes GaN on Si a promising approach for lowering costs. Recently Bridgelux showed cool-white LED efficiencies as high as 160 lm/W at a correlated color temperature of 4350 K, and warm-white GaN on Si LEDs delivered 125 lm/W at a color temperature of 2940 K and CRI of 80.[73] As an alternative to a sapphire substrate, at least one company, Cree, uses SiC. It currently has cost issues similar to those involved in using sapphire but produces the highest-efficiency LEDs. The 231 lm/W record was obtained using an SiC substrate. SiC holds the promise of significantly lower costs[74] if SiC can be made on larger wafers and scaled for larger-volume applications.

The alternating current-direct current (AC-DC) converter and heat sink for high-brightness LEDs also contribute to the high cost. LEDs are DC devices and thus require an AC-DC converter for many applications. Due to parasitic series resistance and other irradiative recombination paths, a substantial amount of heat is produced under high current injection, which degrades the LED performance. In fact, the LED chip itself contributes only 40 percent of the total cost of an LED bulb—more than half of the cost is for converter, heat sink, and light ray design.

The LED efficiency at operating temperatures is still low. The high lumen per watt efficiencies reported by various companies and laboratories were typically measured at room temperature, yet during normal operation, the temperature of the device usually rises to 80-100°C. This reduces the efficiency by 15 to 20 percent. To increase the efficiency, a number of aspects have to be improved.

LED package designs need to focus on heat removal to keep operating temperatures low, even for bulbs producing significant light.

The internal quantum efficiency (IQE) of InGaN/GaN quantum wells (QWs) needs to improve. The IQE is defined as the percentage of electron-hole recombinations in the active region that generates light. GaN materials are grown on a sapphire substrate with a large lattice mismatch, which results in a high defect density (approximately 10^9 cm^{-3}). In addition, there is a greater than 2 percent lattice mismatch between the well and barrier materials for a typical active region, InGaN/GaN QWs, which also leads to defects. These defects present paths for non-radiative recombination of carriers, leading to a reduced IQE and increased heat dissipation in the active region. Improving material quality is thus critical. Many

[73] Nusca, Andrew. 2011. "Bridgelux Shatters LED Efficiency Record with Silicon Tech." Smart Planet. Available at http://www.smartplanet.com/blog/smart-takes/bridgelux-shatters-led-efficiency-record-with-silicon-tech/18252. Accessed December 18, 2011.

[74] Ozpineci, B., and L. Tolbert. 2011. "Silicon Carbide: Smaller, Faster, Tougher." *IEEE Spectrum*. Available at http://spectrum.ieee.org/semiconductors/materials/silicon-carbide-smaller-faster-tougher. Accessed July 24, 2012.

recent publications indicate that defects can be greatly reduced in nanostructures grown on lattice-mismatched substrates.[75,76]

There is a built-in electric field in InGaN QWs grown on a c-plane sapphire, the lowest-cost and most commonly used sapphire substrate. This built-in electric field leads to Quantum-Confined Stark Effect (QCSE) in QWs, causing an even lower IQE and blue-shifted wavelength with current injection. The latter reduces the conversion efficiency of yellow phosphors. Finally, better current-spreading schemes are critical to avoid current crowding, which in turn reduces ohmic heat and Auger recombination. With less heat produced, efficiency droop at high injection current can be suppressed.

FINDINGS

Finding: When compared to today's fossil fuel alternatives, current solar power is not yet cost-effective in most areas for utility-scale application without subsidies. Solar-based electricity generation for utility-scale power is, however, making excellent progress toward the goal of cost parity against fossil-fuel-based electricity generation approaches. It is the opinion of the committee that development of cost-competitive solar power for utility-scale application over the next decade is feasible, and in fact likely. Major solar cell providers have plans to reach cost parity by about 2015. Cost parity will come in stages and initially will occur only in areas of bright sunlight, with expensive alternatives, during peak-load times.

Finding: If cost parity or better is achieved, with solar power becoming the most economically attractive method of building a new electricity generation facility, then much more land will be used for solar power, regardless of the particular solar power approach.

Finding: Optimal solar power concentration approaches are key to effective utility-scale solar power plants in areas with significant direct sunlight.

Finding: The electric grid distribution system can influence the benefits associated with solar power if power can be transferred across time zones, making use of available sunlight in multiple time zones.

Finding: Solid-state lighting is making great progress in efficiency, brightness,

[75] Sburlan, S., A. Nakano, and P.D. Dapkus. 2012. Effect of substrate strain on critical dimensions of highly lattice mismatched defect-free nanorods. *Journal of Applied Physics* 111(5):054907-054907-6.

[76] Zytkiewicz, Z.R. 2002. Laterally overgrown structures as substrates for lattice mismatched epitaxy. *Thin Solid Films* 412(1-2):64-75.

color balance, and cost and is already very attractive for many niche applications. Solid-state lighting is already cost-competitive for general lighting applications in which significant labor is involved in changing lightbulbs.

Finding: There is no suitable green LED available to allow efficient RGB color rendering.

RECOMMENDATIONS AND GRAND CHALLENGE QUESTION

Key Recommendation: The Department of Energy (DOE) should develop a plan for grid parity across the United States by 2020.

"Grid parity" is defined here as the situation in which any power source is no more expensive to use than power from the electric grid. Solar power electric plants should be as cheap, without subsidies, as alternatives. It is understood that this will be more difficult in New England than in the southwestern United States, but the DOE should strive for grid parity in both locations.

Even though significant progress is being made toward reducing the cost of solar energy, it is important that the United States bring the cost of solar energy down to the price of other current alternatives without subsidy and maintain a significant U.S. role in developing and manufacturing solar energy alternatives. There is a need not only for affordable renewable energy but also for creating jobs in the United States. A focus in this area can contribute to both. Lowering the cost of solar cell technology will involve both technology and manufacturing advances.

This key recommendation leads directly to **the fourth grand challenge question:**

4. How can U.S. energy stakeholders achieve cost parity across the nation's electric grid for solar power versus new fossil-fuel-powered electric plants by the year 2020?

The impact on U.S. and world economies from being able to answer this question would be substantial. Imagine what could be done with a renewable energy source, with minimal environmental impact, that is more cost-effective than nonrenewable alternatives. Although this is an ambitious goal, the committee poses it as a grand challenge question, something requiring an extra effort to achieve. Today, it is not known how to achieve this cost parity with current solar cell technologies. Additionally, it is important to recall the statement made in 2010 by the America's Energy Future Panel on Electricity from Renewable Resources in the report *Electricity from Renewable Resources: Status, Prospects, and Impediments.*

> For the time period from the present to 2020, there are no current technological constraints for wind, solar photovoltaics and concentrating solar power, conventional geothermal, and biopower technologies to accelerate deployment. The primary current barriers are the cost-competitiveness of the existing technologies relative to most other sources of electricity (with no costs assigned to carbon emissions or other currently unpriced externalities), the lack of sufficient transmission capacity to move electricity generated from renewable resources to distant demand centers, and the lack of sustained policies.[77]

The second major area in which optics can contribute to energy security in the United States is through solid-state lighting.

Key Recommendation: The DOE should strongly encourage the development of highly efficient light-emitting diodes (LEDs) for general-purpose lighting and other applications.

For example, the DOE could move aggressively toward its 21st-century lightbulb, with greater than 150 lm/W, a color rendering index greater than 90, and a color temperature of approximately 2800 K. Since one major company has already published results meeting the technical requirements for the 21st-century lightbulb, the DOE should consider releasing this competition in 2012. Major progress is being made in solid-state lighting, which has such advantages over current lighting alternatives as less wasted heat generation and fast turn-on time. The United States needs to exploit the current expertise in solid state lighting to bring this technology to maturity and to market.

Recommendation: The DOE should:

1. Develop high-temperature solar cells suitable for use in conjunction with CSP systems. These solar cells can provide sunlight conversion to electricity in addition to the CSP thermal conversion. Moderate-efficiency solar cell electricity generation in addition to thermal electricity generation could drive total efficiency greater than 50 percent;
2. Develop more efficient, and less expensive, methods of solar power concentration for both low and high levels of concentration;
3. Develop methods to reduce LED costs, such as growing LEDs on silicon, or increasing SiC wafer sizes, or other LED substrate cost-reduction approaches; and
4. Develop efficient green LEDs to support quality RGB LED color rendering.

---

[77] NAS-NAE-NRC. 2010. *Electricity from Renewable Resources: Status, Prospects, and Impediments.* Washington, D.C.: The National Academies Press, p. 4.

Recommendation: Governmental land use planning bodies should consider the potential impacts of a significant increase in land use associated with solar electric power plants.

Recommendation: The DOE and electric utilities should evaluate the electric grid for power movement across multiple time zones and evaluate whether changes should be made to support the period of time when grid parity is reached for solar power.

6

Health and Medicine

INTRODUCTION

The U.S. health care industry representing approximately $3 trillion of annual expenditures[1] and employing roughly 15 million people[2] comprises one of the largest sectors of the national economy. Our nation boasts the most technologically advanced, and the most costly, health care system in the world: almost 20 cents of every dollar is spent on health care. To remain the world leader in developing and introducing innovative medical instrumentation while improving and bringing down the cost of health care will require continued investment in research and development. Photonics technology plays a key role in providing the most effective, lowest-cost approaches for diagnosing, treating, and preventing disease and maintaining a healthy U.S. citizenry.[3] In the nearly 15 years since the publication of the National Research Council's (NRC's) *Harnessing Light: Optical Science and Engineering for the 21st Century*,[4] advances in photonics technologies and develop-

[1] Centers for Medicare and Medicaid Services. 2012. "National Health Expenditures 2010 Highlights." Available at http://www.cms.gov/Research-Statistics-Data-and-Systems/Statistics-Trends-and-Reports/NationalHealthExpendData/Downloads/highlights.pdf. Accessed August 1, 2012.

[2] SelectUSA. 2012. "The Health and Medical Technology Industry in the United States." Available at http://selectusa.commerce.gov/industry-snapshots/health-and-medical-technology-industry-united-states. Accessed August 1, 2012.

[3] For a more detailed description of the optics and photonics in the service of health and medicine, see the full description in Appendix C of this report.

[4] National Research Council. 1998. *Harnessing Light: Optical Science and Engineering for the 21st Century.* Washington, D.C.: National Academy Press.

ments in the field of biophotonics have created many opportunities for significant improvements in the quality of health care as well as for substantial reductions in the overall cost.

The discipline of biophotonics deals with the interaction of light, or electromagnetic radiation, with living organisms and biologically active macromolecules: proteins (hemoglobin), nucleic acids (DNA and RNA), and metabolites (glucose and lactose). Light interacts with biological material and organisms in many diverse ways, depending primarily on the energy or color of the photon. At both high (x ray) and low (radio frequency [RF]) energies the body is mostly transparent, thus allowing the non-invasive imaging of the internal structure of organs and bones. In contrast, certain colors or wavelengths in the infrared (IR) and ultraviolet (UV) regions are absorbed strongly by biological tissues. The intense, focused light of lasers with these colors can be used for a wide variety of unique therapeutic interventions: making incisions, cauterizing and sealing wounds, and selectively heating or even vaporizing specific regions of organs and tissues. In the visible region of the spectrum, some biologically active macromolecules naturally absorb specific photon energies or colors. The amount of this intrinsic absorption can be used to determine the physiological health of an organ—for example, whether the tissue is getting sufficient oxygenated blood flow. Non-absorbing macromolecules can be labeled using specifically engineered dyes that selectively bind to macromolecules. These dyes or biomarkers can be used in conjunction with visible and near-infrared light to highlight specific cell and tissue types, such as metastatic cancer cells circulating in the bloodstream. Modern biomedical instrumentation takes advantage of this wide range of interactions between photons and biological materials and provides a remarkably broad set of tools for the physician and life scientist.

HISTORICAL OVERVIEW OF THE IMPACT OF TECHNOLOGY ON MEDICINE

Prior to the modern age of medicine, physicians primarily used their five senses directly to determine the causes of ill health.[5] For example:

- The color of a person's eyes was studied to detect jaundice and possible liver failure.
- The urine of patients was tasted for sweetness, to detect the presence of glucose and diagnose diabetes.
- An ammonia-like odor in urine implied possible kidney failure.

[5] Berger, D. 1999. A brief history of medical diagnosis and the birth of the clinical laboratory. Part 1—Ancient times through the 19th century. *Medical Laboratory Observer* 31(7):28-30, 32, 34-40.

- The chest was struck or thumped and the resulting sound analyzed to identify the presence of fluid in the lungs, implying tuberculosis or pneumonia.
- The abdomen and breasts were palpated to detect cancerous lumps.

Currently biomedical technology extends and enhances the senses of the physician and thus dramatically increases the ability of the physician to diagnose disease.

Considering that the primary sense used by physicians in diagnosis was sight, it is understandable that optics and imaging have played a critical role in improving health care. Over the past 100 years, optics and imaging have allowed the clinician to see the previously unseen. For example, observation of bacteria and microbial parasites led to the development of antibiotics, and direct imaging of skeleton and organs with x ray aided in observing and setting bone fractures and diagnosing traumatic injuries to other organs. For example, laser-based flow cytometers provide detailed quantification of critical blood cell types, which is one of the primary tools for diagnosing and monitoring the treatment of AIDS patients.[6]

In addition to allowing the physician to see what could not be seen unaided, state-of-the-art optical technologies increase the sensitivity and specificity of measurements far beyond the physician's sense of taste, smell, hearing, and touch.[7] Photonics plays a major role in many modern molecular diagnostic instruments. Optical technologies now provide precise measurements of blood serum chemistry for maintaining safe glucose levels in patients with diabetes, replacing urine taste tests.[8] Kidney function tests rely on accurate optical measurements of the glomerular filtration rate (GFR) rather than smelling a patient's urine. Lung diseases such as emphysema, lung cancer, and tuberculosis are detected using computed tomography (CT) and chest x ray imaging. In addition, these imaging modalities provide more complete diagnosis of potential tumors detected by palpation.

During the 20th century, improvements in medical technology have doubled the life expectancy[9] of individuals in high-income countries, changing the primary causes of death for a typical individual. One hundred years ago, infectious diseases often killed most individuals before the age of 50, whereas today the typical individual in a high-income country lives until the age of 80. Optics and photonics have been essential technologies leading to this dramatic increase in life expectancy. For example, the microscope was the key technology allowing discoveries in microbi-

[6] Hazenberg, M.D., S.A. Otto, B.H. van Benthem, M.T. Roos, R.A. Coutinho, J.M. Lange, D. Hamann, M. Prins, and F. Miedema. 2003. Persistent immune activation in HIV-1 infection is associated with progression to AIDS. *AIDS* 17(13):1881-1888.

[7] Bynum, W.F., and Roy Porter, eds. 1933. *Medicine and the Five Senses*. Cambridge, Mass.: Cambridge University Press.

[8] For more information, see Appendix C in this report.

[9] See, for example, the CIA's *World Factbook*, available at https://www.cia.gov/library/publications/the-world-factbook/. Accessed December 1, 2011.

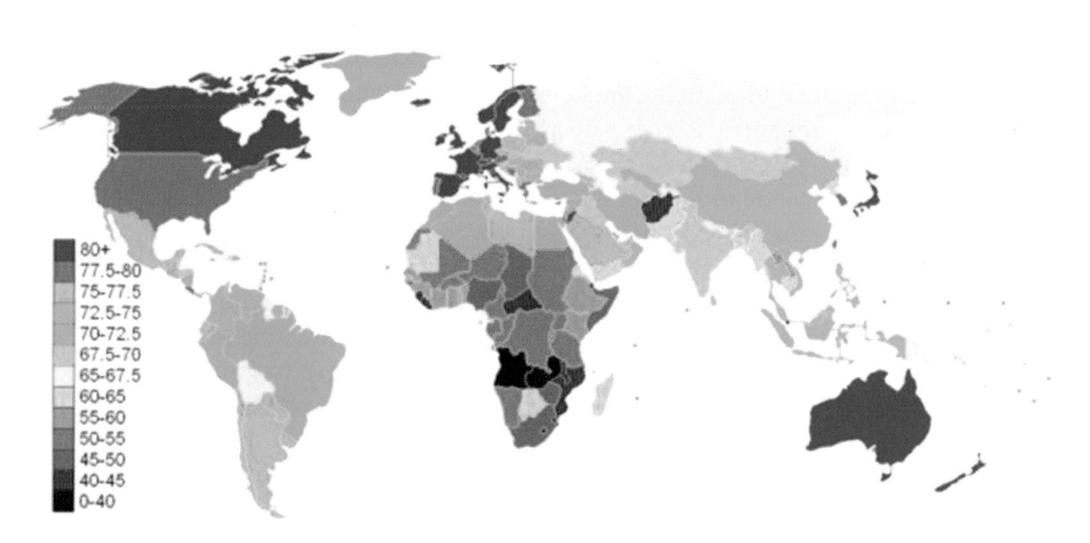

FIGURE 6.1 Life expectancy: 2011 estimates by CIA *World Factbook*. SOURCE: CIA. 2011. *World Factbook*. Available at https://www.cia.gov/library/publications/the-world-factbook/rankorder/2102rank.html.

ology which determined that germs are the underlying cause of most infectious diseases, leading to the development of effective antibiotic drugs.

The primary causes of death today (heart disease, cancer, diabetes, and neurodegenerative diseases) are diseases that are more prevalent at older ages. The success of modern medicine has therefore created a series of new challenges, requiring further innovation.

Many of these challenges are being met using optics and photonics instrumentation, which is providing scientific insights into the underlying molecular biology causing these diseases as well as quantitative new diagnostic instruments to help steer effective interventional therapies.

In low-income countries, infectious diseases, such as tuberculosis, malaria, and AIDS, remain leading causes of death. This is due in large part to the absence of low-cost diagnostic tests that can detect these diseases at an early stage when the infections can be more easily contained and cost-effective intervention strategies can be employed. (See Figure 6.1.)

OPTICS AND PHOTONICS IN MEDICAL PRACTICE TODAY

A patient entering the emergency room (ER) with chest pains or a severe headache almost invariably receives a high-resolution, three-dimensional scan using CT or magnetic resonance imaging (MRI) as an initial diagnostic screening, which can assist in the diagnosis of a heart attack, pulmonary embolism, or stroke. CT and MRI both use photons with wavelengths for which the human body is very transparent, creating comprehensive, high-resolution, three-dimensional images of the anatomy: CT uses x-ray photons and MRI uses RF photons.[10] Stroke and heart attack can lead to sudden death, but effective interventions exist if the conditions can be diagnosed swiftly. If a nonresponsive patient comes to the ER with an unlabeled prescription in his or her pocket, how can the ER attendant determine whether or not the patient's drugs have contributed to his or her condition? New technology using lasers has the potential to provide a fast method for identifying drugs by exposing the pills to laser light and observing the spectra emitted in response to the laser excitation. The colors emitted by the sample provide a unique fingerprint, which can be compared with a database of more than 5,000 common pharmaceuticals to determine the precise makeup of the patient's prescription and to determine quickly whether these drugs contributed to the patient's condition.[11]

In *the surgical suite*, brain damage occurs in less than 5 minutes in anesthetized patients during surgery if sufficient oxygen levels are not maintained. During the early and mid-20th century, the monitoring of blood oxygen levels required taking a blood sample and sending it to the hospital laboratory for analysis. This process typically took tens of minutes to complete, presenting significant hazards to the patient with such slow feedback. These procedures were revolutionized by optical pulse oximeters, developed in the 1980s, which precisely measure the ratio of the absorption levels of the blood at two wavelengths, using convenient, low-cost, disposable optical probes based on light-emitting diodes (LEDs) and inexpensive solid-state detectors.[12] In the past, mortality rates of 1 in 2,000 to 1 in 7,000 were reported in developed countries, and many patients suffered brain damage owing to oxygen deprivation during surgery;[13] today such deaths and injuries have essentially been eliminated.

Surgery almost always results in some unavoidable trauma to the patient. Minimizing the size of the surgical incision reduces this trauma and can dramatically speed recovery. Modern optical endoscopes provide a close-up view of organs and a method for implementing laser surgery, utilizing incisions of less than a few centimeters. In addition, endoscopic visualization is now the standard of care for screening for colon cancer and for diagnosing esophageal cancer. Commonly used today in orthopedic surgery for repairing injuries in almost all of the major joints

---

[10] For more information about developments in CT instruments, see Appendix C.

[11] Gendrin, C., Y. Roggo, and C. Collet. 2008. Pharmaceutical applications of vibrational chemical imaging and chemometrics: A review. *Journal of Pharmaceutical and Biomedical Analysis* 48(3): 533-553.

[12] For more information about O_2 saturation measurements, see Appendix C in this report.

[13] World Health Organization (WHO). 2008. *Global Pulse Oximetry Project.* Background document. First International Consultation Meeting. WHO Headquarters, Geneva, CH. Available at http://www.who.int/patientsafety/events/08/1st_pulse_oximetry_meeting_background_doc.pdf. Accessed August 1, 2012.

(knee, elbow, hip, wrist), this technology has allowed many surgeries to become outpatient procedures, eliminating hospital stays and greatly reducing health care costs.

Besides being ubiquitous in the hospital, optical methods and instruments are also used in monitoring chronic conditions and in many *outpatient surgical procedures.*[14] The most pervasive use of optics and lasers in surgery is in ophthalmology. Laser treatments are standard therapy for treating blindness due to diabetes as well as age-related degenerative disease. One of the most common laser cosmetic surgeries today is the correction of focus of the eye lens by the precise shaping of the cornea, a procedure that has been performed on tens of millions of patients. Lasers and optics are also used in many outpatient elective cosmetic procedures, such as skin resurfacing and hair and tattoo removal.

Almost everyone has had blood drawn and sent to *the clinical laboratory* for blood tests, but few people are aware that these blood samples are analyzed using lasers and optical imaging to provide the measurements characterizing the status of the patient's blood and circulatory system. Many types of blood cells have distinct shapes and unique internal structures, which can be detected by illuminating the cells with a laser beam and analyzing the transmitted and scattered laser light. The laser-based instruments used for these purposes can count many thousands of cells per second and are used to measure with great precision the different types of cells present in the bloodstream.

A doctor detecting a suspicious lump in a patient will often order an exploratory biopsy, which is sent to *the pathology laboratory.* High-resolution imaging of the excised tissue and analysis of the size and shape of the cells provide the most precise diagnosis of tumor malignancy and aggressiveness. These images also can help determine the boundaries between healthy and diseased tumor tissue, providing a surgeon with guidance about how much tissue needs to be removed to fully excise a tumor.

ADVANCES IN TECHNOLOGY PROVIDING THE OPPORTUNITY FOR NEW APPLICATIONS OF PHOTONICS

New optical technology has accelerated the translation of remarkable new capabilities into medical practice. As an example, just 30 years ago the standard technique for sequencing genes involved the radioactive labeling (using a radioactive isotope of phosphorous) of the gene bases, gel electrophoresis to separate the gene fragments by size, and the overnight exposure of an optical film placed in close proximity to the gel. This laborious and time-consuming process allowed the sequencing of only several hundred bases during a typical day. With the introduction of optical methods in the 1980s, including four-color sequencing and optical scan-

[14] For more information, see Appendix C.

ning of the gel, the speed of sequencing increased by a factor of 10. The development of superior separation technologies, including capillary electrophoresis and better optical designs using confocal laser scanning, increased sequencing speed by about another factor of 10. Recently, single molecular fluorescence detection during single-strand DNA synthesis, synthesis-based sequencing (SBS), has replaced electrophoresis. Low-noise, high-resolution charge-coupled device (CCD) imaging devices allow the simultaneous sequencing of millions of individual DNA strands. These technologies have increased the speed of sequencing by a factor of 1,000. (See Figure 6.2.) Not only has the speed increased by 100,000-fold, but the cost of sequencing the genome has decreased from billions of dollars to under $1,000. Data collected during 1 day in a typical sequencing laboratory in 1970 would fill approximately half of a single page in a lab notebook. In 1990, the technology produced enough data to fill a chapter of a lab book, or about 20 pages in a single day. In the year 2000, a single day's data would fill a whole lab book.

Present-day technology, driven by advances in laser sources, nanophotonics, and detectors, generate enough data in 1 hour to fill the contents of 10, 24-volume encyclopedias.

ADVANCES IN TECHNOLOGY PROVIDING THE OPPORTUNITY FOR FUTURE APPLICATIONS OF PHOTONICS

Continuing advances in several critical areas of technology have dramatically increased the capabilities of biomedical optical instrumentation and herald a new era of innovation in biomedical optics, leading to improvements in treating many types of diseases. New optical sources and materials, imaging devices, microfluidic technologies, and detection methods will provide remarkable increases in speed, sensitivity, and precision for biomedical optical instrumentation.

Nucleic Acid Sequence Detection and Mutation Detection

Predilections for many diseases are the result of the specific makeup of an individual's inherited genetic code. Specific alterations in certain genes can dramatically increase the likelihood of cancer, cognitive impairment, and severe allergic reactions to certain types of food. Early identification of these inherited tendencies allows preventive strategies to be in place before the disease causes significant damage. With the advent of rapid and much-lower-cost methods for whole genome sequencing, many of which rely on optical methods, the possibility exists that a wide range of correlations between genetic makeup and disease predilection can be detected earlier, and appropriate interventions put in place.[15]

Today, sequencing a complete human genome costs less than $5,000; it is

[15] For more information, see Appendix C.

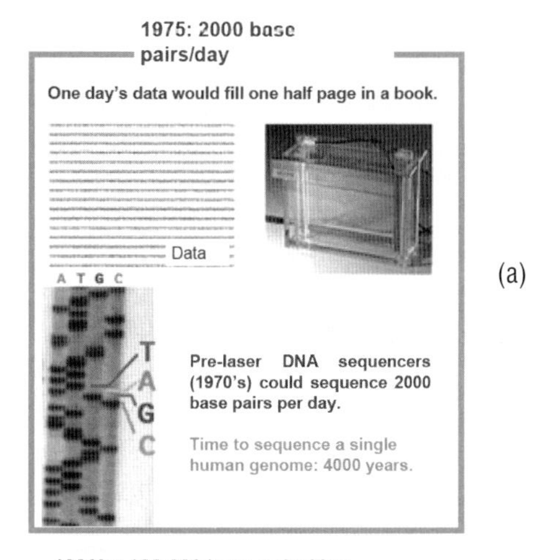

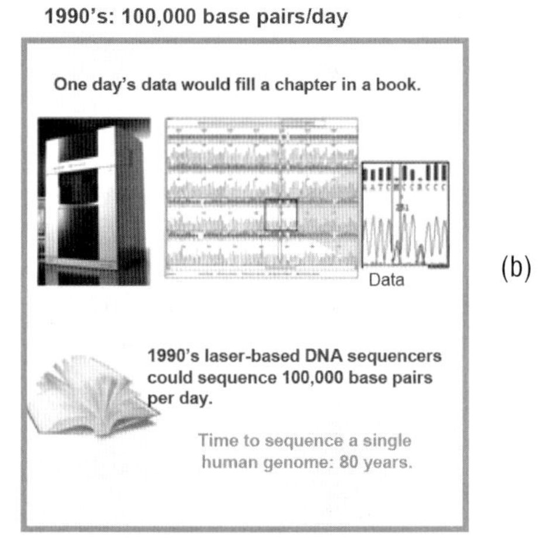

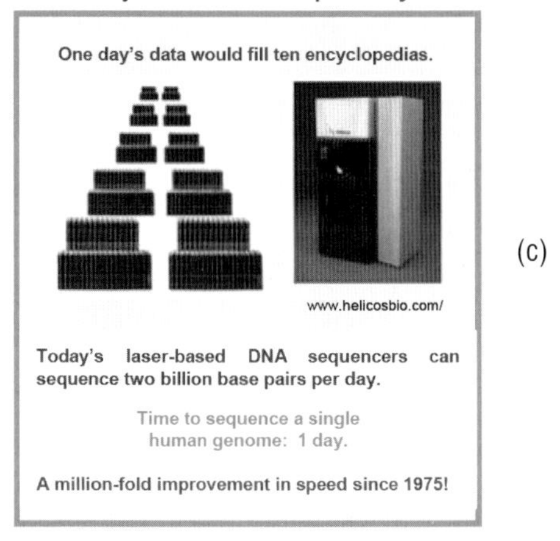

FIGURE 6.2 Technologies have increased the speed of sequencing by a factor of 1,000. Time to sequence a single human genome with (a) pre-laser DNA sequencers (1970s), (b) 1990s laser-based DNA sequencers, and (c) second-generation laser-based DNA sequencers. SOURCE: Congressional briefing by Thomas Baer, Executive Director, Stanford Photonics Research Center, Stanford University, Palo Alto, Calif. November 30, 2010. Available at http://portal.acs.org/preview/fileFetch/C/CNBP_026401/pdf/CNBP_026401.pdf. Accessed November 8, 2012.

projected that the next generation of instruments will reduce the cost to $1,000 over the next few years.[16] The new generation of SBS sequencers (see Box 6.1 and Figure 6.3) combines nanotechnology with photonics to achieve this remarkable increase in speed.[17] To reach their full potential, these instruments will require further investment in research to develop higher-speed and -sensitivity CCD cameras, more efficient labeling dyes, high-speed software to extract quantitative information from large, high-resolution images, and nanophotonic structures optimized to localize the fluorescent signals from the individual DNA strands.

Proteomic Analysis Through Protein and Tissue Arrays

Detecting cancer recurrence early, determining the most appropriate drug therapy to slow tumor growth, and detecting and/or diagnosing a number of deadly infectious diseases are all examples of diagnostic tests that rely on measuring specific proteins in the blood serum. Recent advances in protein detection using photonics technology platforms are providing dramatic increases in sensitivity and specificity.

Microfluidics and robotics combined with optics provide the technology to create arrays of tissue samples on slides that can be automatically laser scanned and analyzed after exposure to fluorescently labeled antibodies which attach to tumor-specific proteins. These tissue arrays are analyzed using laser scanners and automated image analysis of the digital images. Drug interactions and molecular structures can be studied across hundreds of diseased and healthy patient samples all located on a single slide. This technology has the potential for greatly accelerating the drug development process and reducing development costs.

These protein-measuring instruments use automated, high-resolution microscopy, wide-field-of-view, low-noise imaging devices, and quantitative fluorescence microscopy. The performance of these instruments will improve greatly with advances in the areas of optics and photonics technology.

High-Throughput Screening

The development of new drugs based on small molecules is often limited by the rate at which candidate molecules can be screened for their therapeutic effect on target cell cultures. In recent years, optical technology has been combined with

[16] Bio-IT World Staff. 2011. "Illumina Announces $5,000 Genome Pricing." Available at http://www.bio-itworld.com/news/05/09/2011/Illumina-announces-five-thousand-dollar-genome.html. Accessed August 1, 2012.

[17] Margulies, M., M. Egholm, and W.E. Altman, et al. 2005. Genome sequencing in microfabricated high-density picolitre reactors. *Nature* 437:376-380.

BOX 6.1
Nucleic Acid Synthesis-Based Sequencing

The first human genome was sequenced using technology based on electrophoresis, a process by which molecules of different sizes and chemical makeup are separated spatially in special gelatins by running an electric current through the gel. This process allowed different nucleic acid sequences to be isolated and the genetic code to be read out using fluorescent tags and laser scanners.

The newest approaches to gene sequencing use a radically different approach—called synthesis-based sequencing (SBS)—which relies on several key photonics technologies. A sample containing a quantity of DNA molecules with unknown sequences is placed in a special microfluidics chamber. In this chamber separate strands of DNA are captured in different locations and then are copied by special enzymes. As the copy is created, each added base is incorporated into the growing DNA strand along with a fluorescent dye. Each of the four bases is coupled to a different-colored dye. The addition of each new base is detected by exciting the fluorescence dye with a laser and detecting the color of the fluorescence using a high-resolution, high-sensitivity CCD (charge-coupled device) camera. After the dye color is detected and the added base is identified, the dye molecule is enzymatically cleaved from the DNA strand. This process is repeated until the all of the bases have been added and identified and the DNA sequence has been determined.

This synthesis of copies of the various DNA molecules in the sample takes place simultaneously in hundreds of thousands of distinct locations, all of which can be monitored in parallel by high-sensitivity cameras that can detect signals as low as a single photon from each synthesis site. This approach has increased the speed of the sequencing of nucleic acids by many orders of magnitude. Currently, a single instrument can sequence the 3 billion base pairs in a human genome in less than a day. In contrast, the human genome project employed many hundreds of instruments and took more than 10 years to complete.

SOURCE: Fuller, C.W., L.R. Middendorf, S.A. Benner, G.M. Church, T. Harris, X. Huang, S.B. Jovanovich, J.R. Nelson, J.A. Schloss, D.C. Schwartz, and D.V. Vezenov. 2009. The challenges of sequencing by synthesis. *Nature Biotechnology* 27(11):1013-1023.

robotics to provide the ability to screen hundreds of thousands of drug candidates per day, dramatically accelerating the drug discovery process. This high-throughput screening technology relies on robotic sample-handling automation for the precise and rapid parallel processing of multiple samples as well as on optical technology for high-speed quantitative data collection. Optical methods, including fluorescence, bioluminescence, and colorimetry, are used to identify and count viable and nonviable cells affected by the candidate compounds, determine activated molecular pathways in target cells, and detect the overall cellular response to potential small-molecule drugs. An example of an approach that allows very low concentrations of proteins to be detected by actually counting the protein molecules individually is seen in Figure 6.4. More recently, by combining microfluidics with microscopic imaging to enhance protein quantification, researchers have increased sample throughput by 1,000-fold, to an astounding 10 million samples per day,

FIGURE 6.3 The Pacific Biosciences PacBio RS platform is used for single-molecule sequence data. SOURCE: © Pacific Biosciences. Reprinted with permission.

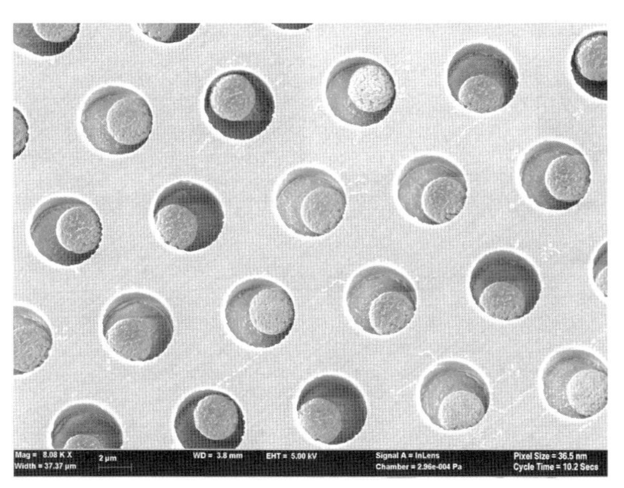

FIGURE 6.4 Showing single occupancy: reaction wells approximately 5 microns in diameter contain a single bead coated with antibodies that trap a single target protein molecule. This single protein in turn binds a fluorescent enzyme that creates a fluorescent signal localized in the well, which is detected by laser scanning the well array. This approach allows very low concentrations of proteins to be detected by actually counting the protein molecules individually. SOURCE: Subbaraman, N. 2010. "Detecting Single Cancer Molecules." *Technology Review.* Available at http://www.technologyreview.com/biomedicine/25462/. Reprinted with permission.

and radically reduced the sample volume and the time required for analysis.[18] These instruments utilize and will benefit from improvements in high-sensitivity, quantum-limited imaging detectors and new, compact, long-lived laser sources.

Flow Cytometry Mass Spectrometry

Although essential to survival, the human immune system also is a key factor in a number of common diseases, such as rheumatoid arthritis, childhood or Type I diabetes, and arteriosclerosis. Understanding the complex processes that make up an immune response requires the simultaneous monitoring of the activity and concentration of dozens of immune-system cell types. Lasers have traditionally been used in flow cytometer instruments to evaluate patient blood samples, identifying and counting the immune-system cell types. These instruments were limited in the past to quantifying only a few different cell types simultaneously. A new generation of flow cytometry instruments combines optical detection with mass spectrometry. This new technology promises to allow a status check of a patient's immune system by simultaneously quantifying all of the major cellular constituents. When combined with other advanced proteomic technologies, including tissue microarrays and protein mass spectrometry, the CyTOF instrument (see Figure 6.5) will provide the most complete understanding of the immune system to date.[19] Identifying immune cell types along with their associated functions in affected tissues—that is arthritic joints, inflammatory tissues, and pancreatic islets—will allow the most complete understanding of the local and systemic processes that underlie many degenerative diseases such as rheumatoid arthritis, diabetes, Alzheimer's disease, and heart disease.[20]

Ophthalmology

Age-related macular degeneration (AMD) and diabetic retinopathy (DR) are two of the leading causes of blindness, particularly in older patients.[21] Laser-based surgical and drug therapies can slow the disease progression, particularly if the disease is detected prior to major damage to the retina. Early detection is difficult because the disease primarily impacts tissue beneath the surface of the retina, which

---

[18] Agrestia, J.J., E. Antipov, A.R. Abate, K. Ahn, A.C. Rowat, J.-C. Baret, M. Marquez, A.M. Klibanov, A.D. Griffiths, and D.A. Weitz. 2009. Ultrahigh-throughput screening in drop-based microfluidics for directed evolution. *Proceedings of the National Academy of Sciences of the United States of America* 107(9):4004-4009.

[19] Cheung, R.K., and P.J. Utz. 2011. Screening: CyTOF—the next generation of cell detection. *Nature Reviews Rheumatology* 7(9):502-503.

[20] For information about use in HIV status, see Appendix C.

[21] For more information, see the section on "Ophthalmology" in Appendix C.

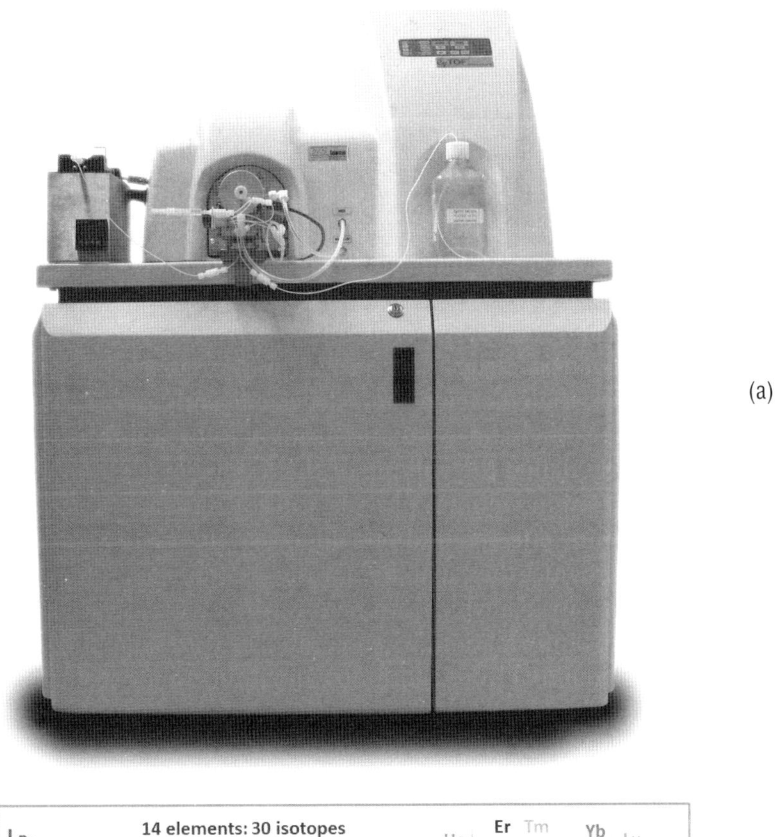

(a)

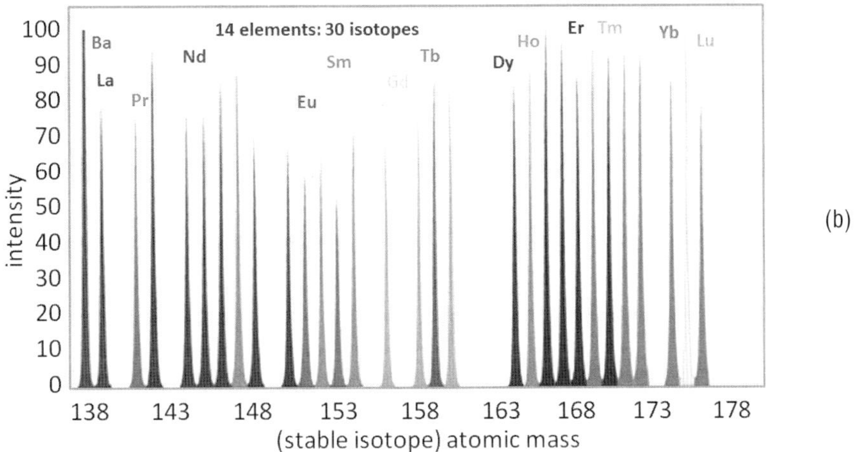

(b)

FIGURE 6.5 (a) A CyTOF instrument, which extends the capability of multi-parameter flow cytometry by atomic mass spectrometry to measure up to 100 biomarkers simultaneously in single cells at a rate of 1,000 cells per second; (b) data: 138-178 segment of mass spectrum for a homogeneous sample of several enriched isotopes of lanthanides. SOURCE: DVS Sciences, Inc. Available at http://www.dvssciences.com/index.html. Reprinted with permission.

is not easily observed in the early stage. New optical methods employing optical coherence tomography (OCT), a type of microscopic laser radar imaging that can probe beneath the surface of the retina, provide a method for precise subsurface imaging. OCT provides a high-resolution, three-dimensional image of the interior of the eye, allowing subsurface structures to be resolved down to a depth of about 1 millimeter below the surface of the retina. This capability allows early diagnosis and early intervention, which can stop or slow down the disease progression, providing the potential for many additional years of visual acuity to affected individuals.

The number of cases of AMD and DR has increased greatly due to the aging of the population and the obesity epidemic.[22] Early detection and intervention with new anti-angiogenesis drugs have proven to be remarkably effective in treating AMD. Moreover, the efficacy of new drugs currently under development can be quantitatively assessed and compared using the three-dimensional OCT images, which accurately define the volume of the lesions in the retina caused by AMD. Changes in the lesion volumes provide a direct measure of the drug efficacy and can help determine effective and safe dosage levels.

OCT also provides the capability for the accurate mapping of the lens and surrounding tissue capsule, which can be measured with great precision in all three dimensions. This information, when combined with laser surgery using ultrafast lasers, has the potential to revolutionize the protocol for treating cataracts. Using OCT guidance, femtosecond lasers can be precisely focused on the capsule and automatically cut close to perfectly round incisions in the capsule. These precise incisions greatly assist in locating and centering the replacement lens. The same ultrafast pulsed laser can also be used to segment the original occluded lens, which can then be much more easily extracted from the patient's eye. This combination of OCT for precise measurement of eye morphology, along with precision femtosecond ultrafast laser machining, is setting a new standard for quality in these ophthalmic procedures, as seen in Figure 6.6.[23]

Image-Guided Surgery

For most solid tumor cancers, surgical excision is often the optimal intervention strategy when it is feasible. Most often it is very important to balance the

[22] AMD Alliance International. 2011. *Increasing Understanding of Wet Age-Related Macular Degeneration (AMD) as a Chronic Disease to ensure that all patients have access to early intervention, regular proactive treatment, and integrated care, and that research is ongoing for improved treatment options.* Available at http://www.amdalliance.org/user_files/documents/AMD_ChronicDiseasePolicy_M03_NoCrops_Low_Res.pdf. Accessed August 1, 2012.

[23] Friedman, N.J., D.V. Palanker, G. Schuele, D. Andersen, G. Marcellino, B.S. Seibel, J. Batlle, R. Feliz, J.H. Talamo, M.S. Blumenkranz, and W.W. Culbertson. 2011. Femtosecond laser capsulotomy. *Journal of Cataract and Refractive Surgery* 37(7):1189-1198.

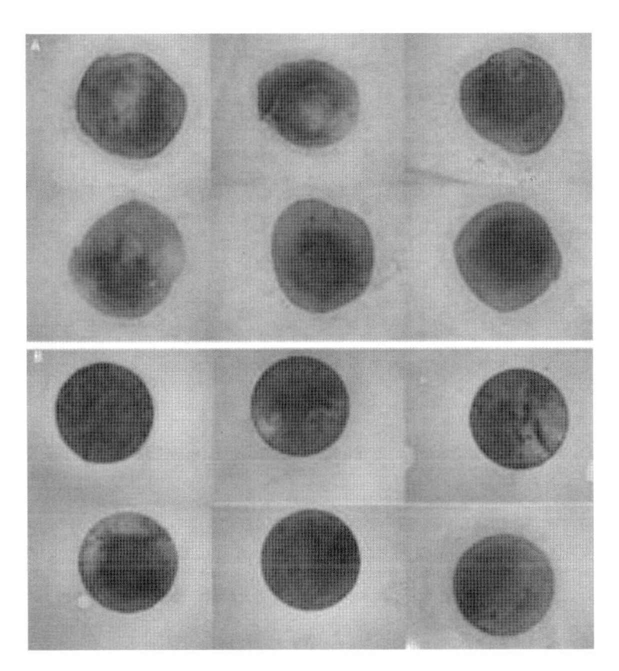

FIGURE 6.6 Excised and stained lens capsule samples from (*top*) a manual capsulorhexis and (*bottom*) a laser capsulotomy with clearly improved boundaries of lens capsule cutting with optical coherence tomography (OCT)-guided femtosecond laser surgery. SOURCE: Reprinted with permission from Friedman, N.J., D.V. Palanker, G. Schuele, D. Andersen, G. Marcellino, B.S. Seibel, J. Batlle, R. Feliz, J.H. Talamo, M.S. Blumenkranz, and W.W. Culbertson. 2011. Femtosecond laser capsulotomy. *Journal of Cataract and Refractive Surgery* 37(7):1189-1198.

need to remove the tumor completely versus the desire to maintain the integrity and function of the surrounding tissue. This is of particular importance in organs such as the brain, liver, and pancreas. Currently many surgical procedures require the excision of a sample portion of the affected organ, which is then sent to the pathology lab for analysis to determine the tumor boundaries, which then help guide the surgeon's decision as to how much tissue to remove. This process is time-consuming and is typically performed while the patient is anesthetized. New optical techniques are being developed[24] that provide real-time images of the tumor boundaries. These techniques employ fluorescent biomarkers (see Figure 6.7), which selectively bind to the tumor cells, providing a clear demarcation between the healthy and diseased tissues that can be visualized directly by the surgeon during the operation. Similar techniques can be used to highlight nearby nerves

[24] For additional information, see Appendix C.

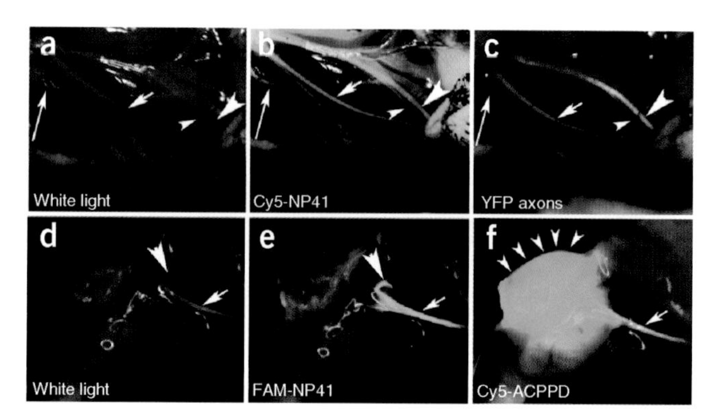

FIGURE 6.7 Comparisons highlighting different fluorescent biomarkers. (a) White light reflectance only used in (a-c); (b) Cy5 fluorescence (pseudocolored cyan, overlaid on reflectance) image highlighting deeply buried nerve (long-stemmed arrow); (c) yellow fluorescent compound (YFP) fluorescence (pseudocolored yellow, overlaid on reflectance) image highlights additional branches (large arrowhead); (d) white light reflectance only used in (d-f); (e) FAM fluorescence (pseudocolored cyan, overlaid on reflectance) image highlighting a stained buried nerve branch (large arrowhead); (f) Cy5 fluorescence (pseudocolored green, overlaid on reflectance) image highlighting a tumor (small arrowheads). SOURCE: Reprinted with permission from Whitney, M.A., J.L. Crisp, L.T. Nguyen, B. Friedman, L.A. Gross, P. Steinbach, R.Y. Tsien, and Q.T. Nguyen. 2011. Fluorescent peptides highlight peripheral nerves during surgery in mice. *Nature Biotechnology* 29:352-356.

and lymph nodes that may need to be carefully avoided or excised as part of the operational procedure.[25]

Dual Energy CT and Quantitative Image Analysis

The primary causes of death in the United States today are heart disease, cancer, and pulmonary disease, such as emphysema.[26] In all of these modern ailments, early detection is the key to effective intervention. Since, in their early stages, these diseases often develop with minor or no symptoms, appropriate routine screening of at-risk populations must be implemented to detect disease in apparently healthy individuals. These screens must be low-cost, minimally invasive, and have low false-positive results in order to prevent unnecessary follow-on procedures. Imaging methods can provide a very effective approach to meeting these criteria.

Recent large-scale clinical trials have demonstrated the ability of low-dose x ray CT scanning as a very effective method to screen for lung cancer tumors,

[25] For information about the use of optic and photonics in biopsies, see Appendix C.

[26] Centers for Disease Control and Prevention (CDC). 2012. "Leading Causes of Death." Available at http://www.cdc.gov/nchs/fastats/lcod.htm/. Accessed August 1, 2012.

cardiovascular disease, and early signs of emphysema. High-resolution CT scans of the chest provide data sets that allow precise measurement of the size of lesions in the lung and allow tracking of changes in the size of these lesions, which is a very specific indication of malignancy. These CT data sets can also provide measures of occlusions or obstruction in cardiac vasculature (early indication of heart attack risk) and assessment of pulmonary function (indicative of emphysema).

CT imaging is a fast (a scan takes less than 10 seconds), non-invasive, and low-cost method for obtaining critical data that can be used to determine effective early intervention, preventing disease progression and greatly reducing the cost of treatment and improving the quality of life for the patient. To achieve the full potential of this technology will require new approaches to data analysis and quantitative feature extraction. For example, to deploy effective screening services, more precise methods for quantifying tumor size, determining the level of calcification in cardiac vasculature, and extracting measures of lung expiration capacity must be developed.

The critical contributions of optical technology to CT instrumentation (and other imaging platforms such as MRI and ultrasound) are often overlooked. Fundamentally, CT devices are optical instruments, employing photons chosen with wavelengths for which the body is partially transparent to precisely image the internal physiology of the patient. The x ray photon sources, the optics for focusing the x rays, and the detectors used to detect the x ray photons are designed employing many of the same techniques developed for the design of imaging instruments using visible light. Similarly, the mathematical methods for analyzing the raw transmitted x ray data, for reconstructing and visualizing three-dimensional models of internal anatomy, are almost identical to comparable techniques employed in other imaging modalities using visible or infrared light. Thus advances in detector technology, image reconstruction models, and techniques for quantitative feature extraction from large three-dimensional data sets will greatly enhance the performance of CT, MRI, and ultrasound imaging platforms.[27]

In general, advances in developing quantitative imaging data analysis procedures are hindered by the inability of the scientific community to access common data sets useful for comparing the performance of automated computerized methods to analyze the data. Establishment of the infrastructure to support public access to large data sets of image data and open-source software tools to extract clinically significant features from these data should be a national priority. Such an infrastructure is vital to accelerating advances in many different imaging modalities, including OCT, CT, MRI, ultrasound, x ray, diffuse optical imaging, and others.

[27] Baer, T.M., J.L. Mulshine, and J.J. Jacobs. 2007. Biomedical Imaging Archive Network. *Skeletal Radiology* 36(9):799-801.

Biomedical Optics in Regenerative Medicine

One of the most active frontiers of modern medicine is in the field of stem cell research. As more is learned about the fundamental processes behind how progenitor cells differentiate and develop into cell types that make up specific tissues and how tissue expands and develops into organs, the potential arises for using this knowledge to repair or replace organs that are damaged due to aging or traumatic injury. Time-lapse microscopic imaging of stem cells as they differentiate into different cells types is playing a key role in identifying specific stages that characterize both normal and abnormal growth pathways. These data can potentially be used to determine which cells are safe to transplant into patients and which may be give rise to tumors.[28]

Biomedical Optics in Research

Advances in technology and the application of new instruments often provide the basis for further insights and discoveries that lead to a deeper understanding of the causes and the molecular basis of diseases. Significant improvements in optical technology have dramatically increased the ability to measure and study biological processes in both in vitro and in vivo environments.

The past few years have witnessed the birth and genesis of a whole new field of biophotonics called optogenetics.[29] (This field was declared the Method of the Year in 2010 by the journal *Nature Methods*.[30]) As defined by Carl Deisseroth, one of its primary developers: "Optogenetics is the combination of genetic and optical methods to achieve gain or loss of function of well-defined events in specific cells of living tissue."[31] This technique has seen primary application in neuroscience, where the function of single neurons or groups of neurons can be monitored and controlled by specific wavelengths of light. Neurons in live, behaving animals can be genetically programmed to fire or be prevented from firing by exposure to different colors of light. Moreover, the active or inactive state of the neuron can be detected by observing fluorescence from the neuron. This provides neuroscientists with tools very analogous to those used by engineers to study electronic circuits: specific neural circuits can be activated or deactivated, and the overall circuit re-

[28] Wong, C.C., K.E. Loewke, N.L. Bossert, B. Behr, C.J. De Jonge, T.M. Baer, R.A. Reijo Pera. 2010. Non-invasive imaging of human embryos before embryonic genome activation predicts development to the blastocyst stage. *Nature Biotechnology* 28:1115-1121.

[29] For additional information, see Appendix C.

[30] *Nature Methods*. 2011. "Method of the Year 2010." Editorial. Available at http://www.nature.com/nmeth/journal/v8/n1/full/nmeth.f.321.html. Accessed July 27, 2012.

[31] Lin, S.-C., K. Deisseroth, and J.M. Henderson. 2011. Optogenetics: Background and Concepts for Neurosurgery. *Neurosurgery* 69(1):1-3.

sponse to these changes and the changes to information flow through the neuronal circuit can be monitored in real time. All of these types of measurements have been performed in a wide variety of alert, active animals.

These remarkable capabilities are made possible by the combination of the development of new forms of bioengineered, light-sensitive proteins and the application of two photon imaging instruments that allow the subsurface probing of neural anatomies. These new bioengineered materials and optical techniques have revolutionized the field of neuroscience and provide, essentially for the first time, the possibility of reverse engineering and modeling of the neural circuitry of complex brains in live animals.

One of the fundamental principles of imaging that has limited the size of objects that can be resolved is the diffraction limit. In essence, objects smaller than the wavelength of light used to illuminate the sample cannot be imaged clearly. Over the past 10 years, several techniques have been developed that allow the precise location of single molecules to be determined to a fraction of a wavelength of light. These approaches allow startlingly vivid images illustrating the locations of proteins and small molecules in organelles and other structures within cells, providing a new capability for gaining insight into the mechanisms and functions of proteins within cells.

Several other areas of biophotonics are in early stages of development but show great promise. These include the use of free electron laser coherent x ray sources to probe the structure of membrane bound proteins in situ. These proteins are often key drug targets, since they control signaling pathways within the cell that are involved with a number of diseases. Laser tweezers and atomic force microscopy are being used to measure what the impact of localized forces on cell membranes is and how these forces can initiate biochemical signaling within the cell. This research has important ramifications on the engineering of tissue structures to support the appropriate growth of specific cell types for organ transplantation.

FINDINGS

Key Finding: Many chronic, debilitating, and often fatal degenerative diseases impacting the aging population are mediated or exacerbated by the patient's own immune system. Understanding and controlling the immune system are thus among the major challenges facing modern medicine today. Optical instrumentation will continue to be the principal enabling technology allowing advances in the understanding of the immune system.

Key Finding: Stem cell science is advancing rapidly, providing great insight into how cells progress from progenitor cells (capable of transforming into any tissue type) to cells of a phenotype characteristic of a specific tissue. Controlling these processes in vivo and developing confidence that once the cells are transplanted into a patient they will continue to develop normally, present a major remaining challenge that must be overcome before stem cells can be used broadly in regenerative medicine. Microscopic imaging technologies will provide key non-invasive methods to monitor the growth process of stem-cell-derived tissues and help ensure their safety and efficacy for transplantation.

Finding: Optical techniques using solid-state light sources and detectors combined with microfluidics are the ideal technology base for automated, low-cost, portable devices that can be operated by personnel without their needing extensive training. In high-income countries the primary causes of death and patient morbidity are degenerative diseases due to longer life expectancy; in contrast, in low-income countries the infectious diseases remain leading causes of death. One of the primary challenges for infectious diseases in low-income countries is to develop low-cost diagnostic methods that can identify disease in its pre-symptomatic and pre-infectious early stage. Additionally, diseases such as malaria and tuberculosis have different phenotypes that can be identified using optically based diagnostic methods and thus help determine the most effective course of therapy.

Finding: The current generation of imaging instruments (CT, MRI, OCT, and ultrasound) provides unprecedented resolution, allowing spectacular three-dimensional, non-invasive images of human anatomy. These data sets contain information that will allow early diagnosis of many potentially fatal diseases, including lung cancer, heart disease, emphysema, and Alzheimer's disease. Early detection often provides the opportunity for the most effective intervention. However, these images contain an enormous amount of information, at times overwhelming the ability of the radiologist or clinician to effectively evaluate and digest all the information available from the raw images. Similar challenges are faced by ophthalmologists interpreting three-dimensional data sets generated by OCT instruments, and likewise by pathologists dealing with large data sets generated by the automated scanning of large tissue sections imaged with subcellular resolution. Automated image-analysis software can provide reliable quantitative measurement of key features from these complex data sets, improving the diagnostic reliability, decreasing the amount of time required, and lowering cost. Clearly all of these new image approaches have many challenges in common and could benefit from a centralized infrastructure for sharing data and image-analysis software algorithms.

Finding: A person's genes determine, in part, that person's tendency to succumb to specific diseases. Developing more cost-effective methods to sequence human genomes could lead to effective identification of an individual's risk factors and potentially to effective early intervention and preventative strategies. Almost all of the most recent generation of high-throughput sequencing instruments are based on optical methods.

RECOMMENDATIONS

Key Recommendation: The U.S. optics and photonics community should develop new instrumentation to allow simultaneous measurement of all immune-system cell types in a blood sample. Many health issues could be addressed by an improved knowledge of the immune system, which represents one of the major areas requiring better understanding.

Key Recommendation: New approaches, or dramatic improvements in existing methods and instruments, should be developed by industry and academia to increase the rate at which new pharmaceuticals can be safely developed and proved effective. Developing these approaches will require investment by the government and the private sector in optical methods integrated with high-speed sample-handling robotics, methods for evaluating the molecular makeup of microscopic samples, and increased sensitivity and specificity for detecting antibodies, enzymes, and important cell phenotypes.

Recommendation: The U.S. health care diagnostics industry, in cooperation with academia, should prioritize the development of low-cost diagnostics for extremely drug-resistant and multi-drug-resistant TB, malaria, HIV, and other dangerous pathogens, and low-cost blood-serum- and tissue-analysis technology to potentially save millions of lives per year.

Recommendation: The U.S. health care industry, in cooperation with academia, should prioritize the development of new optical instruments and integrated incubation technology capable of imaging expanding and differentiating cell cultures in vitro and in vivo, to provide important tools for predicting the safety and efficacy of stem-cell-derived tissue transplants.

Recommendation: The U.S. software and information technology industry, in cooperation with academia, should prioritize the development of new software methods automating the extracting, quantifying, and highlighting of important features in large, two- and three-dimensional data sets to optimize the utility of the latest generation of imaging instruments.

Recommendation: The U.S. life science instrumentation industry, in cooperation with academia and the federal government, should prioritize the development of the next generation of super-high-throughput sequencing devices, required for lowering the cost of sequencing down to the target cost of $1,000 per genome. This will require advances in high-sensitivity, low-noise, high-resolution CCD cameras, high-efficiency laser sources, and nanophotonic devices integrated with microfluidics and automated systematic analysis.

Recommendation: The U.S. government should expand investment in multidisciplinary centers (e.g., at universities with medical and engineering schools) at which critical developments combining medical and engineering discoveries can be efficiently fostered.

Recommendation: The U.S. government, in cooperation with scientific and medical societies, should facilitate the creation of an information technology infrastructure for sharing large amounts of medical and clinical data (e.g., quantitative imaging and molecular data) and open-source analysis tools.

7

Advanced Manufacturing

INTRODUCTION

As of 2009, the United States accounted for 21 percent of global manufacturing value added, measured at 2009 purchasing-power exchange rates.[1] The U.S. share has declined since at least 1990, and the share of producers (including U.S.-owned production facilities) in the industrializing economies in Southeast Asia and elsewhere has grown. Nevertheless, a great deal of cutting-edge innovation and a modest amount of manufacturing activity to support it have remained in the United States, and this generalization applies to photonics and other high-technology industries.

This chapter discusses photonics manufacturing, emphasizing three distinct but closely linked issues. First, it examines the relocation of production of photonics components and products in three key product fields—displays, solar technologies, and optoelectronic components—and the factors behind the offshore movement of much of this production activity. Next, it discusses the relationship between manufacturing and innovation in photonics technologies, highlighting the contrasts and similarities among the three cases in an effort to explain how and where the United States has been able to retain dominance in both production and innovation in selected photonics technologies. Finally, the chapter examines photonics in manufacturing, discussing new advances in manufacturing tech-

[1] United Nations Industrial Development Organization (UNIDO). 2010. *International Yearbook of Industrial Statistics.* Cheltenham, U.K.: Edward Elgar Publishing.

nologies and production capabilities made possible by applications of photonics technologies.

PRODUCTION AND INNOVATION IN PHOTONICS TECHNOLOGIES: THREE CASE STUDIES

This section presents three case studies—displays, solar cells, and optoelectronics components for the converging communications and computing industry—to examine trends in the offshoring of manufacturing and the relationship between manufacturing and innovation. All three deal with optic and photonic applications based on semiconductor technologies that originated in AT&T Bell Laboratories and other large corporate laboratories. While all three cases have similarities, important differences among them have implications for innovative performance, industry structure, and public policy. It is noteworthy that there is a continued need for increased resolution, smaller features, and increased packing density in all cases of production technologies. This need will drive a need for optical sources and imaging tools supporting the increase in resolution.

Displays

Of the three industries, the earliest to move manufacturing overseas from the United States was displays. As discussed by Macher and Mowery in the National Research Council report *Innovation in Global Industries: U.S. Firms Competing in a New World*,[2] although the technological foundations of the display industry were developed in the United States in the 1960s, the industry's production operations quickly migrated to Japan and then to Korea and Taiwan.[3] By 1995, liquid-crystal displays (LCDs) accounted for greater than 95 percent of flat panel display sales by value and thin-film transistor (TFT) LCDs accounted for more than 90 percent of LCD sales, having first found their way into application in calculators, then in cell phones and computers applications, and more recently, as prices continued to decline, largely replacing cathode ray tubes in television receivers.[4] Large TFT LCDs accounted for about 75 percent of the value of TFT LCD sales, although the unit production volume of small and medium-size LCDs was five to six times that of large TFT LCDs.[5] TFT LCDs remain the dominant display technology.

TFT LCD manufacturing and innovation have their roots in the United States

[2] National Research Council. 2008. *Innovation in Global Industries: U.S. Firms Competing in a New World (Collected Studies)*, J.T. Macher and D.C. Mowery, eds. Washington, D.C.: The National Academies Press.

[3] National Research Council. 2008. *Innovation in Global Industries.*

[4] National Research Council. 2008. *Innovation in Global Industries.*

[5] National Research Council. 2008. *Innovation in Global Industries.*

in the late 1960s with a number of research advances by such major firms as RCA, Westinghouse, Exxon, Xerox, AT&T, and IBM.[6] Of those firms, only IBM invested in high-volume manufacturing—through a joint venture with Toshiba in Japan.[7] In contrast, all the major Japanese electronics firms invested in high-volume manufacturing. Manufacturers of TFT LCDs face primarily two strategic decisions: when to invest in the construction of a new fabrication facility and whether to move to the next generation of substrates. In 1996, greater than 95 percent of all TFT LCDs were produced in Japan. By 2005, fewer than 11 percent were made in Japan, and the top two production locations were Korea and Taiwan, each of which produced roughly 40 percent of total output. The main reasons for the shift in production location were the lower engineering and labor costs in Korea and Taiwan and the ability of first Korea and then Taiwan to raise the large amounts of capital needed for investing in state-of-the-art fabrication facilities. The window of opportunity for Korean entry occurred in 1991 when Japanese firms were unable to raise the capital needed for investing. Similarly, the window of opportunity for Taiwanese entry came during the Asia crisis of 1997-1998 when Korean firms experienced difficulties in financing new plants.[8] (See Figure 7.1.)

Although the successful integration of each generation of production equipment depended on investment in high-volume production, new materials and equipment were not necessarily developed in the same countries that invested in manufacturing. Figure 7.1(b) shows the changing proportion of U.S. versus total U.S. Patent and Trademark Office (USPTO) patents in LCDs. Although it might be tempting to focus on the United States' declining share of total worldwide LCD patents, it is important to note both that this represents only LCDs, not the next big thing in display technologies, and that even in LCDs the share of patents fails to show a full picture.[9] Thus, in addition to IBM's joint effort with Toshiba for TFT LCD manufacturing, a number of important U.S. firms participated in the industry—most notably, Corning (in substrate materials), Applied Materials (in chemical vapor deposition equipment), and Photon Dynamics (in test, inspection, and repair equipment). These firms remained key players in the market through their ability to acquire knowledge by working collaboratively with manufacturers outside the United States. At the global level, liquid-crystal materials were devel-

[6] National Research Council. 2008. *Innovation in Global Industries.*

[7] National Research Council. 2008. *Innovation in Global Industries.*

[8] National Research Council. 2008. *Innovation in Global Industries.*

[9] The data described above are based on the U.S. versus other nations' shares of patents in the USPTO database. Past research suggests that U.S. patents are a reasonable measure of unique inventive activity worldwide by internationally competitive companies. Notably, the patents described in this data are in no way weighted by their scientific or market value. Thus, there is no way to tell which patents described in Figure 7-1b may be highly incremental additions to existing knowledge rather than revolutionary.

(a)

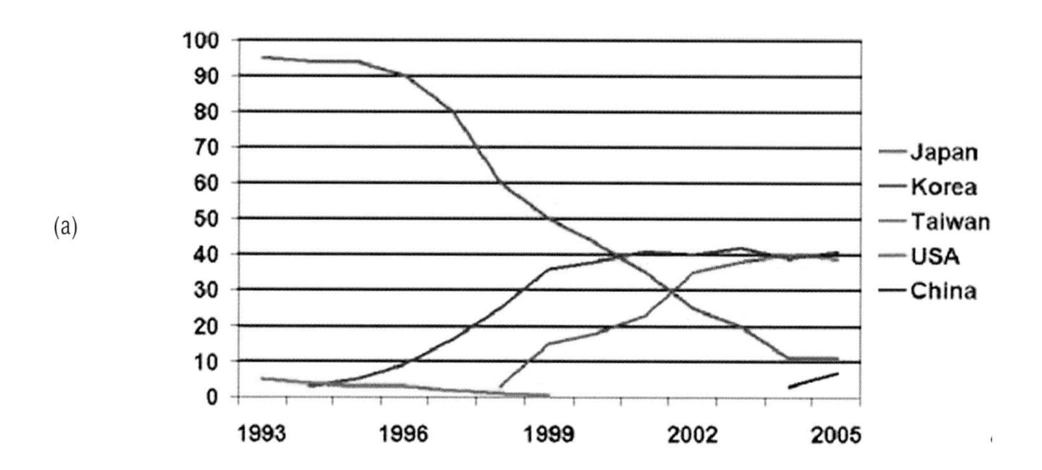

(b)

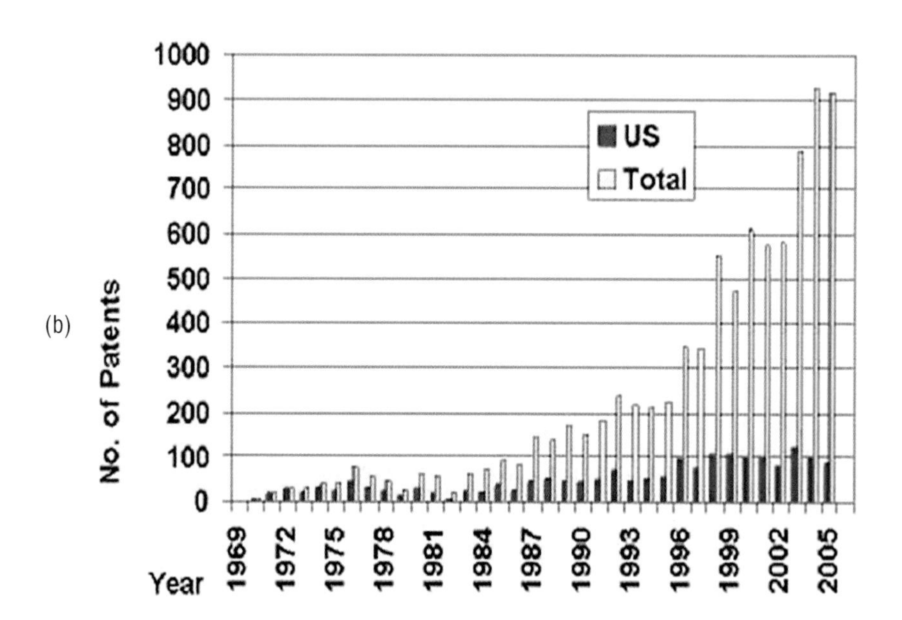

FIGURE 7.1 (a) Percentages of production shares of thin-film transistor (TFT) liquid crystal displays (LCDs). SOURCE: Murtha, T., S.A. Lenway, and J.A. Hart. 2001. *Managing New Industry Creation.* Stanford, Calif.: Stanford University Press. (b) U.S. firms' or laboratories' share of total USPTO LCD patents. SOURCE: U.S. Patent and Trademark Office. 2012. "Patenting in Technology Classes, Breakout by Organization." Available at http://www.uspto.gov/web/offices/ac/ido/oeip/taf/tecasga/349_tor.htm.

oped and fabricated primarily in Western Europe and sold to East Asian producers. Chemical vapor deposition equipment was developed primarily in the United States and Western Europe. Testing equipment was developed mainly in Japan and the United States. Finally, lithographic equipment was developed primarily in Japan, the United States, and Western Europe. Given the time pressures created by the frequent transitions from one generation of production technology to the next, materials and equipment suppliers became more important. Unlike their Japanese counterparts, Korean and Taiwanese firms were generally unable to build their own production equipment and instead had to rely largely on external suppliers.[10]

Although IBM, Corning, Applied Materials, and Photon Dynamics were successful in the display industry, other U.S. firms were less successful. A number of relatively small, niche producers of TFT LCDs engaged in a variety of efforts to catch up with the Japanese, some of which involved financial support from the U.S. government, in particular the Defense Advanced Research Projects Agency (DARPA). Murtha, Lenway, and Hart argue that successful entry by these U.S. firms at this late stage required that they work with partners in East Asia that were experienced in high-volume production.[11] U.S. government policies made it difficult for firms receiving government funding to work closely with manufacturers in Asia, and most of these firms did not recognize the importance of collaborating with high-volume manufacturers.[12]

TFT LCDs dominate today's display markets, but new market opportunities in displays are opening up particularly in flexible displays. Today, the primary sources of innovation in flexible displays are in the United States, and forecasts suggest that this technology's market share in mobile devices will grow in the next few years.

Solar Cells

As in the case of displays, manufacturing and innovation of solar photovoltaics (PVs) have their roots in the United States. As discussed by Colatat, Vidican, and Lester in the paper Innovation Systems in the Solar Photovoltaic Industry: The Role of Public Research Institutions, the first silicon photovoltaic device was invented in 1954 at Bell Laboratories and found a niche market in space satellites.[13] The photovoltaic industry in the 1960s remained very small—the annual market of

[10] National Research Council. 2008. *Innovation in Global Industries.*

[11] Murtha, T.P., S.A. Lenway, and J.A. Hart. 2001. *Managing New Industry Creation: Global Knowledge Formation and Entrepreneurship in High Technology—The Race to Commercialize Flat Panel Displays.* Stanford, Calif.: Stanford University Press.

[12] Murtha et al. 2001. *Managing New Industry Creation.*

[13] Colatat, P., G. Vidican, and R. Lester. 2009. *Innovation Systems in the Solar Photovoltaic Industry: The Role of Public Research Institutions. Massachusetts Institute of Technology Industrial Performance Center Working Paper Series.* Cambridge, Mass.: Massachusetts Institute of Technology.

solar cells was worth $5 million to $10 million, or the equivalent of 50 to 100 kW of capacity—and the U.S. government was the primary customer.[14,15] Throughout the 1950s and 1960s, only five companies produced photovoltaic cells. Two were start-ups: Hoffman Electronics, which acquired National Fabricated Products and its patent license for the Bell Laboratories patents, and Heliotek, founded by Alfred Mann as a spin-off from his previously founded company, Spectrolab.[16] The remaining three entrants were established companies that had diversified into the solar cell market: RCA (which produced radios), International Rectifier (which produced semiconductors), and Texas Instruments (which produced semiconductors). All three of the established companies left the market by the end of the 1960s because it was small and unpredictable.[17]

In 1973, the Arab oil embargo strengthened interest in terrestrial applications of photovoltaics and expanded the market for them.[18] Two well-known U.S. PV firms were founded shortly thereafter by former employees of Spectrolab: Solar Technology International (1975) and Solec International (1976).[19,20] Between the mid-1980s and the mid-1990s, the United States, Japan, and Germany competed for the lead in solar cell production on the basis of the location of production activities. In the mid-1980s, Japan overtook the United States as the number one producer of solar cell modules, with Germany in a distant third place. The United States and Germany then surpassed Japan in world PV module shipments in the mid-1990s. Since the 1995, however, although U.S. production of PVs has risen, the U.S. share of global PV production has fallen from its 43 percent peak in 1995 to an all-time low of 6 percent in 2009.[21] (See Figure 7.2 (b).) As of 2009, Germany is the leader

---

[14] Colatat et al. 2009. *Innovation Systems in the Solar Photovoltaic Industry.*

[15] National Research Council. 1972. *Solar Cells: Outlook for Improved Efficiency.* Washington D.C.: National Academy Press.

[16] Colatat et al. 2009. *Innovation Systems in the Solar Photovoltaic Industry.*

[17] Colatat et al. 2009. *Innovation Systems in the Solar Photovoltaic Industry.*

[18] Colatat et al. 2009. *Innovation Systems in the Solar Photovoltaic Industry.*

[19] Colatat et al. 2009. *Innovation Systems in the Solar Photovoltaic Industry.*

[20] Although both firms are still in operation, both were eventually acquired by non-U.S. companies. Solar Technology International was acquired by Atlantic Richfield Company (ARCO) in 1977 and renamed ARCO Solar. By the time it was acquired by Siemens (a German company) in 1990, ARCO Solar was the largest PV manufacturer in the world. ARCO Solar was later sold by Siemens to an Anglo-Dutch company, Shell, before being sold to Solarworld (a German company). Solec International, which was eventually sold to Sanyo (a Japanese company), is also still in operation today.

[21] Le, Minh, Chief Engineer, Solar Energy Technologies Program, U.S. Department of Energy. 2011. "The SunShot Program—The Great Solar Race: The Apollo Mission of Our Times." Presentation to the NRC Committee on Harnessing Light: Capitalizing on Optical Science Trends and Challenges for Future Research, February 24, 2011.

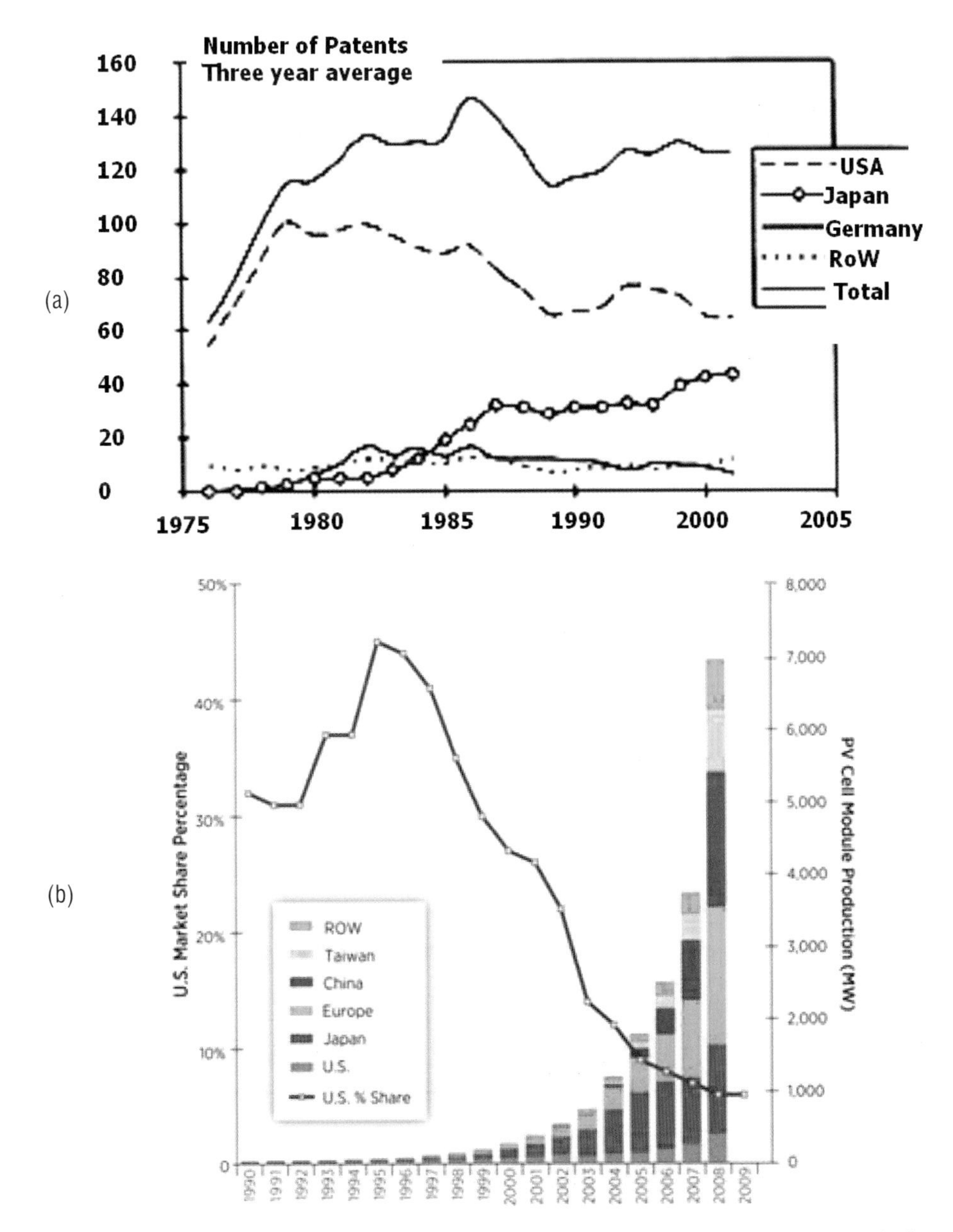

FIGURE 7.2 (a) U.S. share of global patenting. SOURCE: Reprinted, with permission, from Andersson, B.A., and S. Jacobsson. 2000. Monitoring and assessing technology choice: The case of solar cells. *Energy Policy* 28(2000):1037-1049. (b) U.S. share of global PV cell module market and production. SOURCE: Le, Minh, Chief Engineer, Solar Energy Technologies Program, U.S. Department of Energy. 2011. "The SunShot Program—The Great Solar Race: The Apollo Mission of our Times." Presentation to the NRC Committee on Harnessing Light: Capitalizing on Optical Science Trends and Challenges for Future Research, February 24, 2011.

in global production of PV followed by China, with Japan in third place.[22] But the growth of international production means that three of the top five manufacturers of PV modules (measured in megawatts of shipments[23]) in 2010 were Chinese. The firm in second place on the basis of its volume (megawatts) of shipments at the end of the fourth quarter of 2010 is First Solar, a U.S. thin-film PV company that uses cadmium telluride instead of silicon (the material used by all 9 other companies in the top 10) as its semiconductor. First Solar has manufacturing facilities in Perrysburg, Ohio; Frankfurt, Germany; and Kulim, Malaysia. There are no other U.S. companies in the top 10. The Japanese company Sharp is in fourth place, and a Canadian company, Canadian Solar, is in sixth place. The top 10 manufacturers accounted for more than 50 percent of total global PV shipments in 2010.[24]

Although the United States no longer dominates global production of solar modules, it has maintained its lead in patenting in solar technologies,[25] followed by Japan (according to the geographic location reported by the corporate assignees on the patent). As can be seen in Figure 7.2, the United States was the dominant source of solar technology patents during 1975-1995, although Japan's share of global patenting increased significantly after 1980.[26] The United States retains a position of leadership in solar-related USPTO patents, accounting for 52 percent of total solar patents in 2002-2010, followed by Japan at 26 percent and Germany at 6 percent.[27] The 10 leading corporate patentees during 2002-2010 were (in order) Canon (Japanese), Sharp (Japanese), Boeing (U.S.), Sunpower (U.S.), Kanegafuchi (Japanese), Sanyo (Japanese), Emcore (U.S.), Applied Materials (U.S.), Konarka (U.S.), and Rabinowitz (U.S.).[28]

As this discussion suggests, the United States no longer is among the leading

[22] Le, Minh, Chief Engineer, Solar Energy Technologies Program, U.S. Department of Energy. 2011. "The SunShot Program—The Great Solar Race: The Apollo Mission of Our Times." Presentation to the NRC Committee on Harnessing Light: Capitalizing on Optical Science Trends and Challenges for Future Research, February 24, 2011.

[23] The conditions for measuring the nominal power of a photovoltaic module are specified in standards such as IEC 61215, IEC 61646, and UL 1703; the term "power" is also used in describing the size of a shipment or an installation.

[24] *PVinsights*. 2011. "Suntech Lost Championship of Solar Module Shipment to First Solar in 2Q 11." Available at http://pvinsights.com/Report/ReportPMM31A.php. Accessed December 27, 2011.

[25] The above-described data are based on shares of patents in the U.S. Patent and Trademark Office database. Past research suggests that U.S. patents are a reasonable measure of unique inventive activity worldwide by internationally competitive companies. Here, patent locations are assigned on the basis of company assignee location, as reported in the filed patent.

[26] Andersson, B.A. and S. Jacobsson. 2000. Monitoring and assessing technology choice: The case of solar cells. *Energy Policy* 28(14):1037-1049.

[27] Cardona, V. 2011. "Clean Energy Patents—Winners and Losers, Renewable Energy World." Available at http://www.renewableenergyworld.com/rea/news/print/article/2011/03/2010-clean-energy-patents-winners-and-losers. Accessed July 5, 2011.

[28] Cardona, V. 2011. "Clean Energy Patents—Winners and Losers, Renewable Energy World."

sites (by volume) for the manufacturing of solar modules.[29] Notable is the apparent lack of correlation between the nations in which the bulk of solar module manufacturing is sited and the nations that dominate inventive activity (as measured by USPTO patenting). (See Figure 7.2 (b).) Indeed, the United States continues to dominate global inventive activity in solar technologies despite not being the location for the greatest volume of solar module manufacturing output. (See Figure 7.2 (a).) Notably, the largest volume in the world of patents is in cutting-edge solar technologies, such as thin films, which are still produced largely in the United States. In contrast, the largest volume of manufacturing is in crystalline silicon technology modules—a technical field that no longer dominates solar technology patenting in the USPTO.

Innovation in the materials that underpin solar technologies used for energy generation may prove important in affecting the future site of manufacturing activity in this field. Numerous materials and designs can produce photovoltaic effects.[30] Overall, PV technologies can be grouped into four main categories: wafer and thin film (including crystalline and amorphous silicon and cadmium telluride technologies), concentrator, excitonic (including organic polymer and dye-sensitized solar technologies), and novel, high-efficiency technologies (such as plasmonics).[31] Designs based on crystalline silicon have dominated commercial PV technology, accounting for more than 80 percent of the market for commercial modules since the industry's origin.[32] Crystalline silicon may not, however, be the future. Today, thin-film solar technologies hold the second-largest proportion of the commercial market after crystalline silicon, hovering below 20 percent.[33] In the 1980s, both the United States and Japan invested in thin-film amorphous silicon technologies.[34] One report finds that between 1994 and 1998 the number of USPTO patents granted in amorphous silicon exceeded the number granted in crystalline silicon.[35] A more recent report based on National Renewable Energy Laboratory data concluded that the cost per watt of producing thin-film PV was closing the gap with the cost of producing crystalline PV in the late 1990s and early

[29] At the firm level, one U.S.-headquartered firm is among the corporate leaders (by volume) in global manufacturing of solar modules. However, the national headquarters of the corporate entities that dominate solar manufacturing may not correlate with the geographic location of those corporations' manufacturing activities.

[30] Baumann, A., Y. Bhargava, Z.X. Liu, G. Nemet, and J. Wilcox. 2004. *Photovoltaic Technology Review.* Berkeley, Calif.: University of California, Berkeley.

[31] Curtright, A.E., M.G. Morgan, and D. Keith. 2008. Expert assessment of future photovoltaic technology. *Environmental Science and Technology* 42(24): 9031-9038.

[32] Baumann et al. 2004. *Photovoltaic Technology Review.*

[33] *PVinsights.* 2011. "Suntech Lost Championship of Solar Module Shipment to First Solar in 2Q 11." Available at http://pvinsights.com/Report/ReportPMM31A.php. Accessed December 27, 2011.

[34] Baumann et al. 2004. *Photovoltaic Technology Review.*

[35] Andersson and Jacobsson. 2000. Monitoring and assessing technology choice.

2000s.[36] However, even in-depth assessments find it difficult to assess which of the thin-film or the many other PV material and design technologies may be dominant in the future.[37] Inasmuch as the basic materials underpinning solar technologies are likely to undergo considerable change, it is possible that U.S.-based innovation can lead to expanded U.S.-based production of new technologies in this field.

Optoelectronic Components for Communications Systems

Optoelectronic components—which include lasers, modulators, amplifiers, photodetectors, and waveguides produced on semiconductors—are the components necessary to send and receive information in light-based communications systems. The origins of this technology can be traced to the 1960 demonstration of the laser at Hughes Aircraft that followed from research at Columbia University and AT&T Bell Laboratories (see Chapter 2 for further discussion). Further research and development (R&D), much of it at Bell Laboratories, yielded the fabrication methods by which fiber and the other system-critical optoelectronic components could be manufactured economically. In 1970, Corning was the first firm to develop the optical waveguide technology—in particular, low-loss optical fiber that would prove critical to the development of optoelectronics. Corning entered into joint-development cross-licensing agreements with AT&T and cable suppliers in Europe and Japan. By 1986, several other giant corporations had begun production of fiber optics and related components, including DuPont, ITT, Allied Signal, Eastman Kodak, IBM, and Celanese. Large Japanese corporations, including Nippon Electric Company, made similar investments.[38]

The 1984 consent decree that resolved the federal antitrust suit against AT&T produced dramatic changes in the structure of R&D and manufacturing in the U.S. communications industry, and these changes affected the development of optoelectronics in the United States. In 1996, AT&T spun off Bell Laboratories with most of its equipment manufacturing business into a new company named Lucent Technologies. In 2006, Lucent signed a merger agreement with the French company Alcatel to form Alcatel-Lucent. On August 28, 2008, Alcatel-Lucent announced that it was pulling out of basic science, material physics, and semiconductor research to work in more immediately marketable fields.[39]

During the 1990s, many small and medium-size optoelectronic component

[36] Baumann et al. 2004. *Photovoltaic Technology Review.*

[37] Curtright et al. 2008. Expert assessment of future photovoltaic technology.

[38] Sternberg, E. 1992. *Photonic Technology and Industrial Policy: U.S. Responses to Technological Change.* Albany, N.Y.: State University of New York Press.

[39] Ganapati, Priya. 2008. Bell Labs kills fundamental physics research. *Wired Magazine.* August 27. Available at http://www.wired.com/gadgetlab/2008/08/bell-labs-kills/. Accessed November 12, 2012.

manufacturers for communications were founded in the United States.[40] In March 2000, however, the telecommunications bubble burst, throwing the industry into turmoil. By 2002, optical fiber sales had fallen short of monthly projections by more than 80 percent,[41] and competitive survival of producers of fiber and components required that they reduce production costs rather than develop novel technologies.[42]

The collapse of the U.S. telecommunications equipment market led to dramatic change in the location of optical components production. Between 2000 and 2006, the majority of optoelectronic component manufacturers moved manufacturing activities from the United States to developing countries, in particular to developing East Asia.[43] By 2005, five U.S.-based companies (Agilent Technologies, JDSUniphase, Bookham, Finisar, and Infineon) and two Japanese-based companies (Mitsubishi and Sumitomo Electric/ExceLight) accounted for 65 percent of revenues in optoelectronic components.[44] All five of the top U.S. manufacturers had moved assembly activities to East Asia, and all but JDSUniphase had also moved some or all of their fabrication activities to East Asia. The offshore production activities of these U.S. firms did not rely on contract manufacturers, instead producing components in wholly owned foreign subsidiaries. Only a few U.S. entities, mainly start-ups relying on funding from venture capitalists or Small Business Innovation Research (SBIR), chose to keep all manufacturing in the United States. Included among these start-ups were Infinera, Kotura, and Luxtera.

Despite these changes in the location of manufacturing, as can be seen in Figure 7.3, the United States has maintained a dominant role in total optoelectronics patent production as measured by USPTO data, with Japan a close second. In the 4-year period (2001-2004) after the bursting of the telecommunications and Internet bubble, U.S. patents filed annually by assignees in the United States fell while U.S. patenting during the same period by assignees in Japan continued to rise.

Although those data depict trends in issued patents, which have to pass a formal review for novelty and non-obviousness by USPTO examiners, they do not adjust for the fact that the economic or technological importance of individual patents varies widely. Nor do these patent data indicate the location of the inventive activity

---

[40] Yang, C., Nugent, R., and Fuchs, E. 2011. *Gains from Other's Losses: Technology Trajectories and the Global Division of Firms.* Carnegie Mellon University Working Paper. Available at http://papers.ssrn.com/sol3/papers.cfm?abstract_id=2080595. Accessed November 12, 2012.

[41] Fuchs, E.R.H., E.J. Bruce, R.J. Ram, and R.E. Kirchain. 2006. Process-based cost modeling of photonics manufacture: The cost competitiveness of monolithic integration of a 1550-nm DFB laser and an electroabsorptive modulator on an InP platform. *Journal of Lightwave Technology* 24(8):3175-3186.

[42] Fuchs et al. 2006. Process-based cost modeling of photonics manufacture.

[43] Yang et al. 2011. *Gains from Other's Losses.*

[44] Fuchs et al. 2006. Process-based cost modeling of photonics manufacture.

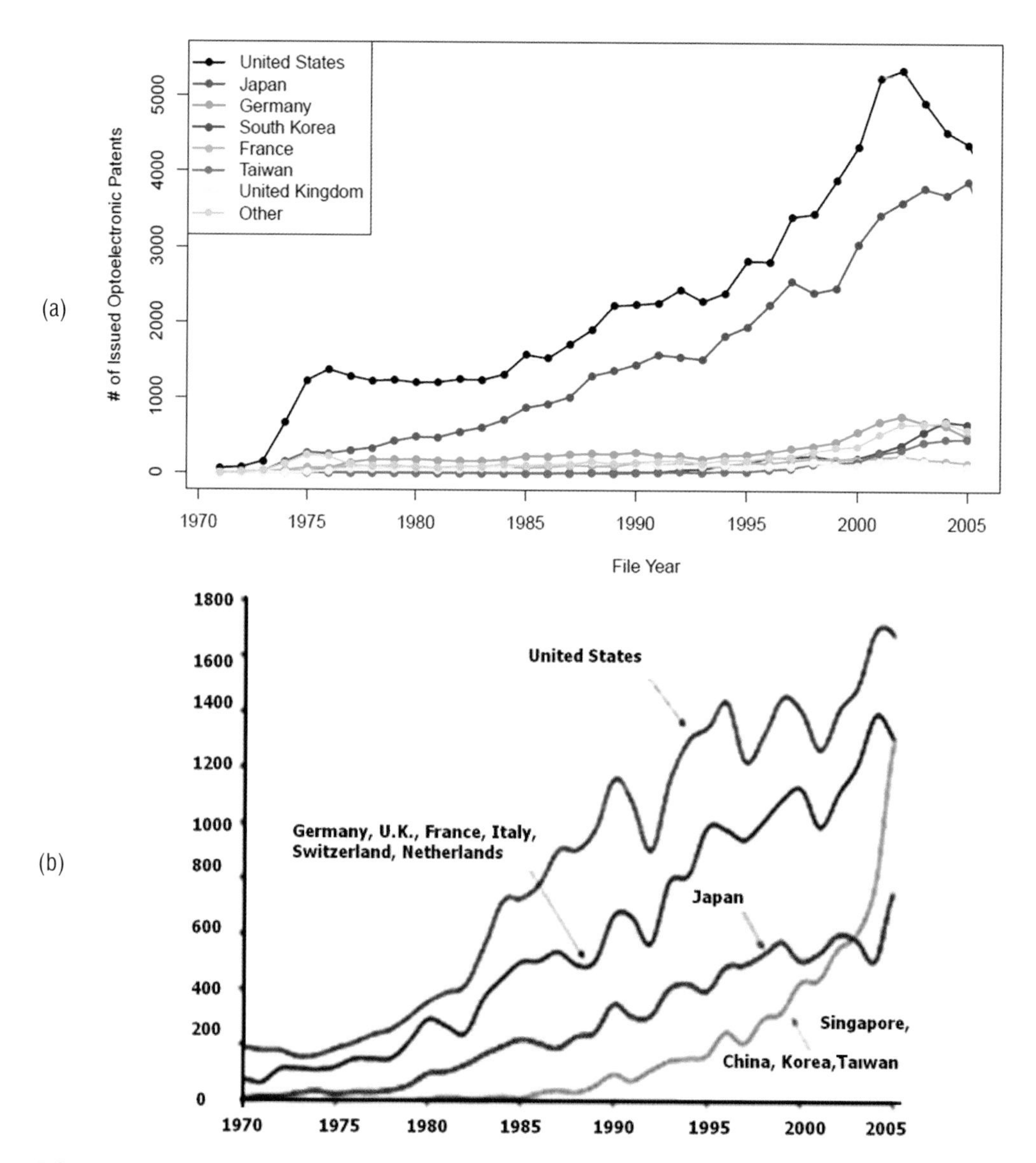

FIGURE 7.3 (a) The United States and Japan dominate U.S. Patent and Trademark Office (USPTO) patents in optoelectronics. NOTE: Country assigned by patent assignee's filed location. SOURCE: Reprinted, with permission, from Ventura, S., R. Nugent, and E. Fuchs. 2012. *Methods Matter: Revamping Inventor Disambiguation Algorithms with Classification Models and Labeled Inventor Records.* Working paper. Abstract available at http://ssrn.com/abstract=2079330. (b) Optoelectronic publications of some major countries (1970-2005). Historically, the United States dominates publications in optoelectronics, followed by Japan and then various countries in Europe; in recent years, however, publication rates from East Asia have risen dramatically. SOURCE: Reprinted, with permission, from Doutriaux, T., 2009. "The Resiliency of the Innovation Ecosystem: The Impact of Off-shoring on Firms Versus Individual Technology Trajectories." Work toward a master's thesis. Advisor: E. Fuchs. Pittsburgh, Pa.: Carnegie Mellon University.

that led to the patents that were issued in the United States. The trends in patenting thus provide some information on the location of inventive activity, but they also reflect inventor (or assigned) perceptions about the location of market opportunities for the exploitation of the inventions. And the United States retains a dominant position as a focus of global patenting activity in optoelectronics.

Publication records covering the 1970-2005 period in optoelectronic components reveal a slightly different picture. For the first three decades of this period, the United States accounts for the largest number of publications, followed by key countries in Europe and Japan. Since roughly 2000, however, an almost exponential rise can be seen in publications from developing East Asia as the total number of publications from China, Korea, Taiwan, and Singapore surpassed the total number of publications from Japan by 2003 and reached parity with the number of publications from Germany, France, Italy, the Netherlands, Switzerland, and the United Kingdom by 2005. Those publication data are not normalized for quality and therefore can be best interpreted as measures of overall research activity in optoelectronics within each country, in contrast with the patent data described above. Arguably, the publications data indicate a significant rise in optoelectronics research activity in the industrializing economies of East Asia since 2000.

To elucidate the relationship between manufacturing location and the rate and direction of innovation in optoelectronic components better, the following discussion addresses the technical details of the innovations that U.S. optoelectronic component manufacturers were pursuing when they began to move their production activities offshore. As was true of the electronics industry 30 years ago, three competing design technologies are used in optoelectronic components. Older, more established designs use discrete components that are wire-bonded together. The most recent designs, however, rely on the capability to do "monolithic integration"—the fabrication and integration (using semiconductor techniques) of multiple functions on a single chip. Between those extremes are designs that use intermediary "hybrid integration" techniques with a variety of methods to bond components to one another. Monolithic integration holds considerable promise for product innovation in the photonics industry, because it significantly reduces size, packing alignment, complexity, and cost while potentially improving the reliability of photonic components, thereby enabling their application to markets outside telecommunications, such as computer chips (especially optical buses), biosensors, and other small-scale sensing and information processing applications.[45]

During the 1980s and 1990s, when optoelectronics was being used extensively

[45] Akinsanmi, W., R. Reagans, and E.R.H. Fuchs. 2011. *Economic Downturns, Technology Trajectories, and the Careers of Scientists.* Presented June 2 at Atlanta Conference on Science and Innovation Policy. Available at http://smartech.gatech.edu/bitstream/handle/1853/42529/527-1641-1-PB.docx?sequence=1. Accessed November 12, 2012.

in telecommunications, an optoelectronic firm's competitiveness depended on the speed with which it could bring the latest innovations to market.[46] During that period, firms manufacturing optoelectronics components for applications in telecommunications focused their process-technology development on monolithic integration (Figure 7.4) to improve component reliability and reduce cost.[47] The transfer by many U.S. optoelectronic component manufacturers of their assembly and fabrication activities offshore after the bursting of the telecommunications bubble shifted their focus in process innovation away from monolithic integration to discrete-technology solutions. That shift largely reflected the different manufacturing-cost environment of the 1990s in the industrializing East Asian economies within which the U.S. firms were operating their offshore facilities and the shift in the competitive environment to favor low-cost production over rapid introduction of new component designs.[48]

Those firms found that the lowest-cost option was to manufacture the discrete technologies in developing countries and abandon U.S. production of monolithically integrated technologies.[49] It is not clear, however, that the apparent declines in monolithic-integration patenting in the firms that have moved production activities offshore will necessarily lead to a decline in overall innovation by U.S. firms in monolithic integration. Several start-up firms that focus on monolithic-integration technologies have emerged in the United States since 2000. It is also possible that established U.S. optoelectronic component manufacturers that have kept fabrication in the United States will increase R&D and patenting in monolithic integration for optoelectronics. Finally, and perhaps most important, firms outside telecommunications and data communications, such as computing firms, may find it in their interest to develop monolithic-integration design and fabrication capabilities for communications applications, as evidenced by Intel's recent establishment of a silicon photonics design and fabrication facility at the University of Washington.

The case of optoelectronics components illustrates a strong relationship between the location of production activities by U.S. firms and the direction of these firms' innovative efforts. But the evidence presented in this case suggests that the movement of optoelectronics-component production to non-U.S. locations has thus far not resulted in the "loss" by the U.S. economy of innovative capabilities

[46] Fuchs, E.R.H., and R.E. Kirchain. 2010. Design for location: The impact of manufacturing offshore on technology competitiveness in the optoelectronics industry. *Management Science* 56(12): 2323-2349.

[47] Fuchs et al. 2010. Design for location.

[48] Yang et al. 2011. *Gains from Other's Losses.*

[49] Recent research shows that moving assembly activities offshore is associated with a decrease in monolithic integration patenting activities, although overall patenting other than monolithic integration increases. U.S. firms that move both assembly and fabrication activities offshore, however, display declines in all optoelectronics patenting. Yang et al. 2011. *Gains from Other's Losses.*

(a)

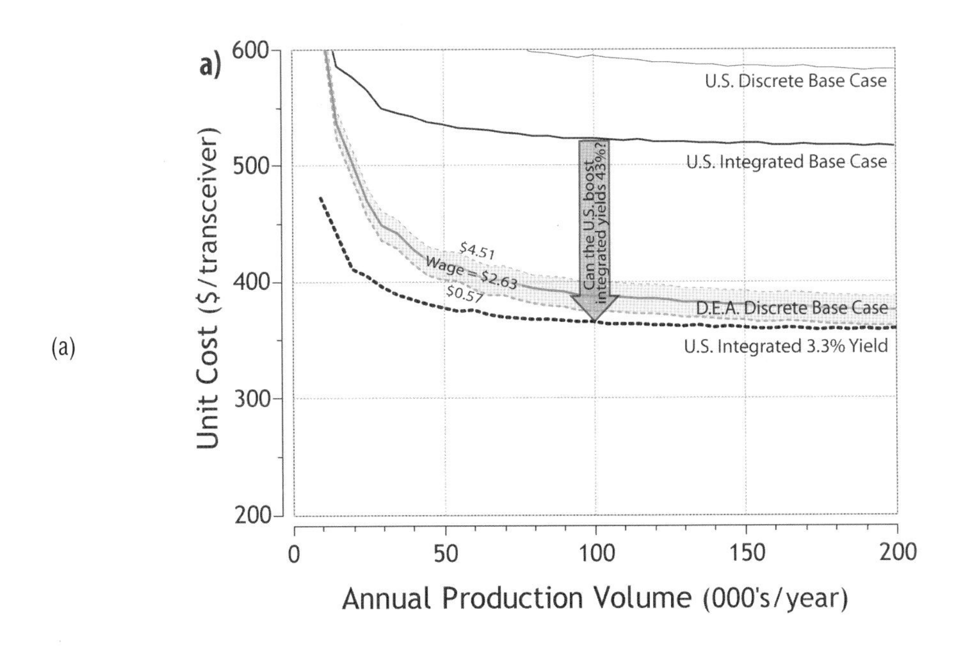

(b)

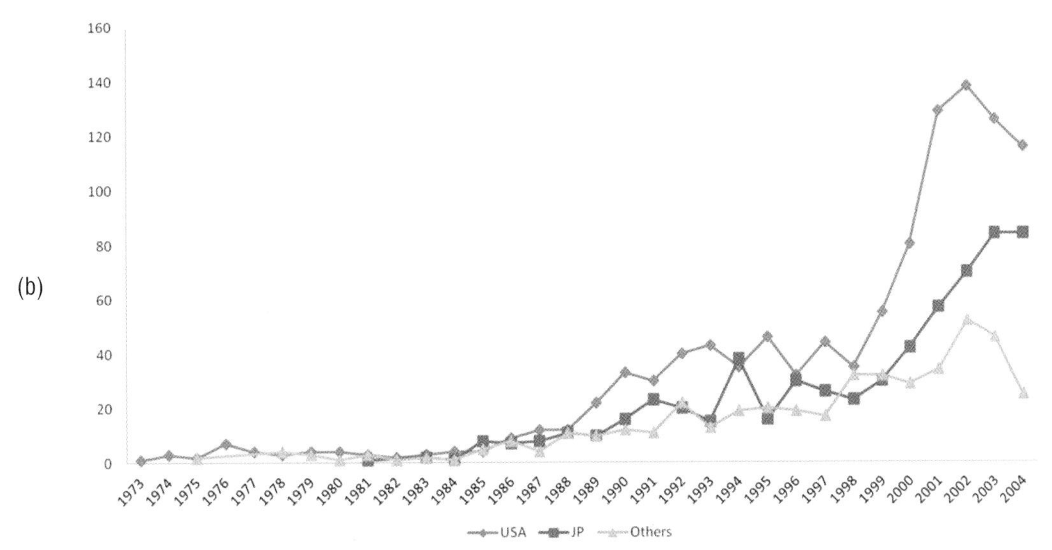

FIGURE 7.4 (a) U.S. monolithically integrated design produced in the United States cannot cost-compete with the discrete design produced in developing East Asia (D.E.A.). SOURCE: Reprinted, with permission, from Fuchs, E., and R. Kirchain. 2010. Design for location: The impact of manufacturing offshore on technology competitiveness in the optoelectronics industry. *Management Science* 56(12):2323-2349. (b) U.S. firms maintain cumulative dominance in USPTO patents in optoelectronic integration; however, Japanese firms have equal or higher numbers of patents in some years. SOURCE: Reprinted, with permission, from Yang, C., R. Nugent, and E. Fuchs. 2011. *Gains from Others' Losses: Technology Trajectories and the Global Division of Firms.* Carnegie Mellon University Working Paper. Available at http://papers.ssrn.com/sol3/papers.cfm?abstract_id=2080595.

in monolithic integration. Instead, the committee observes that a different set of U.S. corporations (and universities) now have become active in this technological field.[50,51]

Similarities and Differences Among the Three Cases

The displays, solar, and optoelectronic communications components cases demonstrate important similarities and differences. All three cases involve semiconductor processing technologies in which R&D and manufacturing activities are linked. In all three, technologies originated in the 1950s and 1960s in corporate R&D laboratories in the United States—for example, AT&T Bell Laboratories, IBM, and Corning. In all three cases, manufacturing (often in U.S.-owned facilities) has moved overseas and the United States has lost its position as the leading site for production activities. In the case of displays, manufacturing moved first to Japan, then to Korea and then Taiwan. In the case of solar, the dominant site of manufacturing moved to Germany and China. In the case of optoelectronic components for communication systems, the dominant manufacturing position moved to developing East Asia.

In all of those cases, manufacturing moved overseas, but the primary offshore manufacturing site did not always become the leading source of innovation. Indeed, in all three cases, U.S.-based firms and inventors are leading technological advances, at least some of which could displace established and dominant producers. The potentially disruptive technologies include flexible displays, thin-film and other technologies in the solar field, and monolithically integrated design and fabrication technologies in optoelectronic communications systems.

In contrast with the pattern of innovation, entry, and early-stage growth of many of these technologies in the 1950s and 1960s, the new technological possibilities are being pursued by start-up firms, often in collaboration with U.S. government laboratories or universities (see Figure 7.5, which shows the growing role of U.S. universities in optoelectronics patenting). Moreover, the U.S. defense market often is a less central source of demand for innovative technologies. The new approach to technology development that relies more heavily on universities and small and medium-size firms for innovation, in which venture-capital funding plays a more important role, may increase the importance of mechanisms to support cross-industry and cross-institutional coordination in helping the United States to maintain leadership in photonics innovation.

The three cases also highlight important differences among the technologies. In displays and optoelectronic components, the vast majority of manufacturing has

[50] Akinsanmi et al. 2011. *Economic Downturns, Technology Trajectories, and the Careers of Scientists.*
[51] Yang et al. 2011. *Gains from Other's Losses.*

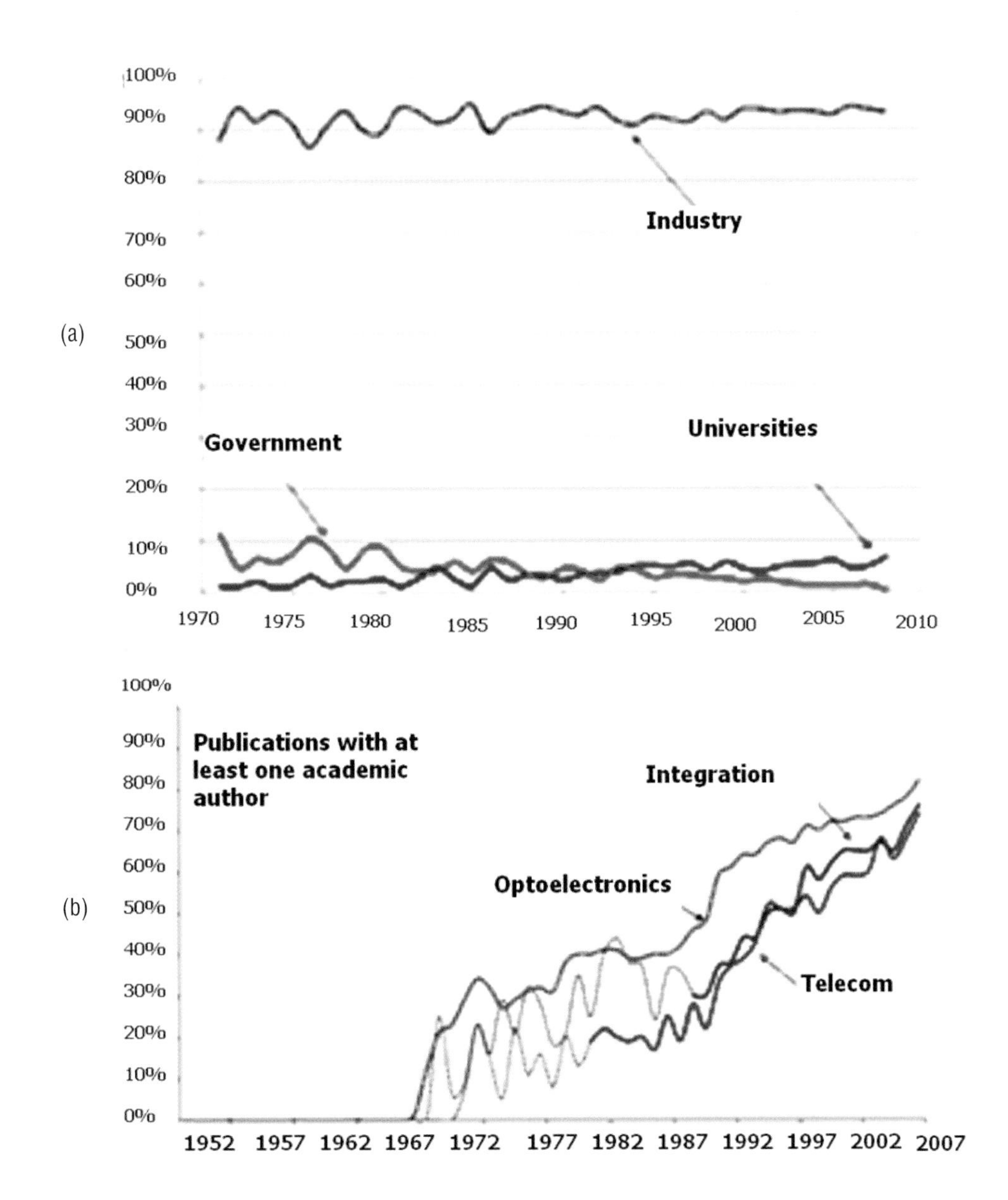

FIGURE 7.5 The institutional locus of innovation in optics. An example from optoelectronic components for the communications industry. (a) Over the last nearly four decades, firms have remained the predominant source of U.S. Patent and Trademark Office patents in optoelectronics. NOTE: The industry, government, or university label is according to patent assignee. (b) Over the last four decades, universities have contributed to a growing percentage of optoelectronic publications. SOURCE: Reprinted, with permission, from Doutriaux, T. 2009. "The Resiliency of the Innovation Ecosystem: The Impact of Offshoring on Firms Versus Individual Technology Trajectories." Work toward a master's thesis. Advisor: E. Fuchs. Pittsburgh, Pa.: Carnegie Mellon University.

moved to developing countries where labor and engineering are cheaper. In solar, however, the leading producer, Germany, is a developed country in which direct-line and engineering wages are as high as or higher than those in the United States.

The close tie between R&D and manufacturing also has different consequences in the three cases. In the case of displays, the close tie led U.S. firms to collaborate with foreign manufacturers to remain at the cutting edge of innovation. In the case of solar, the close tie between R&D and manufacturing means that R&D occurs largely in the same country as manufacturing and has enabled U.S.-based R&D and manufacturing of thin-film technologies to remain dominant. Finally, in the case of optoelectronic components for communications systems, the close tie between R&D and manufacturing led firms that moved manufacturing overseas to abandon monolithically integrated technologies. As a result, monolithic integration continues to be dominated by private firms that remain in the United States.

Those differences yield different policy implications for each industry. In the case of displays, government policies preventing firms from collaborating with foreign enterprises (such as might exist as stipulations for certain work with the Department of Defense [DOD] could have a negative impact on innovation. In the case of solar, the dominant position of German-based production suggests that developing countries may not be the site for cost-competitive manufacturing. The dominance of patenting and production in thin-film technologies by U.S.-based firms suggests that the United States may have the opportunity to be a leader not only in innovating but also in the next generation of advanced solar manufacturing. Finally, in the case of optoelectronic components for communications systems, government and venture funding of small and medium-size enterprises pursuing next-generation monolithically integrated technologies may be critical to overcome the gap between current market demands and longer-term markets in computing and biotechnology that may require the low-power and smaller form-factor performance offered by monolithically integrated solutions. Those differences highlight the importance of avoiding a single-blanket policy for all of photonics. Instead, it is essential to engage technical and industry experts in policy development to exploit their contextual understanding of implications and possible outcomes.

ADVANCED MANUFACTURING IN OPTICS

This section discusses advances in manufacturing technologies in optical component and optical systems production, which have experienced considerable progress over the last decade. Technologies that were considered innovative a dozen years ago have undergone significant evolution and are now found in operation in many optics manufacturing firms. Improvements in generation, finishing, assembly, and metrology technologies are being leveraged to generate higher-performing optical systems and push the upper end of the precision scale. Many of these systems include aspherical lenses, which have advantages over spherical lenses but are more difficult to produce. This section briefly describes the advantages of aspherical lenses and then addresses some of the recent improvements in their production (for a much more detailed description, see Appendix C). The push toward the upper end of the precision scale will drive the need for improvements in optical sources and imaging tools to support the increase in resolution.

Optical Surfaces

Spherical lenses have been the workhorse of optical systems for centuries. The curved surfaces of a lens cause rays of light from a point on a distant object to come to a focus. A single lens with spherical surfaces forms an image that is not a perfect point. (See Figure 7.6.) Optical design has traditionally been a search for combinations of spherical-surfaced components.

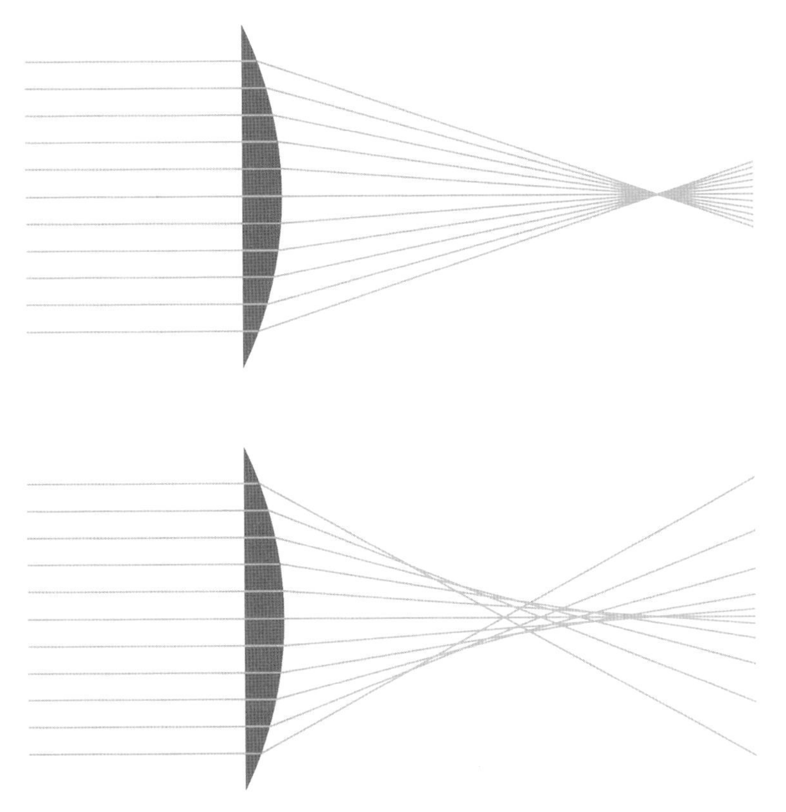

FIGURE 7.6 Light rays are focused to a single focal point with an ideal lens (*top*). A lens with a spherical surface has an associated spherical aberration, and therefore has no distinct focal point (*bottom*). (Image is exaggerated).

Spherical components are relatively inexpensive, partly because of improvements in lapping processes that allow the simultaneous processing of multiple lenses. Although time-consuming, lapping is capable of producing high-quality polished surfaces that deviate from the designer's specifications by as little as a few hundredths of a wavelength. The result is high-precision, cost-effective lenses.

Aspherical Lenses

Aspherical lenses allow an expansion in the optical designer's solution set. In an aspherical lens, the surfaces can be nonspherical. The addition of high-order curvature to an otherwise spherical surface in an aspherical lens permits independent correction or balancing of spherical aberration. That leads to a reduction in the number of lens surfaces needed for aberration-corrected imagery.

Because they can reduce lens surfaces, the use of aspherical lenses can improve transmission and reduce the weight, space requirements, and cost of optical systems. Although their potential for improving performance of an optical design had been known for many years, their use had been limited by the inability to produce them reliably and accurately. Advances in manufacturing process technologies—including deterministic computer numerically controlled (CNC) manufacturing equipment and processes, single-point diamond turning and grinding, polymer molding, glass molding, and precision metrology—have made possible the manufacture and use of these important surface geometries.[52] Among the diverse applications of high-precision aspheric lenses (aspheres) are military aerospace systems, optical data storage, photolithography, and astronomy. Lower-precision aspheres have a wider range of application, including photography and video imaging (especially zoom lenses), such medical instruments as endoscopes, telecommunications, and document scanners and printers. At the low end of the market, aspheres find use in such applications as condenser elements for illumination. Asymmetrical aspheres are also becoming important, especially in conformal applications, in which the outer surface of an optical component must conform to the aerodynamic shape of an aircraft or missile.

Fabrication Processes and Equipment

Processes and equipment available to fabricate optical surfaces, particularly aspherical surfaces, have undergone notable improvement during the last decade. Improvements have been made in both the ability to produce and the ability to measure precision optical components. Improvement has included nearly all as-

[52] For more information on the techniques discussed in this section, see Appendix C of this report.

pects of fabrication from surface generation to coating and the ability to measure these surfaces.

Today, CNC grinding and polishing allow dynamically adjustable cutting paths for tool wear and can be programmed for edging, beveling, sagging, concave, and convex surface grinds. Polymer molding has also become common in fabrication of lenses for consumer and commercial products. Mobile phones, DVD players, digital cameras, and conferencing systems have all incorporated polymer lenses. Glass-molding technology also has become increasingly available in the last decade, as has magnetorheological finishing (MRF) in the fabrication of optical components. In MRF, ferrous-laden fluid passes through an electromagnetic field, where its viscosity is increased, allowing the creation of a precise and repeatable polishing tool. Single-point diamond turning (SPDT) has also grown in popularity in fabricating optical components. It is now routinely used to produce mold inserts for polymer lenses, mold inserts for glass molding, and finished optical elements. The machining process uses single-crystal diamond cutting tools and nanometer-precision positioning to generate spherical surface geometries and more complex geometries, such as toroids, aspheres, and diffractives.

Optical thin-film coatings technology has advanced in response to requirements in diverse markets, including telecommunications, health and medicine, biometrics, and defense. Evaporation deposition processes, in which materials are deposited by way of a transformation from solid to vapor and back to solid, are the most widely used in the optics industry in spite of problems stemming from the porosity of coatings and their sensitivity to humidity and thermal conditions. Metrology is an important enabling technology in the optics industry. There is an old saying, "If you can't measure it, you can't make it;" over the last decade, advances in interferometry have improved the ability to measure increasingly challenging optics, particularly aspheres.[53,54,55]

APPLICATIONS OF PHOTONICS IN MANUFACTURING

This section discusses photonics-enabled advances in process technologies with potentially broad applications in numerous manufacturing industries. As noted below, many of these process-technology innovations hold considerable potential

[53] PalDey, S., and S.C. Deevi. 2003. Single layer and multilayer wear resistant coatings of (Ti,Al)N: A review. *Materials Science and Engineering A* 342(1-2):58-79.

[54] Kelly, P.J., and R.D. Arnell. 2000. Magnetron sputtering: A review of recent developments and applications. *Vacuum* 56:159-172.

[55] Svedberg, E.B., J. Birch, I. Ivanov, E.P. Munger, and J.E. Sundgren. 1998. Asymmetric interface broadening in epitaxial Mo/W (001) superlattices grown by magnetron sputtering. *Journal of Vacuum Science and Technology A: Vacuum, Surfaces, and Films* 16(2):633-638.

to enable the United States to retain manufacturing capability in the face of intensifying international competition.

Photolithography

For the last several decades, photolithography has been the dominant printing technology used by integrated circuit (IC) manufacturers, and it is a key factor in increasing the transistor density per silicon area and lowering the cost per transistor as described by Moore's law.[56] It has played an important role in fabricating high-volume ICs, microelectromechanical systems (MEMS), and other microdevices and nanodevices. Since the publication of the National Research Council report *Harnessing Light: Optical Science and Engineering for the 21st Century*, photolithography in IC manufacturing has remained dominant and has continued to achieve impressive technical advances.

Photolithography is similar to photography in that both use imaging optics and a photosensitive film to record an image. In photolithography, the surface of a semiconductor wafer is coated with a light-sensitive polymer known as a photoresist. Light passing through a mask that contains the desired pattern is focused on the photoresist-coated wafer. The material properties of a photoresist change when it is exposed to light, and the changed material can be selectively removed from the wafer surface. The wafer is then chemically treated to engrave (etch) the exposure pattern in it. The process is repeated many times with different masks to form billions of complicated three-dimensional structures (such as transistors and interconnections) on the wafer. A generic lithographic process is schematically shown in Figure 7.7.

Photolithography has enabled manufacturers to increase transistor density (and thus the complexity of advanced chips) while lowering the cost per transistor; historically, the cost per transistor has decreased by close to 30 percent per year. An important aspect in the advancement of photolithography is the minimum feature size (resolution) provided by the optical projection system that projects the mask image onto the wafer. The minimum feature size is generally formulated by the Rayleigh scaling equation and is directly related to λ, the exposure wavelength, and inversely related to the numerical aperture (NA) of the projection optics, by k_1, a scaling constant that is a function of the process, the photoresist, the type of feature being printed, and other factors. From Rayleigh's equation, resolution can

[56] Brunner, T.A. 2003. Why optical lithography will live forever. *Journal of Vacuum Science and Technology B* 21(6):2632-2637.

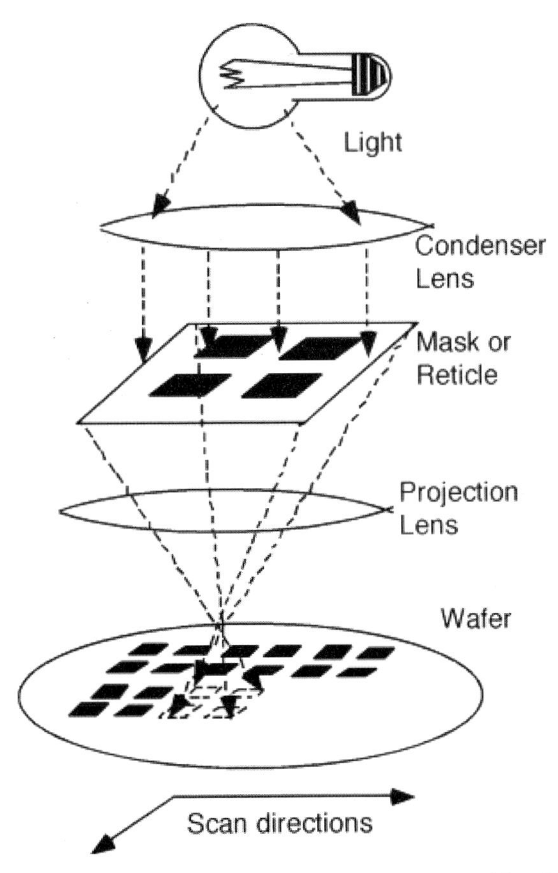

FIGURE 7.7 Generic schematic diagram of the lithographic process. SOURCE: Bill Wilson. 2007. "Photolithography." *Connexions*. Available at http://cnx.org/content/m1037/2.10/. Reprinted with permission.

be improved by (1) reducing the wavelength of the light source, (2) increasing the NA of the lens, and (3) decreasing the value of k_1.[57]

One common approach to reducing the feature size is to reduce the wavelength of light. In the early days of photolithography, the mercury arc lamp "G-line" optical source with a wavelength of 436 nm was used; this resulted in a feature size down to about 700 nm. The transition to "I-line" sources at a 365-nm wavelength enabled resolution below 400 nm, and the use of KrF excimer lasers at a 248-nm wavelength

[57] Mack, C.A. 2007. *Fundamental Principles of Optical Lithography: The Science of Microfabrication*. Hoboken, N.J.: Wiley.

allowed resolution down to 150 nm. Current state-of-the-art lithography technology uses light with a 193-nm wavelength to fabricate feature sizes below 32 nm.

Feature size could also be reduced by increasing the NA of the projection system. The first projection lithography tools had an NA of 0.16 in 1973, and the NA has steadily increased to its current value of 1.35 with the use of immersion imaging; note that at a wavelength of 193 nm, water has a refractive index of 1.44 relative to air and is quite transparent.

Small features can also be achieved by decreasing k_1, at least for a single exposure, and this method has been demonstrated with resolution enhancement technology (RET).[58] RET approaches include optical proximity correction (OPC), off-axis illumination (OAI), and phase-shifting masks (PSM). However, in the printing of dense patterns of lines and spaces, there is a 0.25 lower limit of k_1.

The demise of photolithography was predicted so often that John Sturtevant remarked in 1997 that "the end of optical lithography is always 7 years away;" this statement has come to be known as Sturtevant's law.[59] Nonetheless, photolithography has remained the most desirable technology among IC manufacturers. A main reason for this preference has to do with productivity. Over the last several decades, lithography tools have progressed to the state-of-the-art production line lithography with an exposure wavelength of $\lambda = 193$ nm for a feature size of 22 nm.[60] Over the same period, the cost of a state-of-the-art lithography tool has grown from $100,000 to greater than $50 million. Fortunately, such dramatic increases in tool cost have been accompanied by equally dramatic increases in tool throughput, so the cost of printing a square centimeter of silicon has remained roughly constant.

Continued improvement in resolution while maintaining cost-effectiveness cannot be taken for granted. At the 193-nm wavelength, increasing NA above 1.35 is a serious challenge and will require the development of higher-index lens materials and immersion fluids. Moreover, k_1 is nearing its theoretical minimum of 0.25. Possible approaches to decreasing k_1 below 0.25 include the use of double patterning, in which the original design is split into two masking layers, each patterned by a single exposure. Various approaches have been taken to implement double pat-

[58] Brunner, T.A. 2003. Why optical lithography will live forever. *Journal of Vacuum Science and Technology B* 21(6):2632-2637.

[59] Mack, C.A. 2007. The future of semiconductor lithography: After optical, what next? *Future Fab International* 23.

[60] Sivakumar, S. 2011. EUV lithography: Prospects and challenges. *Proceedings of the 2011 16th Asia and South Pacific Design Automation Conference.* January 25-28, 2011, Yokohama, Japan.

terning schemes, for example, litho-etch-litho-etch (LELE), litho-litho-etch (LLE), litho-freezing-litho-etch (LFLE), and self-aligned double patterning (SADP).[61,62,63]

Lowering the wavelength below 193 nm offers the potential for continued progress in photolithography, but challenges remain. The industry is working diligently to enable extreme ultraviolet (EUV) lithography, with a wavelength of 13.5 nm for high-volume production of 10-nm resolution by 2015. One of the challenges faced by EUV is the light source at the 13.5-nm wavelength. Two types of light sources that are being developed to help meet that challenge are laser-produced plasma (LPP) sources and discharge-produced plasma (DPP) sources. The LPP sources shown by Cymer and Gigaphoton have produced 10-20-W EUV at the intermediate focus. The DPP sources produced by Xtreme Technologies can project 15 W of EUV at the intermediate focus. For comparison, the output of a modern of a 193 nm laser is as much as 90 W in high-volume production.

Table 7.1 shows the top semiconductor equipment suppliers in 2011; the United States has an important but not dominant position. Recently, the Netherlands' Advanced Semiconductor Materials Lithography (ASML) has begun using a preproduction EUV tool (NXE:3100) that has shown the ability to print lines with a periodicity of 18 nm.[64] Using this machine, Imec, a Belgium-based leading nanoelectronics research company, expects to improve the technology to produce 16-nm features by 2013.[65] Yet, Intel, the world's largest semiconductor chip manufacturer, is expected to use EUV for 10-nm features and 193-nm immersion lithography for its 14-nm features.[66]

Although lithography has seen significant technical advances in feature-size

---

[61] Lucas, K., C. Cork, A. Miloslavsky, G. Luk-Pat, L. Barnes, J. Hapli, J. Lewellen, G. Rollins, V. Wiaux, and S. Verhaegen. 2008. Interactions of double patterning technology with wafer processing. *Proceedings of SPIE—The International Society for Optical Engineering Conference on Optical Microlithography XXI*, February 26-29, 2008, San Jose, Calif. Bellingham, Wash.: The International Society for Optical Engineering.

[62] Pan, D.Z., J. Yang, K. Yuan, M. Cho, and Y. Ban. 2009. Layout optimizations for double patterning lithography. *Proceedings of IEEE 8th International on Application Specific Integrated Circuit (ASIC)*, October 20-23, 2009, Changsha, Hunan, China.

[63] Hori, M., T. Nagai, A. Nakamura, T. Abe, G. Wakamatsu, T. Kakizawa, Y. Anno, M. Sugiura, S. Kusumoto, Y. Yamaguchi, and T. Shimokawa. 2008. *Proceedings of SPIE—The International Society for Optical Engineering Conference on Advances in Resist Materials and Processing Technology XXV*, February 25-27, 2008, San Jose, Calif. Bellingham, Wash.: The International Society for Optical Engineering.

[64] Advanced Semiconductor Materials Lithography. 2012. "EUV Questions and answers." Available at http://www.asml.com/asml/show.do?ctx=41905&rid=41906. Accessed June 5, 2012.

[65] Advanced Semiconductor Materials Lithography. 2012. "EUV Questions and answers." Available at http://www.asml.com/asml/show.do?ctx=41905&rid=41906. Accessed June 5, 2012.

[66] LaPedus, M. 2011. "Intel EUV late for 10-nm milestone." *EETimes*. Available at http://www.eetimes.com/electronics-news/4213628/Intel—EUV-misses-10-nm-milestone. Accessed August 3, 2012.

TABLE 7.1 Top Semiconductor Equipment Suppliers in 2011

2011 Rank	Area of the World	Company Name	2011 Sales ($ Millions; 2011 exchange rates)
1	Europe	ASML	7,877.1
2	North America	Applied Materials[a]	7,437.8
3	Japan	Tokyo Electron	6,203.3
4	North America	KLA-Tencor	3,106.2
5	North America	Lam Research	2,804.1
6	Japan	Dainippon Screen Mfg. Co.	2,104.9
7	Japan	Nikon Corporation	1,645.5
8	Japan	Advantest[b]	1,446.7
9	Europe	ASM International	1,443.0
10	North America	Novellus Systems	1,318.7
11	Japan	Hitachi High-Technologies	1,138.7
12	North America	Teradyne	1,106.2
13	North America	Varian Semiconductor Equipment[c]	1,096.3
14	Japan	Hitachi Kokusai Electric	838.4
15	North America	Kulicke and Soffa	780.9
Total, Top 15			40,347.7

[a]Applied Materials includes Varian revenues for November 1 to December 31, 2011.
[b]Advantest includes Verigy's revenues from July 1 to December 31, 2011.
[c]Varian includes revenue as an independent company from January 1 to October 31, 2011.
SOURCE: Ha, P., and R. Puhakka. 2012. "2011 Top Semiconductor Equipment Suppliers." *VLSIresearch.* News release. Available at https://www.vlsiresearch.com/public/cms_pdf_upload/706001v1.0.htm.

reduction,[67] there remain several challenges, including (1) brighter EUV light sources to enable high throughput and low per wafer cost, (2) photoresists that can operate at low exposure doses while providing both high resolution and small line-edge roughness, (3) the ability to manufacture and inspect defect-free masks at the smaller resolutions, and (4) overcoming line-edge roughness, which is caused by photon and chemical stochastic effects that begin to dominate at feature sizes under 30 nm and might be mitigated by directed self-assembly of molecules.[68,69] However, the economic issues are also significant. For example, it is possible to make smaller features but not necessarily while lowering the manufacturing cost per transistor. Thus, efforts to improve resolution while maintaining cost-effectiveness are of great interest. It is noted that even greater resolution enhancements are possible in the future if soft x-ray sources (wavelengths shorter than EUV) with reasonable brightness and cost would be available.

[67] Brueck, S.R.J. 2005. Optical and interferometric lithography—Nanotechnology enablers. *Proceedings of the IEEE* 93(10):1704-1721.

[68] Mack, C.A. 2006. *Field Guide to Optical Lithography.* Bellingham, Wash.: SPIE Press.

[69] Galatsis, K., K.L. Wang, M. Ozkan, C.S. Ozkan, Y. Huang, J.P. Chang, H.G. Monbouquette, Y. Chen, P. Nealey, and Y. Botros. 2010. Patterning and templating for nanoelectronics. *Advanced Materials* 22(6):769-778.

Lasers in Manufacturing

Since the first carbon dioxide (CO_2) laser demonstrations in the 1960s, the use of laser systems in manufacturing has grown rapidly. Expanded applications of lasers throughout manufacturing have been driven by continual innovation in this technology. Laser systems have been transformed from tools applicable only to highly specialized processes to commonplace tools that are used extensively in shop floor operations, such as cutting, drilling, piercing, and welding. In 2011, lasers used to process materials, including applications in shop floor operations, accounted for 26 percent of the lasers in use. (See Figure 7.8.) Many of these lasers were CO_2 lasers. Fiber lasers, however, are replacing CO_2 lasers in some applications, in part because of their superior operating economics, low upfront cost, high energy efficiency, and lower maintenance cost. The laser marking industry is one of the largest users of fiber lasers.

Precision lasers that are used to process materials by cutting, welding, drilling, and piercing can provide advantages over conventional processes, including improvements in the ability to hold tight tolerances, reduction in downtime associated with setups, reduction in part cleaning and deburring, and reduction in distortion of parts during processing. Those attributes lend themselves well to complex machining operations. (See Figure 7.9.) Lasers not only provide the ability to cut materials but provide the ability to produce high-quality components and assemblies precisely and repeatedly.

Lasers have done much to enable micromachining. In particular, solid-state and excimer lasers have been deployed to produce microstructures that would have been either impossible to produce or too costly with conventional machin-

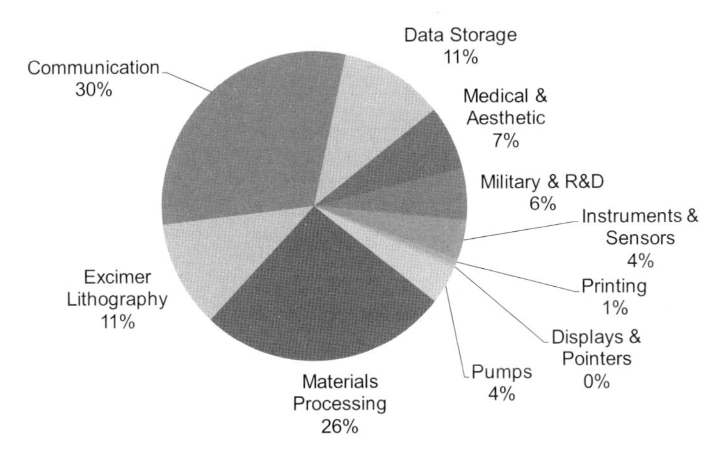

FIGURE 7.8 Review of lasers by application by sales, 2011. SOURCE: Courtesy of Strategies Unlimited and Laser Focus World (2012).

FIGURE 7.9 Laser cutting is used to produce exact components. SOURCE: Image by Bystronic.

ing processes. Diode-pumped solid-state lasers operating in the near infrared (IR) and excimer lasers operating in the UV to the near IR are most frequently used. Laser selection is a function of a variety of factors, including process scalability, the characteristics of the materials being machined, and operating cost. An important feature of micromachining lasers is their high-pulsed, non-continuous output, with pulse repetition rates that range from a few hertz to several hundred hertz and pulse durations between a few picoseconds and microseconds. The nature of the pulsing aids in the management of the heat load in the material being machined. Figure 7.10 shows slots cut into in stainless steel as small at 75 μm in width.

ADDITIVE MANUFACTURING

"Additive manufacturing," three-dimensional printing, describes a group of technologies that are used to create parts by building up layers to, in effect, "grow a part." Additive processes are fundamentally different from traditional subtractive processes in which material is removed from a block to create a part. One of the big advantages of additive processes is that the amount of waste material is greatly minimized because only as much source material as is needed is used to build the part. First developed in the mid-1980s, additive manufacturing has grown in ac-

FIGURE 7.10 Slots cut into stainless steel are 75 μm wide. It should be pointed out that lasers are going to be needed in the future not only for machining but for interferometry for precision manufacturing. Laser interferometry is already being used in IC manufacturing for controlling etching and deposition on a nanometer scale and is going to be even more important in future high-resolution three-dimensional additive manufacturing. SOURCE: Courtesy of Potomac Photonics.

ceptance and practice to the point that today it is an effective development and shop floor tool. The improvements in performance and cost-competitiveness associated with additive manufacturing reflect advances in a number of enabling technologies, many of which are based on photonics.

Although not yet a complete replacement for conventional machining or fabrication processes, additive manufacturing has several key advantages. Most notable of the advantages are the short time from computer-aided design (CAD) file to "part complete" and the cost-effectiveness of low-volume production, which reflect the elimination of time to design and produce custom tools or fixtures. Low-volume cost-effectiveness comes from the elimination of high-cost capital tools that would be required to be leveraged over the production of a low number of parts.

Combined advances in three-dimensional design tools, CNC technologies, and lasers have enabled a steady growth in additive-manufacturing capability. Whereas in the past the technology was confined to simple parts for experimental purposes, today complex geometries with a high degree of precision are produced in the laboratory and on the shop floor. The improved capability leveraged with the advantages described earlier makes additive manufacturing a good fit for a class of products that can be produced effectively in the United States. They are prototypes, products with a high degree of customization and complexity, and products produced in low volume.

The following are brief descriptions of a few of the additive-manufacturing processes that illustrate the use of laser technology to produce parts.

Stereolithography

In contrast with photolithography, stereolithography is used in larger-scale products. Developed in 1988, it was the first of the rapid prototyping processes. As shown in Figure 7.11, the stereolithographic process deposits layers of approximately 0.002-0.003 in. thick and uses a UV laser to cure the resin only where the

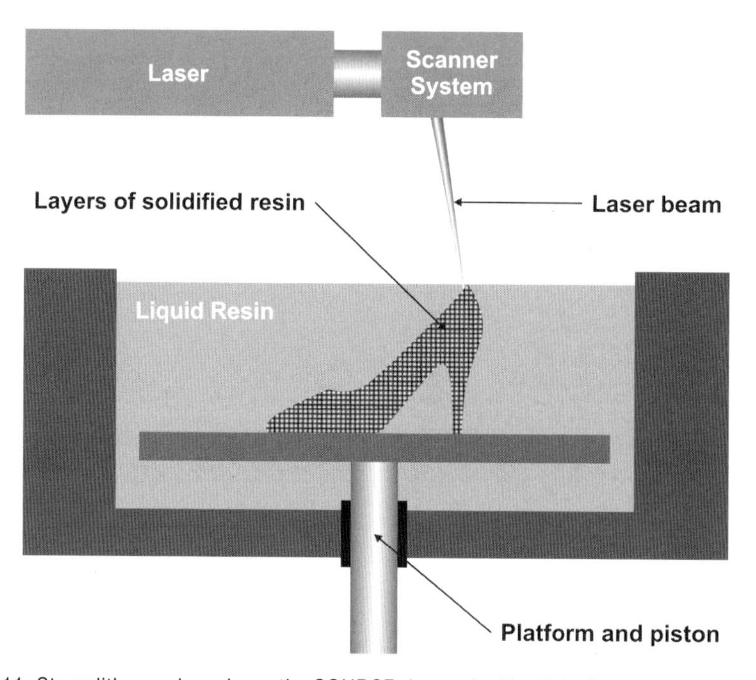

FIGURE 7.11 Stereolithography schematic. SOURCE: Image by Usdabhade.

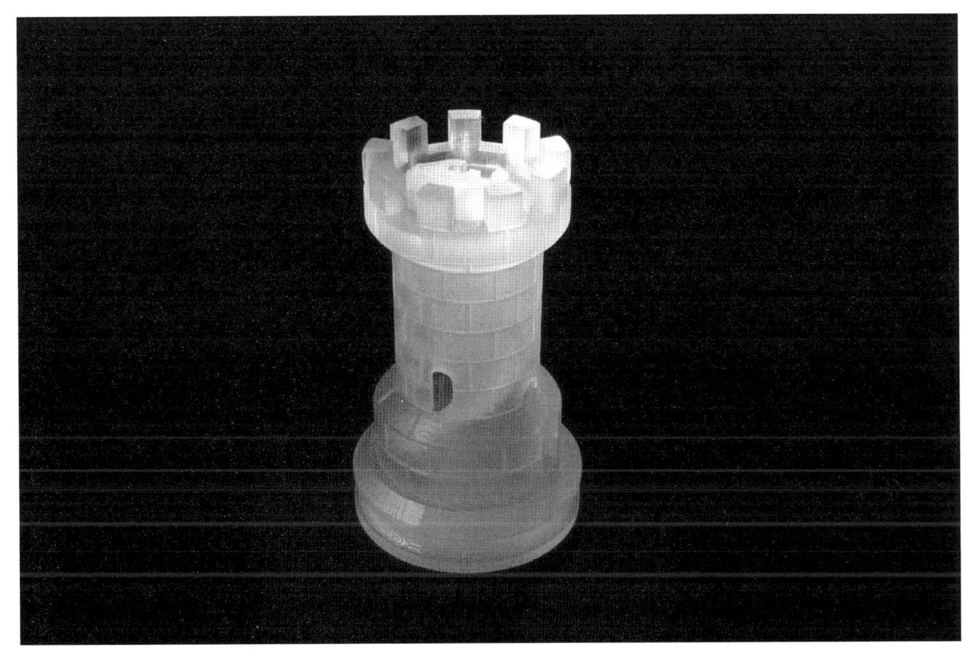

FIGURE 7.12 A stereolithographed chess piece. SOURCE: Courtesy of Potomac Photonics.

material is needed for the part. The laser wavelength and power can vary between manufacturers but is approximately 325 nm from a low-power He-Cd source. The part is built up in layers until the final geometry is completed.

On completion of the layering process, the part is subjected to high-intensity UV light for the postcuring process, which fully hardens the resin. A chess piece fabricated with stereolithography is shown in Figure 7.12.

Selective Laser Sintering

Selective laser sintering (SLS) was developed in the mid-1980s and is capable of producing parts from thermoplastics, ceramics, or metals. Like stereolithography, SLS, shown in Figure 7.13, is a layering process that builds a part from a powder based on a three-dimensional CAD model. In the SLS process, a laser fuses the layers of powder in localized areas to create the final part geometry. Although systems vary between manufacturers, the laser used is approximately a 50-W CO_2 laser. The process can yield very accurate parts with tolerances of ±0.05-0.25 mm. Components fabricated with SLS require no postprocessing. Figure 7.14 shows a replica of a violin produced with the SLS process.

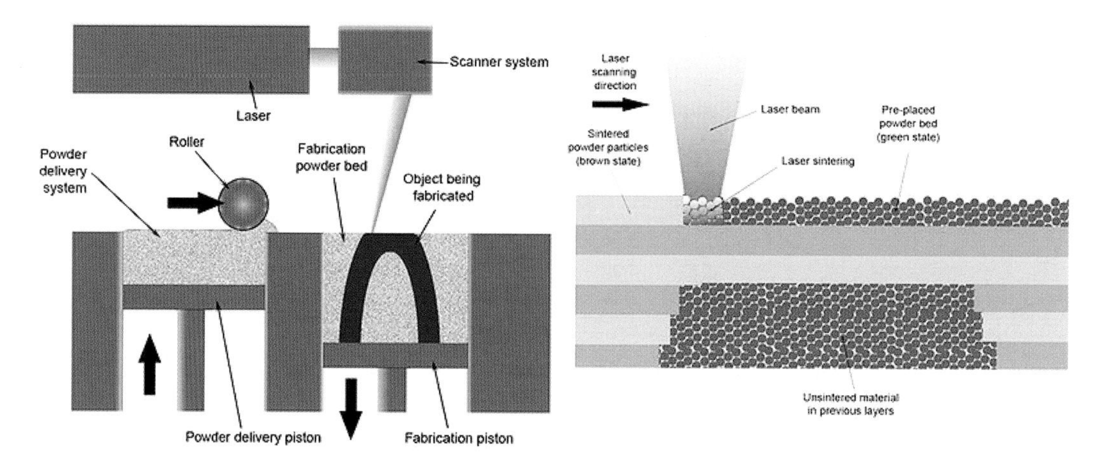

FIGURE 7.13 Selective laser sintering (SLS) schematic. SOURCE: Image by Materialgeeza.

FIGURE 7.14 Martha Cohen, of the Hochschule für Musik und Theater, München, plays a replica of a Stradivarius violin fabricated with SLS during the Kleine Zukunftsmusik der Photonik event. SOURCE: Erik Svedberg, National Research Council.

Laser Engineered Net Shaping

Laser Engineered Net Shaping (LENS™) was developed in the mid-1990s at the Sandia National Laboratories. The process, shown in Figure 7.15, uses a neodymium-doped yttrium aluminum garnet (Nd:YAG) laser operating at 500-600 W that is enclosed in an argon gas environment. The laser creates a molten pool into which powdered metal is injected. Parts have been produced from stainless-steel alloys, nickel-based alloys, tool-steel alloys, titanium alloys, and other specialty materials, including composites. As in other additive-manufacturing processes, parts originate from three-dimensional CAD models, and material is built up in layers to create the final part. The significant difference between LENS and other additive processes is that the parts obtain the same density as the metal used to fabricate them. Figure 7.16 shows a tool produced with the LENS process.

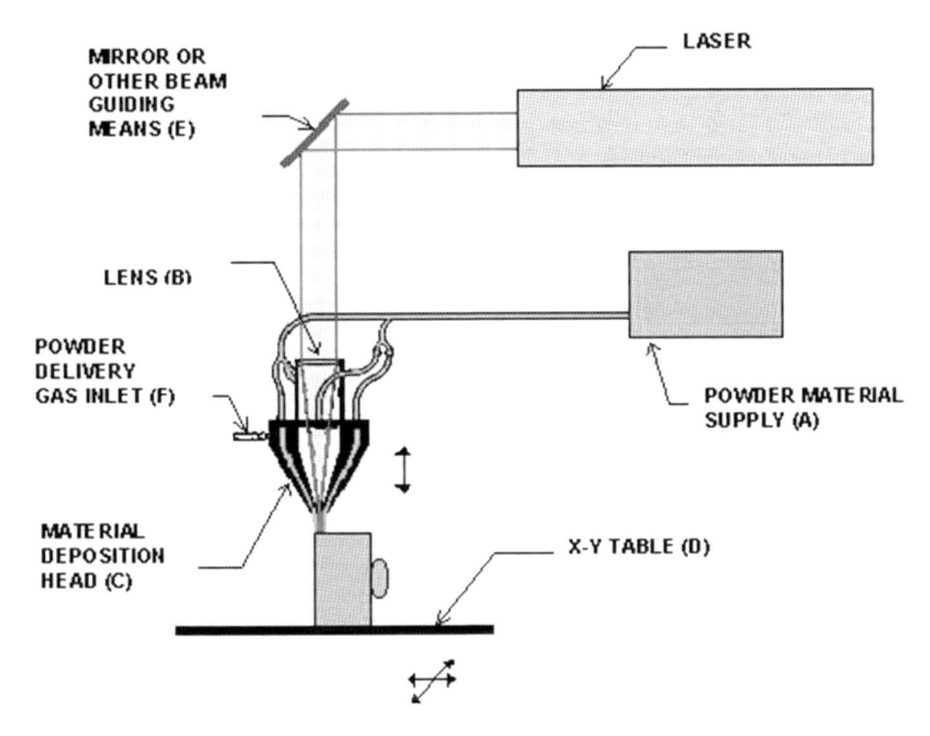

FIGURE 7.15 Laser Engineered Net Shaping (LENS™) process schematic. SOURCE: Worldwide Guide to Rapid Prototyping Website, © Copyright Castle Island Co., all rights reserved. Available at http://www.additive3d.com/len_int.htm. Reprinted with permission.

FIGURE 7.16 H13 tooling created with LENS™ process. SOURCE: Courtesy of Sandia National Laboratories.

A common opportunity exists among these techniques to increase the precision of three-dimensional manufacturing. If shorter-wavelength lasers and imaging were available, it would be possible to reduce the scale of the smallest-possible three-dimensional voxel (a three-dimensional pixel).

In general, one important part of additive manufacturing is an increased emphasis on in situ metrology that uses coherent optics (interference) for feedback and control, especially when the dimensions of parts shrink. Pattern-placement metrology, used ordinarily for lithographic purposes, can rely on phase-coherent fiducial gratings patterned by interference lithography.[70] Potential uses include measuring process-induced distortions in substrates, patterning distortions in pattern-mastering systems, and measuring field distortions and alignment errors in steppers and scanners. For example, spatial-phase-locked electron-beam lithography has been implemented to correct pattern-placement errors at the nanometer level.[71]

[70] Schattenburg, M.L., C. Chen, P.N. Everett, J. Ferrera, P. Konkola, and H.I Smith. 1999. Sub-100 nm metrology using interferometrically produced fiducials. *Journal of Vacuum Science and Technology B: Microelectronics and Nanometer Structures* 17(6):2692-2697.

[71] Hastings, J.T., F. Zhang, and H.I. Smith. 2003. Nanometer-level stitching in raster-scanning electron-beam lithography using spatial-phase locking. *Journal of Vacuum Science and Technology B: Microelectronics and Nanometer Structures* 21(6):2650-2656.

PHOTONICS AND THE FUTURE OF U.S. MANUFACTURING

As noted earlier, U.S.-based manufacturing now accounts for a smaller share of global manufacturing value added than 20 years ago. The drive for competitiveness and increased shareholder value has caused corporations in nearly every area of photonics to search for alternative manufacturing locations. The exception may be products that have substantial defense-related markets and applications and that therefore are subject to controls over their export (International Traffic in Arms Regulations, ITAR). But government licensing has allowed "offshoring" for some components in this product field as well. During the same period, however, some components continue to be manufactured in the United States and have remained competitive. What distinguishes the components and final assemblies whose production has remained in the United States from those now produced mainly offshore?

A critical factor affecting the location of production is volume. Typically, high-volume production operations are more sensitive to labor and capital cost differentials, and these activities have been among the most likely to move offshore from the United States in photonics and other high-technology products. In photonics, as in other high-technology industries, high-volume production operations are most common in consumer products, and low-volume operations range from the production of test lots to the manufacture of specialized systems.

Advances in optical materials and processing have enabled the manufacture of precision optical components for very low cost with sufficient volume to amortize the required tooling. One example is the mass production of molded polymer aspheres. Their unit costs can be very low as long as the volume is sufficiently high. Low-volume production of these components, however, tends to be expensive because of the large amounts of labor and time required to manufacture and test precision tools. Even when advanced capabilities are used in a highly automated manufacturing process, the cost of the equipment coupled with low volume drives production costs up significantly. Recent advances in several manufacturing capabilities, such as different methods of additive manufacturing, hold out considerable promise for the development of low-cost machines capable of providing precision optics, with surface figures not restricted to the narrow range of surfaces possible with current grinding and finishing techniques. In addition to providing a new set of potential optical surface figures and the associated capabilities, these advances may enable low-cost precision optics even for low-volume applications and thereby remove much of the benefit of moving optics manufacturing overseas by minimizing the impact of labor costs on the optics. Photonics-enabled advances in manufacturing technology thus could slow the erosion, or perhaps support renewed growth, in U.S.-based manufacturing activity.

High-Volume Products

High-volume products, particularly consumer retail products, generally have cost as a high priority and tend to use manufacturing processes focused on production cost minimization. Manufacturing processes that minimize labor and infrastructure costs are preferred. U.S. manufacturers have, over the years, had difficulty in competing in the high-volume market and have seen much of this work move offshore. That move has been driven in large part by the cost of labor in the United States, which is reflected in raw materials, operations, and overhead. In an effort to compete in at least a portion of the high-volume market (the lower-volume portion of the high-volume market), U.S. manufacturers have used a variety of strategies. For example, manufacturers have changed from commodity components and moved toward precision components and subassemblies in their U.S.-based operations. The high-volume sector tends to be very cost-sensitive, but the lower-volume end of this sector is somewhat less so.

In addition to focusing on products that are less cost-sensitive, manufacturers have reduced the amount of direct labor in their U.S.-based manufacturing processes. In optics grinding and polishing operations, such as lens centering, use of CNC equipment capable of running unattended, as shown in Figure 7.17, has

FIGURE 7.17 Automated lens handling. SOURCE: Courtesy of Rochester Precision Optics.

helped to reduce the amount of labor at the component level. In polymer lens molding, the use of automation to remove parts from molds, degate those parts, and the use of molds with higher cavitation, has reduced per part cost and improved competitiveness. Those cost-saving strategies can be capital-intensive, and manufacturers must evaluate the associated economics case by case. The purchase, installation, and maintenance costs of labor-saving equipment must be offset by actual labor savings to justify expenditures.

Low-Volume Products

Low-volume products are generally high-precision and complex or subject to export restrictions linked to national security concerns, as in the case of ITAR. ITAR thus has offsetting effects on the location of innovation and production for U.S. firms. On the one hand, restrictions on export of ITAR-controlled products may lead U.S. firms to site their self-financed product development activities for these products offshore to avoid the restrictions; at least some types of innovation may move offshore from the United States as a result of ITAR. On the other hand, the production of ITAR-controlled products, especially products based on R&D that draw on defense-funded programs or products that are sold in large part to federal agencies, may be less likely to move offshore because of ITAR restrictions on procurement of such products from foreign producers or foreign production sites.

Products not subject to ITAR in the low-volume sector are often in early-stage development and require prototypes or are products in the medical industry, such as complex Lasik surgery equipment. A common characteristic of these products is the requirement for tightly specified precision optical components; emphasis is placed on consistently and reliably satisfying difficult specifications. Although cost is always an important element with all products, it often falls behind the requirement for precision and reliability.

U.S. manufacturers have excelled in the production of low-volume, high-precision optical components and devices. Manufacturers have successfully pushed legacy technologies and adopted newer technologies to satisfy the requirements of this segment. Precision CNC equipment capable of producing components repeatedly and accurately is in wide use. Assembly processes that often require active alignment are used to enable compliance with the requirement for tight tolerances. Complex precision metrology is used for testing components and assemblies to ensure that specifications are met.

THE U.S. MANUFACTURING WORKFORCE

It is the judgment of this committee that advances in photonics provide clear potential for growth in U.S.-based manufacturing by (1) expanding the applications of photonics throughout manufacturing and thereby improving the cost competitiveness of processes used to produce high-volume products, (2) accelerating the commercial exploitation of new photonics-based technological opportunities, and (3) improving the cost-competitiveness of U.S. manufacturing of photonics products and components.

Realizing that potential, however, requires a well-trained manufacturing workforce that includes both advanced-degree holders and skilled operators and craftspeople. As was discussed in previous chapters, because of the nature of photonics as a technology rather than an industry on which data are collected by U.S. government statistical agencies, the committee was not able to assemble estimates of current employment in photonics-enabled production activity, nor was it able to forecast growth for various occupations within this sector.[72] The collection and reporting of better employment data in photonics will be important for any future federal initiative in this field.

Nonetheless, it is the judgment of the committee that U.S. holders of advanced degrees from university-based programs in photonics, optics, and related disciplines too often pursue careers in R&D or academia rather than pursue opportunities in design or manufacturing within industry. It is important that U.S. firms develop more attractive career paths for advanced-degree holders to pursue careers in photonics manufacturing and in the applications of photonics technologies throughout manufacturing.

Some committee members cited the example of the U.S. semiconductor industry as one that has developed attractive employment opportunities in manufacturing for engineering and science advanced-degree holders. That industry seems to have experienced fewer problems in recruiting top graduates into manufacturing process development, perhaps because leading firms recognize its significance and are willing to make manufacturing jobs attractive by acknowledging that manufacturing is a key ingredient in their competitive advantage.

Similarly, the committee concluded that improvements in technical education are needed to increase the quality of skilled blue-collar workers in optics and photonics. Although U.S. community colleges provide abundant opportunities for students to pursue technical education, there are fewer opportunities for apprenticeship-based training in the U.S. optics industry than in similar industries

[72] Although Chapter 2 presents rough estimates of employment in U.S. firms that are active in photonics, based on their membership in professional or trade organizations, the committee notes in that discussion that such estimates do not represent measures of "photonics-dependent" employment, nor do they enable an analysis of employment prospects or current shortages of skilled workers, scientists, or engineers in photonics-related production or applications. Chapter 4 discusses concerns about the effects of impending retirements and the small pool of U.S. nationals with advanced degrees available for employment in defense-related production and R&D activities, although here, too, precise data on the implications of these trends for photonics applications are lacking.

in Germany. The difficulties of expanding a skilled technical workforce in the United States are considerably exacerbated by the declining performance of U.S. primary and secondary education.[73,74] It is, of course, true that weaknesses in the quality and quantity of the skilled blue-collar workforce in the United States can be partially compensated for by expanded investment in automation. But even the intelligent deployment of more highly automated manufacturing processes will be hampered by weaknesses in the skilled workforce in U.S. manufacturing.

FINDINGS

Finding: Production of many photonics applications—such as TFT displays, solar modules, and optoelectronic components for communications systems, most of which were first developed and commercialized by U.S. firms in the U.S. economy—now is dominated by foreign production sites, even when these production activities are still controlled by U.S.-based firms. The effects of this offshore movement of manufacturing on innovation in photonics, however, vary considerably among different sectors and technologies within photonics. Indeed, the United States remains the leading source of USPTO patents in two key sectors of photonics (solar and communications components) and is the leader in potential next-generation technologies, such as flexible displays, "paint-on" and other thin-film solar cells, and monolithically integrated optoelectronic devices.

Key Finding: To enable the United States to be productive in manufacturing photonics goods, a capable and fully trained workforce must exist at all levels, including shop floor associates, technicians, and engineers. Because photonics is not yet recognized as an industry and data are not tracked in a way that facilitates analysis, it is difficult to evaluate the extent of personnel shortages in photonics manufacturing and in applications of photonics elsewhere in the manufacturing industry in general. It seems that it would be beneficial if industry and government did more to increase training and employment opportunities in photonics manufacturing.

Key Finding: Additive manufacturing, which uses significant optics and photonics technologies, has become important in manufacturing, and its position is expected to increase. Additive manufacturing tends to require low labor use and is therefore advantageous in regions with high labor costs, such as the United States.

[73] National Research Council. 2010. *Standards for K-12 Engineering Education?* Washington, D.C.: The National Academies Press.

[74] National Research Council. 2011. *Successful STEM Education: A Workshop Summary.* Washington, D.C.: The National Academies Press.

RECOMMENDATIONS AND GRAND CHALLENGE QUESTION

Key Recommendation: The United States should aggressively develop additive-manufacturing technology and implementation.

Current developments in the area of lower-volume, high-end manufacturing include, for example, three-dimensional printing, also called additive manufacturing. With continued improvements in manufacturing tolerances and surface finish, additive manufacturing has the potential for substantial growth. The technology also has the potential to allow three-dimensional printing near the end user no matter where the design is done.

Key Recommendation: The U.S. government, in concert with industry and academia, should develop soft x-ray light sources and imaging for lithography and three-dimensional manufacturing.

Advances in table-top sources for soft x rays will have a profound impact on lithography and optically based manufacturing. Therefore, investment in these fields should increase to capture intellectual property and maintain a leadership role for these applications. The committee views development of soft x-ray light sources and imaging as an appropriate field for expanded federal R&D funding under the sponsorship of a national photonics initiative undertaken with the advice and financial support of U.S. industry. This chapter indicates the need for an order-of-magnitude or greater increase in resolution in manufacturing.

The above two key recommendations help to inform **the fifth and last grand challenge question:**

5. How can the U.S. optics and photonics community develop optical sources and imaging tools to support an order of magnitude or more of increased resolution in manufacturing?

Meeting this grand challenge could facilitate a decrease in design rules for lithography, as well as providing the ability to do closed-loop, automated manufacturing of optical elements in three dimensions. Extreme ultraviolet is a challenging technology to develop, but it is needed in order to meet future lithography needs. The next step beyond EUV is to move to soft x rays. Also, the limitations in three-dimensional resolution on laser sintering for three-dimensional manufacturing are based on the wavelength of the lasers used. Shorter wavelengths will move the state of the art to allow more precise additive manufacturing that could eventually lead to three-dimensional printing of optical elements.

Recommendation: Industry and public (both federal and state) sources should expand financial support for the training of skilled workers in photonics production and in applications of photonics-based technologies in manufacturing. The photonics industry also should enhance incentives for holders of advanced degrees in photonics, optics, physics, and related fields to pursue employment opportunities in manufacturing. One potential vehicle for such expanded support and for needed improvements in data collection on photonics employment trends at all levels is the federal initiative in photonics discussed in the recommendations of Chapter 2.

8

Advanced Photonic Measurements and Applications

INTRODUCTION

Advances in sensing, imaging, and metrology over the last decade have been critically dependent on optics and photonics, and precision sensing has moved progressively to optically based measurements. Optical techniques are already at the core of some of the most precise measurements. For example, the NIST-F1 cesium time standard in use in the United States since 1998, around the time of the publication of the National Research Council's (NRC's) report *Harnessing Light: Optical Science and Engineering for the 21st Century*,[1] exploits laser cooling of cesium atoms, optical monitoring of fluorescence, and various other optical techniques to lock in the microwave frequency of the atomic clock, and a second generation of such a system is under construction.[2] This chapter describes the advances made in these technologies since 1997.

Precision metrology is important for advances in the following: fundamental research that relies on precision measurements, communication that relies on precision timing for high data rates and long ranges, and the Global Positioning System (GPS), which relies on precision timing. GPS devices were just becoming commercially available in 1998, and now they are in nearly every cell phone. The advent of octave-spanning optical frequency combs allows a small table-top appa-

---

[1] National Research Council. 1998. *Harnessing Light: Optical Science and Engineering for the 21st Century.* Washington, D.C.: National Academy Press.

[2] Jefferts, S.R., T.P. Heavner, T.E. Parker, and J.H. Shirley. 2007. NIST cesium fountains—Current status and future prospects. *Acta Physica Polonica A* 112(5):759-767.

ratus to provide a direct link between radio frequency (RF) and optical standards, which took several rooms to perform at only a few laboratories around the world a decade ago. Now this capability is commercially available. Since the NRC's 1998 study, miniature atomic clocks on a chip have been developed to provide precise local measurements. Quantum cascade lasers on the market extend the range of chip-scale laser sources for near and remote sensing applications into the middle-infrared wavelength range of the electromagnetic spectrum (3-30 μm). The field of terahertz imaging has matured to the point of deployable systems in airports and other points of entry into our nation for the secure and efficient passage of trade goods. New construction—such as bridges, tunnels, dams, skyscrapers, pipelines, railroad tracks, and power plants—and renovation of civilian and military infrastructure around the world routinely have many kinds of active and passive optical sensors (for example, of vibration, temperature, strain, displacement, and cracks) embedded for the real-time monitoring of operation and for the forecasting of hazardous conditions before disaster strikes. Optical sensors are also common in cars, trucks, airplanes, and ships.

Optics and photonics advances have enabled advances in precision manufacturing, which have enabled further improved sensors. Low-cost, high-resolution cameras in cell phones now make advanced digital imaging available to a substantial fraction of the world's population with capabilities comparable with the best high-end cameras of a decade ago. Those components will enable a new wave of secondary niche markets that have the potential to have a significant impact on the U.S. economy and job pool. This broad growth of optical sensing and metrology—from the most precise scientific applications to universal consumer devices—makes the next decade an exciting time for optics and photonics in sensing and measurement, in research, and in consumer and industrial applications and offers significant opportunities for U.S. leadership.

IMPACT OF OPTICS AND PHOTONICS ON SENSING, IMAGING, AND METROLOGY

Advanced photonic measurements and applications have had a profound impact on our daily lives. For example, GPS has had a significant impact on navigation. In the late 1990s, consumer GPS devices were only beginning to enter the market. Now this capability is a commonplace consumer item found in cell phones, car navigation equipment, and even pet identification tags. GPS relies on precision timing to enable high-resolution positioning, which also enables high data rates and long-range communications. That timing is enabled by several advances in photonics, such as compact atomic clocks on a chip (see Figure 8.1). Sensing and metrology have enabled a new level of integrated-circuit (IC) manufacturing, which has driven the entire consumer electronics industry. Those advances have

FIGURE 8.1 Schematic (*left*) and photograph (*right*) of a microfabricated atomic clock. The total volume of the device is less than 1 cm^3, making it practical for use in handheld, battery-powered electronics. (See source for detailed image labels.) SOURCE: Reprinted, with permission, from Knappe, S., L. Liew, V. Shah, P. Schwindt, J. Moreland, L. Hollberg, and J. Kitching. 2004. A microfabricated atomic clock. *Applied Physics Letters* 85:1460.

also enabled the incorporation of low-cost, high-resolution imaging sensors in a broad range of consumer devices (such as cell phones and tablets). The proliferation of low-cost sensors connected by a high-bandwidth data transfer capability will enable the rapid growth of applications that would not have been economically viable without this large technology base. One example will be low-cost medical sensing devices that leverage consumer electronics components.

Since the NRC's 1998 study, advances in octave-spanning optical combs have enabled a small table-top apparatus that provides a direct link between RF and optical frequency and time standards—apparatus that used to take several rooms full of specialized equipment. Such advances have narrowed the gap in measurement capabilities between premium laboratories with specialized equipment and those with modest funding, and this will be a game changer in advancing both basic and applied research.

Photonic measurement and application advances have enabled improvements in manufacturing (for example, in lithography, machining, cutting, and welding), which have provided improved devices that are used to make improved sensors. That spiral threading of improvements feeds itself. Although the United States tends not to compete well in high-volume manufacturing, there is now a market opportunity for leveraging the application of these improved capabilities, as in the examples above, from consumer devices to address lower-volume niche sensor markets.

There has been a steady progression from RF to optically based sensing, which has advanced significantly since the *Harnessing Light* appeared in 1998. One example is in synthetic aperture imaging. Synthetic aperture radar (SAR) has been used since the 1950s; however, only in the last decade have advances in photonics enabled simultaneously agile and stable optical sources that have made SAR viable at optical wavelengths. The move to optically based sensing is partially due to the potential for improved resolution made possible by the much shorter wavelength. However, in many systems the resolution requirements are modest. In those cases, the primary motivations are to achieve easily interpreted imaging and improve illumination efficiency. The shorter wavelength enables a smaller illumination area because of diffraction, and the reflectivity at optical wavelengths closely matches what we are accustomed to viewing with our eyes. In contrast, typical SAR images require significant training for interpreting the resulting data.

Since the NRC's 1998 study, there have been significant advances in emitter and detector materials for practical sources and sensors at new wavelengths. One example is the substantially improved capability at wavelengths near 2 μm, which is important for atmospheric research and military sensing. Significant advances in devices have also enabled photon-counting detectors to be extended to Geiger-mode detector arrays and to photon-number-resolving Geiger-mode detectors. Such advanced photon-counting techniques need to be expanded not only to higher count rates but to exploitation of novel quantum states of light in advanced optical sensors that are likely to come onto the horizon in the next decade or so.[3,4,5] Moreover, current research will potentially provide a true linear-mode single-photon detector that will open new doors for sensing, imaging, and metrology.

---

[3] An example is the planned incorporation of squeezed quantum states of light in the advanced Laser Interferometer Gravitational Wave Observatory (LIGO). Johnston, Hamish. 2008. Prototype gravitational-wave detector uses squeezed light. *Physics World*. Available at http://physicsworld.com/cws/article/news/33755. Accessed August 1, 2012.

[4] More information on the Laser Interferometer Gravitational-Wave Observatory (LIGO) is available at http://www.ligo.caltech.edu/. Accessed August 1, 2012.

[5] More information is available at LIGO Scientific Cooperation, http://www.ligo.org/. Accessed August 1, 2012.

TECHNOLOGY OVERVIEW

Since the issuance of *Harnessing Light: Optical Science and Engineering for the 21st Century* 14 years ago,[6] the role of optics in advanced photonic measurements and applications has undergone a revolution. New fields have blossomed, such as the advent of carrier-envelope mode locking (which earned the 2005 Nobel Prize in physics for John L. Hall and Theodor W. Hänsch),[7] which enables highly coherent pulse trains and precisely spaced lines of optical frequency (about 1 cycle per second, 1 Hz) that span more than an octave in spectrum from middle-infrared to deep blue. That enables a direct link between RF and optical standards in a small table-top apparatus; such precision makes possible extremely precise spectroscopy for metrological applications, which would have been impossible when *Harnessing Light* was published. Moreover, the availability of mass-market optical imagers (such as fairly high-resolution cell phone cameras) is making possible personalized sensing and imaging applications. Such applications are likely to be not only affordable but highly precise in sensitivity and resolution because of the tight linking that exists between the optics and the sophisticated onboard electronic signal processing tools. We have also seen significant new technological opportunities emerge as nanotechnology has increasingly enabled new kinds of optical and optoelectronic structures, some without precedent in the classical optical world. Nanophotonic structures that are patterned or fabricated on sub-wavelength scales open new or enhanced functions for almost any application in which tailoring the properties of light is important.

While the new scientific developments are breathtaking and will continue to spawn new directions in advanced photonic measurements and applications in laboratories worldwide, it is the category of transitioning to mass-market devices that might have a much greater impact on the economy and people's daily quality of life. Imagine an optics-enabled attachment to one's cell phone that allows monitoring of blood glucose by simply inserting a finger into an orifice in the attachment, thereby avoiding pricking one's finger several times a day. What if the same attachment had sensing elements that recorded other vital signs at the same time, to keep track of the user's general well-being and issue an early warning when a trend in some vital measure was spotted?

Much has happened in science in the years since *Harnessing Light* appeared. Many scientific breakthroughs that were in their infancy in 1998 have matured,

---

[6] National Research Council. 1998. *Harnessing Light.*

[7] More information on the Nobel Prize is available at http://www.nobelprize.org/nobel_prizes/physics/laureates/2005/. Accessed August 18, 2011.

have penetrated the marketplace, and are having an impact on our lives.[8] Of course, many other breakthroughs are just beginning to be understood. The following are some the exciting areas of science and technology that are being pursued aggressively today:

- Development of coherent sensing and imaging techniques;
- Emergence of highly coherent optical pulse trains (carrier-envelope mode locking made possible by highly nonlinear and novel microstructure optical fibers);
- Development of attosecond pulse trains by means of high-harmonic generation;
- Table-top availability of extreme intensities by means of chirped pulse amplification;
- Terahertz and middle-infrared sources of radiation (for example, quantum-cascade lasers);
- High-power fiber lasers;
- Advances in non-linear optics, quasi-phase matching, photonic bandgap fibers, and magneto-optics;
- Nano optics and plasmonics, negative index materials, and transformation optics;
- Advances in controlled generation of quantum light states and their manipulation and detection;
- Advances in detector technologies, wider wavelength coverage, pixel count, quantum limited operation, and single-photon and photon-number resolved counting;
- Advances in adaptive optical techniques, guide stars, deformable mirrors, and turbulence control; and
- Computational imaging and sensing.

Some of these areas are expected to mature technologically and lead to new applications that will penetrate the marketplace or make existing applications work better in the coming years. The next section presents a few of the major advances with an eye toward the technologies that might have the most impact on society in the future.

It should be noted that the list of scientific advances above only briefly touches on subjects pertaining to quantum information science and technology. Light plays

[8] Optical coherence tomography is one example. More information is available at *Optical Coherence Tomography News*, http://www.octnews.org/. Accessed October 26, 2011. The ubiquitous social networking enabled by massive wavelength division multiplexing (WDM) optical communications is another example.

an important role in almost all implementations of quantum processing, not just quantum communications and the so-called linear-optics paradigm of quantum computing. Those subjects are aptly covered in the National Research Council report *Controlling the Quantum World: The Science of Atoms, Molecules, and Photons.*[9] In a similar vein, scientific advances in astronomy are only briefly touched on where adaptive optics and photon-counting arrays are playing a transformative role in Earth-based telescopes and photon-counting arrays are likely to play a similar role in space-based telescopes, such as the James Webb Space Telescope.[10] The focus of this chapter is the advances that may have direct applications in sensing, imaging, and metrology systems.

CHANGES SINCE *HARNESSING LIGHT*

There have been significant changes in advanced photonic measurements and applications since the publication of *Harnessing Light.*[11] The changes have created new capabilities, improved the resolution and precision of measurements, and provided capabilities to modest facilities that were previously available in only a few locations around the world. Some of the significant changes are highlighted here.

Changes in SI Definitions

Around the time that *Harnessing Light* was published, the Système International (SI, or International System of Units) definition of the second was changed from the 1967 definition—the duration of 9,192,631,770 periods of the radiation corresponding to the transition between the hyperfine levels of a cesium-133 atom—to include the stipulation of ground state at a temperature of 0 K. The change is made practical by the extremely low temperature that is available from the use of optical cooling of collections of cesium atoms to temperatures as low as 1.3 μK. Combined with the 1983 change in the definition of the meter (defined as the path traveled by light in vacuum in 1/299,792,458 second), the change in the definition of the second reflects the continued importance of optics and photonics in precision measurements. The definition of the kilogram is also undergoing a fundamental change: the current definition defines the kilogram as the mass of the

[9] National Research Council. 2007. *Controlling the Quantum World: The Science of Atoms, Molecules, and Photons.* Washington, D.C.: The National Academies Press.

[10] More information on NASA's James Webb Space Telescope is available at http://www.jwst.nasa.gov/index.html. Accessed May 28, 2012.

[11] National Research Council. 1998. *Harnessing Light.*

international prototype kilogram, and the new definition relates it to the equivalent energy of a *photon* by means of Planck's constant.[12]

Development of Attosecond Pulse Trains by Means of the Generation of High Harmonics

When light passes through a medium, such as glass, its wavelength usually is not affected; such transmission of light through matter is termed linear optics. However, when the strength of the light is high, nonlinear optical phenomena occur, one of which is harmonic generation. Consider what happens when we turn the volume up too high in a loudspeaker. Instead of clean, pure tones, we get distortion, which consists of higher harmonics of the pure tones and other frequencies produced by mixing the tones. In similar fashion, light is a wave—just like a sound wave, but made of electromagnetic (EM) energy. When the light passing through a material gets too intense, harmonics of the light wave can be created. Blue light, for example, is the second harmonic (one-half the wavelength and twice the frequency) of near-infrared light and can be created by a non-destructive change in the response of the medium to the intense lightwave. Such phenomena are captured by the field of nonlinear optics.

By using the techniques of laser mode-locking and chirped-pulse amplification, scientists in the United States, Europe, and Japan have learned to create compact, cost-effective table-top sources of highly intense pulses of light. When such pulses of light are focused on inert gases, extreme nonlinear optical phenomena occur.[13] Generation of these high-harmonics leads to extremely short pulses of light at a very short wavelength (the second harmonic is one-half the wavelength, the third harmonic is one-third the wavelength, and so on). Scientists at JILA (University of Colorado, Boulder) have created table-top sources of coherent x rays[14] by such methods of extreme nonlinear optics.[15] Such x-ray light sources are likely to have a revolutionary impact on such applications as imaging and lithography on the nanoscale. (See Box 8.1.)

---

[12] Mohr, P. 2010. "Recent Progress in Fundamental Constants and the International System of Units." White paper. Third Workshop on Precision Physics and Fundamental Physical Constants. Available at http://physics.vniim.ru/SI50/files/mohr.pdf. Accessed January 17, 2012.

[13] As the second and third harmonics are generated, which themselves can become very intense, this can cause generation of harmonics of the harmonics, which generate further harmonics, and so on.

[14] Popmintchev, T., M.-C. Chen, P. Arpin, M.M. Murnane, and H.C. Kapteyn. 2010. The attosecond nonlinear optics of bright coherent x ray generation. *Nature Photonics* 4:822-832.

[15] Kapteyn, H.C., M.M. Murnane, and I.P. Christov. 2005. Extreme nonlinear optics: Coherent x rays from lasers. *Physics Today* 58:39-44.

BOX 8.1
Table-Top, Lensless, Soft-X ray Microscope

The optical microscope has contributed greatly to our understanding of the world around us. Unfortunately, the smallest object that can be imaged is determined—and limited—by the wavelength of the light used. To visualize much smaller objects on the nanoscale, x-ray microscopes are needed. A team led by the Kapteyn–Murnane research group at JILA (University of Colorado, Boulder) has recently demonstrated a table-top, lensless, soft-x-ray microscope with a resolution that is very close to the wavelength of the extreme ultraviolet light used. A lensless microscope uses a computer algorithm to analyze the scatter patterns produced from the illuminated sample. Figure 8.1.1 shows imaging of a test sample with 13-nm coherent light. A resolution of 92 nm is obtained. Higher-repetition-rate ultrafast lasers currently under development will significantly reduce image capture time and thus improve resolution toward the wavelength-limited value. This table-top soft-x-ray diffraction microscope should find applications in biology, medicine, nanoscience, and materials science.

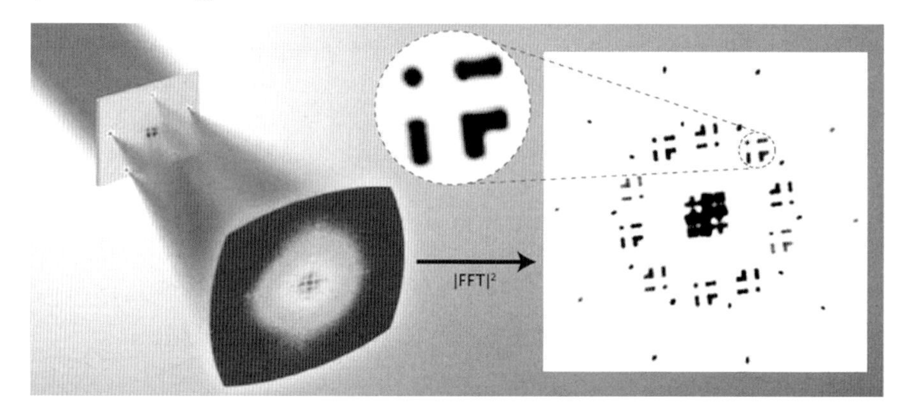

FIGURE 8.1.1 Lensless diffractive imaging combined with multiple-reference fast Fourier transform holography. The spatial autocorrelation of the object can be retrieved. Further refinement of the image to a resolution of 50 nm is possible with phase-retrieval algorithms to recover the spatial frequency information scattered at high angles. SOURCE: Reprinted with permission from McKinnie, I., and H. Kapteyn. 2010. High-harmonic generation: Ultrafast lasers yield x rays. *Nature Photonics* 4(3):149-151.

SOURCE: McKinnie, I., and H. Kapteyn. 2010. High-harmonic generation: Ultrafast lasers yield x rays. *Nature Photonics* 4(3):149-151.

Table-top Availability of Extreme Intensities by Means of Chirped-Pulse Amplification

Among the many attributes of laser light (monochromaticity, directionality, polarization purity, and brightness), the brightness or intensity (power density) is the most used property. Applications include cutting, welding, printing, data storage, and many more. Much has been accomplished in the technology of boosting laser beams to high, and sometimes lethal, power.[16] However, amplifying laser light without affecting its other attributes presents several challenges.

Nano-optics and Plasmonics, Negative-Index Materials, and Transformation Optics

Modern nanofabrication techniques allow us to control the size, shape, and structure of the material used with features on deeply sub-optical-wavelength scales, thereby opening up a broad range of technical opportunities. Such controlled fabrication means that optical properties can be tailored by the size, shape, or structure rather than by just the natural properties of materials themselves. Structures with controlled dimensions from tens to hundreds of nanometers fabricated in dielectrics, semiconductors, and metals[17] allow a broad range of new optical possibilities, such as photonic crystal structures, metamaterials,[18] compact high-quality-factor micro-ring resonators, and other nanometallic and plasmonic structures.[19] Those approaches offer new ways of concentrating or manipulating light[20] for enhancing or controlling sensing of various kinds,[21] such as chemical sensors, or such techniques as Raman scattering, and allow us to tailor optical response, such as spectral sensitivity, in ways beyond conventional optics. The science and basic technology of many such opportunities have been increasingly explored in research over the last decade as various nanofabrication tools have become more available.

Sensing with surface plasmon phenomena,[22] in which light is concentrated very near the surface of a metal, has been exploited in commercial biochemical sensing devices since the 1990s. Small changes in refractive index resulting from specific biochemical activity can be detected in very small detection volumes. The

---

[16] More information on chirped-pulse amplification is available at http://www.rp-photonics.com/chirped_pulse_amplification.html. Accessed January 17, 2012.

[17] von Freymann, G., A. Ledermann, M. Thiel, I. Staude, S. Essig, K. Busch, and M. Wegener. 2010. Three-dimensional nanostructures for photonics. *Advanced Functional Materials* 20:1038-1052.

[18] Chen, H., C.T. Chan, and P. Sheng. 2010. Transformation optics and metamaterials. *Nature Materials* 9:387-396.

[19] Brongersma, M.L., and V.M. Shalaev. 2010. The case for plasmonics. *Science* 328:440-441.

[20] Schuller, J.A., E.S. Barnard, W. Cai, Y.C. Jun, J.S. White, and M.L. Brongersma. 2010. Plasmonics for extreme light concentration and manipulation. *Nature Materials* 9:193-204.

[21] Richens, J.L., P. Weightman, W.L. Barnes, and P. O'Shea. 2010. "*In Vivo* Spectroscopic Imaging of Biological Membranes and Surface Imaging for High-Throughput Screening." Chapter 17 in *Nanoscopy and Multidimensional Optical Fluorescence Microscopy*, A. Diaspro, ed. Boca Raton, Fla.: CRC Press.

[22] Homola, J. 2008. Surface plasmon resonance sensors for detection of chemical and biological species. *Chemical Reviews* 108:462-493.

use of nanometallic particles is expected to improve such sensitivities further.[23] Another example of a nanometallic approach, recently commercialized, uses sub-wavelength holes in metals to allow optical detection of individual nucleotides in DNA sequencing.[24]

Nanophotonic techniques with dielectrics or metallic nanostructures show promise for making extremely compact spectrometers. Quantum mechanical properties can also be tailored once dimensions can be controlled on about a 10-nm or smaller scale. For example, quantum-dot (QD) fluorescent tags for biological experiments allow the fluorescent color to be controlled by choice of the size of the quantum dots.[25]

Advances in Controlled Generation of Quantum Light States and Their Manipulation and Detection

An ideal laser—and many practical lasers approach this ideal limit—emits light in the form of what is called a coherent state, so termed by Roy J. Glauber[26] in the early 1960s. In this quantum state, the light quanta (photons) exit the laser at random times, forming a Poisson distributed stream of photons even though the emitted light beam has constant power in the case of a continuous-wave laser. At the macroscopic level, the EM field associated with the emitted light wave approaches a sinusoid much like that seen on a string when it is repetitively shaken. Microscopically, however, the same randomness causes the wave to possess an uncertainty in its amplitude (height of the crests and troughs) and phase (zero-crossing points of the wave amplitude), but in this wave picture the uncertainty can be tied to the fluctuations in the vacuum EM field that permeates all space. The fundamental uncertainty caused by the vacuum field cannot be removed, but its effect can be manipulated in judicious ways to bypass its degrading effect on precise measurements in some situations. For example, the uncertainty in the amplitude can be traded at the expense of the uncertainty in the phase and vice versa, whereas the uncertainty product remains unchanged, as dictated by the Heisenberg uncertainty

[23] Offermans, P., M.C. Schaafsma, S.R.K. Rodriguez, Y. Zhang, M. Crego-Calama, S.H. Brongersma, and J. Gómez Rivas. 2011. Universal scaling of the figure of merit of plasmonic sensors. *ACS Nano* 5:5151-5157.

[24] More information is available at Pacific Biosciences, http://www.pacificbiosciences.com/. Accessed August 1, 2012.

[25] Alivisatos, P. 2004. The use of nanocrystals in biological detection. *Nature Biotechnology* 22:47-52.

[26] Roy J. Glauber shared the 2005 Nobel Prize in physics "for his contribution to the quantum theory of optical coherence." More information is available at http://www.nobelprize.org/nobel_prizes/physics/laureates/2005/. Accessed November 14, 2011.

principle,[27] a fundamental law of quantum physics. Such novel quantum light states have been called squeezed states, and tremendous progress has been made in the development of sources of squeezed light in the last couple of decades.[28]

One current grand challenge in the scientific world of sensing and precision measurement is the quest for the detection of gravity waves predicted by Einstein's theory of general relativity. Even though these waves in the fabric of space-time continuum were predicted almost a century ago, their direct observation has eluded scientists. National-scale efforts are underway in different parts of the world to detect gravity waves, and the most advanced sensor is in the Laser Interferometer Gravitational-Wave Observatory (LIGO).[29,30,31,32] It turns out that the strain sensitivity achieved in the current generation of LIGO does not reach a level that is high enough to ferret out the faint signatures of the gravity waves. The ultimate barrier to improving the strain sensitivity of the LIGO further is the above-discussed fundamental noise on the waves of light that bounce between the arms of LIGO's giant interferometer. The use of squeezed light can lead to enhanced performance, and a prototype demonstration of the expected enhancement has been made (see Figure 8.2).[33,34] It is expected that the use of this novel quantum state of light will play a pivotal role in the ultimate detection of gravity waves and in the opening of a new window on the universe. Continued development of highly efficient sources of squeezed light motivated by the grand challenge of detecting gravity waves, particularly those in the telecommunications wavelength bands prevalent in today's

[27] More information on Heisenberg's uncertainty principle is available at AIP Center for the History of Physics, http://www.aip.org/history/heisenberg/. Accessed August 1, 2012.

[28] Vahlbruch, H., M. Mehmet, S. Chelkowski, B. Hage, A. Franzen, N. Lastzka, S. Goßler, K. Danzmann, and R. Schnabel. 2008. Observation of squeezed light with 10 dB quantum noise reduction. *Physical Review Letters* 100:033602-033606.

[29] Johnston, H. 2008. "Prototype Gravitational-Wave Detector Uses Squeezed Light." *Physics World*. Available at http://physicsworld.com/cws/article/news/33755. Accessed August 1, 2012.

[30] A more advanced interferometer is also being planned. More information on the Laser Interferometer Space Antenna (LISA) is available at http://lisa.nasa.gov/. Accessed August 1, 2012.

[31] More information is available at Laser Interferometer Gravitational-Wave Observatory (LIGO), http://www.ligo.caltech.edu/. Accessed August 1, 2012.

[32] More information is available at LIGO Scientific Cooperation, http://www.ligo.org/. Accessed August 1, 2012.

[33] Goda, K., O. Miyakawa, E.E. Mikhailov, S. Saraf, R. Adhikari, K. McKenzie, R. Ward, S. Vass, A.J. Weinstein, and N. Mavalvala. 2008. A quantum-enhanced prototype gravitational-wave detector. *Nature Physics* 4(6):472-476.

[34] The LIGO Scientific Collaboration. 2011. A gravitational wave observatory operating beyond the quantum shot-noise limit. *Nature Physics* 7(12):962-965.

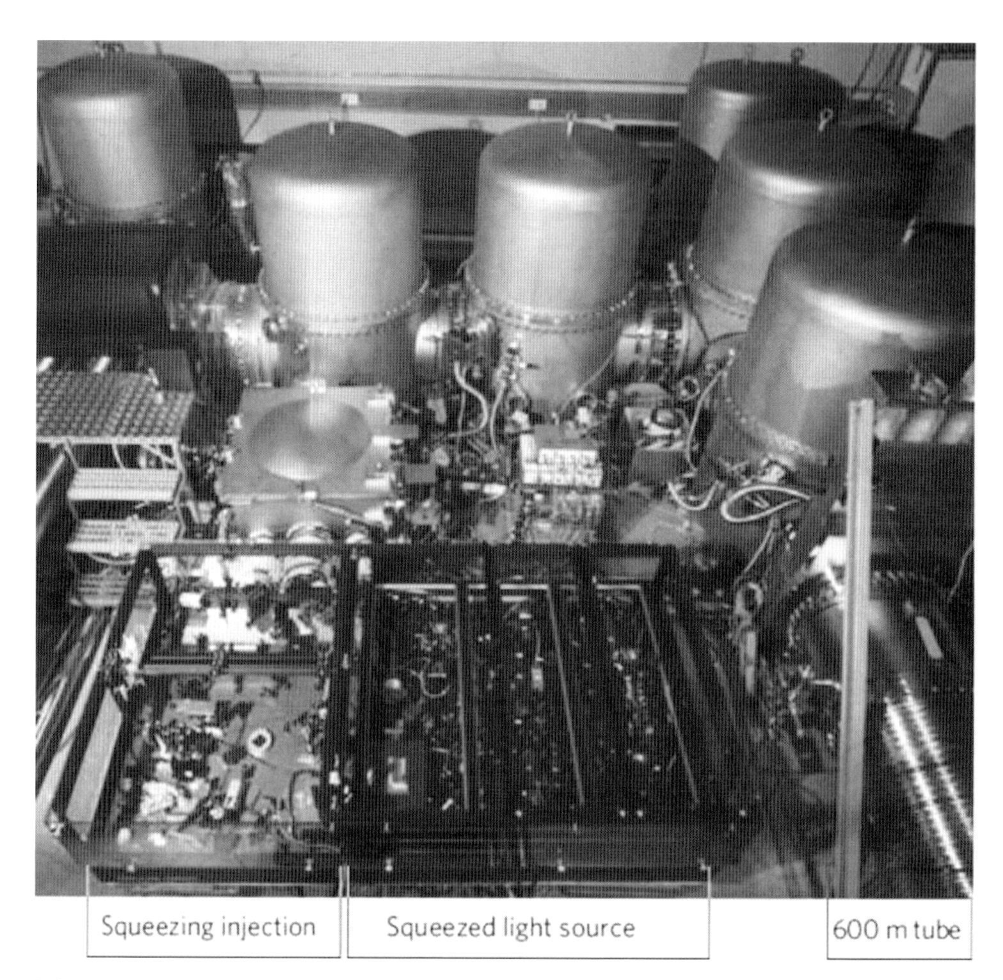

FIGURE 8.2 View into the GEO600 central building in Schäferberg, Germany. In the front, the squeezing bench containing the squeezed-light source and the squeezing injection path is shown. The optical table is surrounded by several vacuum chambers containing suspended interferometer optics. SOURCE: Reprinted with permission from The LIGO Scientific Collaboration. 2011. A gravitational wave observatory operating beyond the quantum shot-noise limit. *Nature Physics* 7(12):962-965.

communication and sensing systems,[35] is likely to spawn new sensing applications, such as quantum-enhanced laser radar (LADAR) imagers as recently proposed.[36]

---

[35] Mehmet, M., S. Ast, T. Eberle, S. Steinlechner, H. Vahlbruch, and R. Schnabel. 2011. Squeezed light at 1550 nm with a quantum noise reduction of 12.3 dB. *Optics Express* 19:25763-25772.

[36] Dutton, Z., J.H. Shapiro, and S. Guha. 2010. LADAR resolution improvement using receivers enhanced with squeezed vacuum injection and phase-sensitive amplification. *Journal of the Optical Society of America B* 27:A63-A72.

High-Resolution Remote Sensing with Optical Synthetic Aperture Radar

Although high-resolution remote sensing with optical synthetic aperture radar is discussed in Chapter 4 on defense, the ability to do high-resolution imaging at long ranges can have applications in areas other than for the military. *Planet Earth* videos[37] used defense-developed equipment for what at the time was long-range imaging, so animals could be remotely observed in their natural habitat without the observation changing the animals' behavior. In disaster scenarios, long-range imaging can help to plan relief activities.

Advances in Adaptive Optical Techniques

The performance of astronomical telescopes and free-space laser communication systems is severely limited by the effects of atmospheric distortion. Similarly, in microscopy and retinal imaging, optical aberrations can prevent one from achieving diffraction-limited resolution. "Adaptive optics (AO) is a technology that is used to improve the performance of optical systems by reducing the effect of wavefront distortions. It works by measuring the distortions in a wavefront and compensating for them with a device that corrects the errors, such as a deformable mirror or a liquid-crystal array."[38] Tremendous advances continue to occur in the technology and applications of adaptive optics.[39] For example, in the not-too-distant future, a patient may, after having cataract surgery, be able to have a personalized aberration-corrected lens implanted that would give the person better vision than she or he had been born with.[40]

Identification of Technological Opportunities from Recent Advances

The ultimate technical challenge in sensing is to be able to detect something even at very low levels or with very high specificity, such as trace concentrations of toxic pollutants in the atmosphere, a specific biochemical structure, vibrations on the fuselage or wings of an airplane in order to gain early indications of crack formation, or variations in Earth's gravity to facilitate a search for oil or other hidden objects. Imaging is sensing as a function of location to obtain a spatial rendering of whatever is being sensed. The goal of metrology is to ensure that the output

---

[37] More information on the British Broadcasting Corporation's (BBC's) *Planet Earth* series is available at http://www.bbc.co.uk/programmes/b006mywy. Accessed August 1, 2012.

[38] *BBC News*. 2011. "'Adaptive Optics' Come into Focus." Available at http://www.bbc.co.uk/news/science-environment-12500626. Accessed May 29, 2012.

[39] *BBC News*. 2011. "'Adaptive optics' Come into Focus."

[40] Chris Dainty, Professor of Applied Optics, National University of Ireland, Galway. Communication to the committee. May 15, 2011.

of a sensing device can be accurately tied to the sensed quantity, such as "This many units of the sensor reading correspond to this many grams of the pollutant per liter" in the first example above. Therefore, harnessing light for ever-more-advanced and reliable applications in advanced photonic measurements and applications is intimately tied to our basic understanding of how light interacts with matter and how we can manipulate and detect light at the very fundamental level. The technological advances since the publication of the NRC's 1998 *Harnessing Light* report[41] have already enabled new measurement capabilities and narrowed the gap between "high-end" laboratories and more modest facilities in terms of measurement capabilities. Those advances will be a significant catalyst for the next wave of advances in both fundamental and applied research. The proliferation of high-resolution sensors in consumer devices has enabled a market opportunity to leverage these new measurement capabilities for applications that would otherwise not be economically viable. Below are some examples of technological opportunities enabled by recent advances in sensing, imaging, and metrology.

Cost-Effective Biomedical Sensing Devices

The general field of nanophotonics is likely to remain promising and active in research in coming years for biochemical and biomedical sensing. Because many nanopatterning and nanofabrication tools (such as optical lithography developed for IC fabrication and other novel techniques, such as nanoimprint lithography[42]) are capable of mass manufacture of precisely controlled nanostructures, there is significant potential for implementing novel practical applications. Research focused on those application possibilities will be increasingly important. Highly chemical-specific and low-cost biochemical sensing will be a particularly important application.

Such devices as cell phone cameras already offer a ubiquitous optical sensing platform that is networked. Mobile phone subscriptions worldwide have passed 5 billion.[43] Extensions of such technology—possibly with the addition of light sources to excite fluorescence, novel microscopy approaches, or more sophisticated spectral detection capabilities—may allow widely available remote medical or

[41] National Research Council. 1998. *Harnessing Light*.

[42] Osborne, M. 2005. "Enhanced Nanoimprint Process for Advanced Lithography Applications." White paper. Available at http://www.fabtech.org/white_papers/_a/enhanced_nanoimprint_process_for_advanced_lithography_applications/. Accessed January 17, 2012.

[43] Associated Press. 2010. "Number of Cell Phones Worldwide Hits 4.6B." Available at http://www.cbsnews.com/stories/2010/02/15/business/main6209772.shtml. Accessed December 5, 2011.

physiological monitoring or diagnostic techniques[44] with major impact on global health.

Exploiting the Quantum Detection and Manipulation of Light

At the macroscopic level, such as experienced when one is sitting in a lighted room, one perceives light to vary in a continuous, classical manner. For instance, a dimmer switch can control the brightness of light in a room and can be continuously varied from daylight conditions to the extreme darkness of nighttime. At the microscopic level, however, light consists of quantized packets of energy. A beam of light can be thought of as a flux of photons. When faint light is detected, instead of a detector output changing continuously, the detector observes random clicks corresponding to the absorption of specific photons by the detector. A familiar analogy is watching sand flow through an hourglass. When viewed from a distance, the falling of sand appears to be a smooth continuous flow. However, when viewed close up, it can be seen as the granular dropping of the sand particles. If one were to count the number of sand particles crossing the neck of the hourglass per second, one would obtain a randomly varying number from one second of counting to the next, and the flow rate would only seem to be constant. The same applies to the measurement of light by a detector. The light that one would want to detect after it interacts with the transducer in the sensor would have random variations (usually called noise) in the measured photon count, yielding uncertainty or error in the value of the sensed quantity. That kind of noise is called the shot noise, and the resulting error is a fundamental property of the process because it is related to the elementary nature of light. Thus, it would appear that the error due to shot noise would set ultimate limits on the sensitivity of sensing, imaging, and metrology systems. That is, the very basic granular nature of light would in general prevent us from sensing extremely weak signals.[45]

The quantum manipulation of the generation and detection of light, however, offers new opportunities. Research in the last couple of decades has shown that the arrangement of quanta in a beam of light can be manipulated. For example, instead

[44] Zhu, H., S. Mavandadi, A.F. Coskun, O. Yaglidere, and A. Ozcan. 2011. Optofluidic fluorescent imaging cytometry on a cell phone. *Analytical Chemistry* 8(17):6641-6647.

[45] For example, it is possible to reduce shot noise by means of squeezed light injection in the LIGO; this is leading to enhanced sensitivity in the quest for the detection of the gravity waves.

of being a random flow,[46] the photons in a light beam can be regularized (photon antibunching[47]) so that on detection the uncertainty in measurement would be reduced. Similarly, many other types of manipulations of photons in light beams can be made, such as creating paired photons that maintain their intimate quantum mechanical phase-coherent correlation (entanglement[48]) no matter how far apart they are.[49,50] Such novel photonic quantum states of light are already proving to be extremely potent. For example, there is the possibility of using entangled photons for creating shared secrets between remote users for the purpose of communicating securely.[51] Such techniques of quantum cryptography have been demonstrated and are being commercialized,[52,53] and there is much potential for ensuring the privacy of communications in ways that are tamperproof.[54] However, much more research and technology development need to happen before the promise of global-scale, highly secure communications protected by the fundamental laws of quantum physics can be realized. For example, the current systems have limited reach owing to the lack of a suitable quantum repeater technology—unlike the ubiquitous optical amplifiers in the case of conventional optical communications—and are slow owing to poor quantum efficiency and low speed of single-photon detectors. Many promising paths of research and technology development are being pursued worldwide, but the United States is consistently losing ground in this field for lack

[46] It turns out that ordinary lasers at their best emit light beams in the form of random flow of photons characterized by the so-called Poisson distribution. When such light is detected, the shot-to-shot variation in the photon count (standard deviation) in a unit time interval equals the square root of the average photon count in that time interval. Detection of light is thus very uncertain when the irradiance is weak enough (low-light-level illumination) for the detector to see only a few photons over its response time.

[47] Teich, M.C., and B.E.A. Saleh. 1990. Antibunched light. *Physics Today* (43)6:26-34.

[48] Zeilinger, A. 2010. *Dance of the Photons: From Einstein to Quantum Teleportation*. New York, N.Y.: Farrar Straus Giroux.

[49] Ursin, R., F. Tiefenbacher, T. Schmitt-Manderbach, H. Weier, T. Scheidl, M. Lindenthal, B. Blauensteiner, T. Jennewein, J. Perdigues, P. Trojek, B. Ömer, M. Fürst, M. Meyenburg, J. Rarity, Z. Sodnik, C. Barbieri, H. Weinfurter, and A. Zeilinger. 2007. Entanglement-based quantum communication over 144 km. *Nature Physics* 3:481-486.

[50] Dynes, J.F., H. Takesue, Z.L. Yuan, A.W. Sharpe, K. Harada, T. Honjo, H. Kamada, O. Tadanaga, Y. Nishida, M. Asobe, and A.J. Shields. 2009. Efficient entanglement distribution over 200 kilometers. *Optics Express* 17:11440-11449.

[51] Gisin, N., G. Ribordy, W. Tittel, and H. Zbinden. 2002. Quantum cryptography. *Reviews of Modern Physics* 74:145-195.

[52] More information on the products offered by ID Quantique is available at http://www.idquantique.com. Accessed August 1, 2012.

[53] More information on the products offered by NuCrypt, LLC, is available at www.nucrypt.net. Accessed August 1, 2012.

[54] Scarani, V., H. Bechmann-Pasquinucci, N.J. Cerf, M. Dušek, N. Lütkenhaus, and M. Peev. 2009. The security of practical quantum key distribution. *Reviews of Modern Physics* 81:1301-1350.

of adequate support for basic science and technology development. For instance, Europe, Japan, and China have roadmaps for breaching the distance limit by way of low-Earth-orbit satellite terminals, but the U.S. funding agencies, once a leader in free-space quantum cryptography and communications, so far have announced no plans.[55]

The fundamental quantum nature of light is such that our ability to *produce* light beams with prearranged photonic structure (light of a specified quantum state) is intimately tied to our ability to *measure* the arrangement of photons in a light beam. Although tremendous progress has been made in the last couple of decades in "seeing" photons,[56] it remains a technical challenge to detect light at single-photon resolution with a high degree of confidence and precision and certainly in a cost-effective manner. This is despite the widely accepted belief that the human eye is capable of resolving single or very small numbers of photons[57] and that photomultiplier tubes capable of detecting light at the single-photon level have been around for over a half-century. Instead of measuring light with single-photon resolution, the current generation of instruments puts out either no click with high probability when no photons arrive or one click no matter how many photons arrive in the detector's response time. In addition, when the photons do arrive, the probability of detection is very limited (about 70 percent for visible to near-infrared light and about 20 percent in the telecommunications waveband).[58,59] Nonetheless, progress is being made; devices and instruments with arrays of single-photon detectors for imaging applications are beginning to appear on the market, and technologies based on superconducting devices have been demonstrated in research laboratories.

In addition to diagnosing the photonic structure of light beams, the technology of detecting light efficiently and reliably at the single-photon level will open a host of other opportunities because such technology will revolutionize how we quantify light. Measuring light level (brightness) is typically an analog measurement that is notoriously hard to make precise and accurate. Counting photons will turn such measurements into an inherently digital form by basing the measurements on fun-

---

[55] Hughes, R., and J. Nordholt. 2011. Refining quantum cryptography. *Science* 333:1584-1586.

[56] National Research Council. 2010. *Seeing Photons: Progress and Limits of Visible and Infrared Sensor Arrays.* Washington, D.C.: The National Academies Press.

[57] Wolpert, H.D. 2002. "Life lessons: Photonic Systems in Nature Can Offer Technical Insights to Designers of Optical Systems and Detectors." SPIE Newsroom. Available at http://spie.org/x25379.xml?ArticleID=x25379. Accessed August 1, 2012.

[58] More information on the products offered by ID Quantique is available at http://www.idquantique.com. Accessed August 1, 2012.

[59] More information on the products offered by NuCrypt, LLC, is available at www.nucrypt.net. Accessed August 1, 2012.

damental unit of energy.[60] Because the precision resulting from counting increases with the count rate, the ability to count photons at a high rate would spawn new metrological applications of light.

Manufacturing

Although many advances that originated in the United States address optical manufacturing capabilities, there is almost no high-volume manufacturing of sensors and imagers within the United States. However, the proliferation of devices developed for consumer products presents a significant marketing opportunity. Many niche sensor markets could not be addressed without the capabilities enabled by these devices. One example is in biomedical sensing. There are capabilities in microscope systems costing more than $400,000 that could be partially addressed in a small device costing less than $10,000 that leverages capabilities provided by high-volume consumer device components. Because the resulting sales could be about 1,000 per year, these markets would not be efficiently addressed by a large microscope manufacturer. However, a small company could profitably address such a market. These niche markets rely on moving research advances into the market efficiently while exploiting the capabilities of components developed and priced for high-volume markets. A small company could keep most of the created jobs within the United States by leveraging the manufacture of low-cost devices that have steadily moved overseas. To address this market opportunity efficiently, an efficient coupling between basic and applied research in optics- and photonics-related technologies with industrial application partners is critical. An efficient partnership in this field could significantly add to U.S.-based jobs at all levels.

U.S. GLOBAL POSITION

For many years, the United States has benefited from a leadership position in research in optics and photonics. However, the research capabilities of many countries have been steadily improving, and the gap is rapidly narrowing. As discussed earlier, several advances over the last decade have hastened that narrowing, and cutting-edge measurement capabilities are now available to a much broader set of researchers. While continued research in fundamental optical sciences will be critical in maintaining a leadership position, it will also be critical for the U.S. economy to move those advances into the market efficiently to capture the financial benefit of generated intellectual property. Although high-volume manufacturing is not typically done within the United States, there is a significant market oppor-

[60] Migdall, A. 1999. Correlated photon metrology without absolute standards. *Physics Today* 52: 41-46.

tunity for leveraging high-volume consumer components with research advances to address low-volume markets. Capitalizing these niche markets efficiently could have a significant impact on U.S.-based jobs.

FINDINGS

Key Finding: Optics and photonics have been critically important to advances in precision metrology, which has had a significant impact since publication of the NRC's 1998 study *Harnessing Light: Optical Science and Engineering for the 21st Century* (for example, GPS, communications, and manufacturing). The importance of optics and photonics is now reflected in the adoption of optics-based SI definitions of the second and the meter.

Key Finding: There is a significant opportunity for the U.S. economy to exploit niche sensor markets that leverage consumer components and cutting-edge research applications. One example is in biomedical sensing in which low-volume manufacturing of devices could efficiently be maintained within the United States by leveraging high-volume consumer components, such as the high-resolution networked imagers now almost universally available in the form of cell phone cameras. Exploiting this advanced technology could enable portable and/or remote health monitoring and diagnosis.

Key Finding: Techniques of extreme nonlinear optics that promise table-top, coherent sources of extreme ultraviolet (EUV) and x-ray light have been developed. If this promise becomes real, it will profoundly affect such applications as sub-nanometer-scale lithography and determination of the structure of complex matter (biological proteins, for instance) on the atomic scale, further enabling advances in fields such as optical machining that rely on progressively shorter illumination wavelengths to improve manufacturing tolerances. This increased precision will be important for maintaining advances consistent with Moore's law of ICs.

Key Finding: The ultimate sensitivity of any advanced photonic measurement and application system is fundamentally tied to the intrinsic photonic granularity of light. Measuring light with single-photon resolution and accuracy at high speeds will therefore improve the performance of such systems tremendously in analogy to how counting cycles of light waves for shorter and shorter wavelengths is paving the way for more accurate and precise measurements of time (first key finding above).

Finding: Precision metrology has improved and become more widely available because of the significant technological advances since the NRC's *Harnessing Light* study was published in 1998. One example is octave-spanning optical combs, which provide a direct link between RF and optical standards within a small table-top apparatus that is now commercially available. At the time of the 1998 study, linking between RF and optical standards took instrumentation that filled several rooms and was performed at only a few locations around the world.

Finding: Several countries around the world have made significant advancements in photonics research capabilities in the measurement area, and the research leadership gap between these countries and the United States has significantly narrowed in many disciplines.

Finding: Progress in nanophotonics, plasmonics, metamaterials, and other related fields of science and technology is opening a broad range of possibilities for the enhanced sensitivity, greater specificity, lower size, and lower cost of sensors. These possibilities will have significant impacts in various fields, including biochemical sensing.

RECOMMENDATIONS AND GRAND CHALLENGE QUESTION

Entangled photons and squeezed states are new subjects of research and development in the optics and photonics field and allow sensing options never previously considered.

Key Recommendation: The United States should develop the technology for generating light beams whose photonic structure has been prearranged to yield better performance in applications than is possible with ordinary laser light.

Prearranged photonic structures in this context include generation of light with specified quantum states in a given spatiotemporal region, such as squeezed states with greater than 20-dB measured squeezing in one field quadrature, Fock states of more than 10 photons, and states of one and only one photon or two and only two entangled photons with greater than 99 percent probability. These capabilities should be developed with the capacity to detect light with over 99 percent efficiency and with photon-number resolution in various bands of the optical spectrum. The developed devices should operate at room temperature and be compatible with speeds prevalent in state-of-the-art sensing, imaging, and metrology systems. U.S. funding agencies should give high priority to funding research and development—at universities and in national laboratories where such research is carried out—in this fundamental field to position the U.S. science and technology base at the forefront of applications development in sensing, imaging, and metrology. It is believed that this field, if successfully developed, can transfer significant technology to products for decades to come.

Key Recommendation: Small U.S. companies should be encouraged and supported by the government to address market opportunities for applying research advances to niche markets while exploiting high-volume consumer components. These markets can lead to significant expansion of U.S.-based jobs while capitalizing on U.S.-based research.

Recommendation: U.S. funding agencies should continue to support fundamental research in optics and photonics. Important subjects for future research include nanophotonics, extreme nonlinear optics, and number-resolving photon counters for a truly linear-mode single-photon detector. Support should be provided for applying advances to devices for market application.

The fifth grand challenge question is partially supported by the discussion in this chapter and is thus repeated here with some supporting information.

> How can the U.S. optics and photonics community develop optical sources and imaging tools to support an order of magnitude or more of increased resolution in manufacturing?

Meeting this grand challenge could facilitate a decrease in design rules for lithography, as well as providing the ability to do closed-loop, automated manufacturing of optical elements in three dimensions. Extreme ultraviolet is a challenging technology to develop, but it is needed in order to meet future lithography needs. The next step beyond EUV is to move to soft x rays. Also, the limitations in three-dimensional resolution on laser sintering for three-dimensional manufacturing are based on the wavelength of the lasers used. Shorter wavelengths will move the state of the art to allow more precise additive manufacturing that could eventually lead to three-dimensional printing of optical elements.

9

Strategic Materials for Optics

INTRODUCTION

Materials are playing an increasingly important role in the technological evolution of photonic and optical applications. Whether the applications are related to imaging of cellular functions, the development of new types of sensors and solar cells, or the integration of materials for optoelectronics, the study of techniques to alter how light interacts with materials has become an important element in the advancement of various applications. Moreover, defense applications require an assured and secure manufacturing source of key materials. While materials were not called out in the National Research Council's (NRC's) 1998 report *Harnessing Light: Optical Science and Engineering for the 21st Century*,[1] the role of materials in optics and photonics has become much greater since its publication. Engineered materials, including photonic crystals, have come of age. We have seen key optoelectronic materials play an important role in negotiations between countries. Material development has always been a slow and expensive process, but with the advent of engineered materials and the rise of awareness of the importance of certain materials, materials have taken on a strong enough strategic importance to warrant their own chapter in the present report.

One specific challenge that optics faces in most industries where it is being used is that optics remains an enabling technology, supporting another area such

[1] National Research Council. 1998. *Harnessing Light: Optical Science and Engineering for the 21st Century.* Washington, D.C.: National Academy Press.

as cancer treatment or welding. The principal value of a product is not always attributed to the optical technologies associated with the product's applications. For example, developing new biological materials, such as fluorescent proteins,[2] that have new optical characteristics is associated more with new advances in biotechnology than with optical imaging.

This chapter outlines the role that strategic materials play in the development of new optical phenomena in specific categories of applications, outlines a few key technological problems that need to be solved to enhance the impact of these materials on the evolution of those applications, and, finally, identifies a set of challenges that need to be addressed by policy makers to support the research needed to solve the technological problems.[3]

ENERGY APPLICATIONS

The Sun has been identified as one of the primary sources of alternative energy as the United States transitions from a fossil-fuel-driven energy infrastructure in the next 20 years. However, for solar energy to become a viable and cost-effective source, its price needs to drop significantly. As discussed in Chapter 5, energy critical elements (ECEs) were identified as extremely important for the development of thin-film photovoltaics (TFPV), which will be instrumental in meeting cost targets. The ECEs are critical because of their limited supply in the United States. Gallium, germanium, indium, selenium, silver, and tellurium are all critical elements for development of TFPV. A challenge for the United States in connection with the ECEs is to determine whether there is a need to produce them in the United States as opposed to relying on a foreign supply.[4] For example, most of the easily extractable lithium in the world is localized in South American countries. Lithium is present in most of the promising battery technologies being produced and developed, and access to this material could be restricted by government intervention.

Several of the exotic elements required by emerging solar technologies rely on joint production methods, which pose another potential risk to scale-up. These

[2] The 2008 Nobel Prize in chemistry was awarded jointly to Martin Chalfie, Osamu Shimomura, and Roger Y. Tsien "for the discovery and development of the green fluorescent protein, gfp." More information is available at http://www.nobelprize.org/nobel_prizes/chemistry/laureates/2008/. Accessed May 31, 2012.

[3] For further discussion of strategic materials, such as erbium and other rare earths, see National Research Council. 2012. *The Role of the Chemical Sciences in Finding Alternatives to Critical Resources: A Workshop Summary.* Washington, D.C.: The National Academies Press.

[4] American Physical Society (APS) and Materials Research Society (MRS). 2011. *Energy Critical Elements: Securing Materials for Emerging Technologies.* A report by the APS panel on public affairs and the MRS. Available at http://www.aps.org/policy/reports/popa-reports/loader.cfm?csModule=security/getfile&PageID=236337. Accessed February 10, 2011.

materials are currently produced as by-products of refinement or extraction of more common materials and are bought from suppliers as waste products. Increased demand for these materials may surpass the amount extracted with current production methods, thus placing a risk on the scaling up of systems reliant on these elements. Elements that are known to pose joint-production risks are gallium, tellurium, and selenium. Gallium, an essential component of many high-efficiency thin-film and multi-junction photovoltaic cells, is obtained as a by-product of aluminum refining, and tellurium and selenium are most often obtained as by-products of copper refining. The locations of these common production metals will determine production of these ECEs and are shown in Figure 9.1. Cadmium is usually produced from zinc processing but is refined out because of its toxicity and thus is commercially available at relatively low and stable prices and not qualified as an ECE.[5]

The use of light-emitting diodes (LEDs) for lighting is another example of how optics is affecting energy conservation in several industrial nations. Several countries around the world have initiatives to phase out the use of incandescent lightbulbs by 2020. LEDs have been identified as the leading replacement technology for this mass market. One promising approach to obtain white light is to use gallium-nitride (GaN)-based blue LEDs in conjunction with fluorescent phosphors, which convert part of the blue light into yellow and red. Many of the efficient phosphors contain rare earth elements (REEs). The two key REEs that provide color LED lighting are europium and terbium. Currently, the majority of the REEs are produced in China, and the Chinese government has imposed export restrictions.[6] The United States will have to develop a novel strategy either to develop new materials for LED applications or to encourage research toward cost-effective and environmentally friendly purification, mining, and production. Finally, the United States should consider reclamation from waste to extract rare materials inasmuch as U.S. waste management derives most of its profit from processing waste to remove precious and rare metals and other substances.

NOVEL STRUCTURES: SUB-WAVELENGTH OPTICS, METAMATERIALS, AND PHOTONIC CRYSTALS

In addition to new core materials, there is much promise in tailoring existing materials in novel ways to produce innovative results. These new materials, known as metamaterials or nanophotonic materials, are materials that can be developed

[5] APS and MRS. 2011. *Energy Critical Elements.*

[6] Bradsher, K. 2010. "China Said to Widen Its Embargo of Minerals." *New York Times.* October 19. Available at http://www.nytimes.com/2010/10/20/business/global/20rare.html?pagewanted=all. Accessed August 29, 2011.

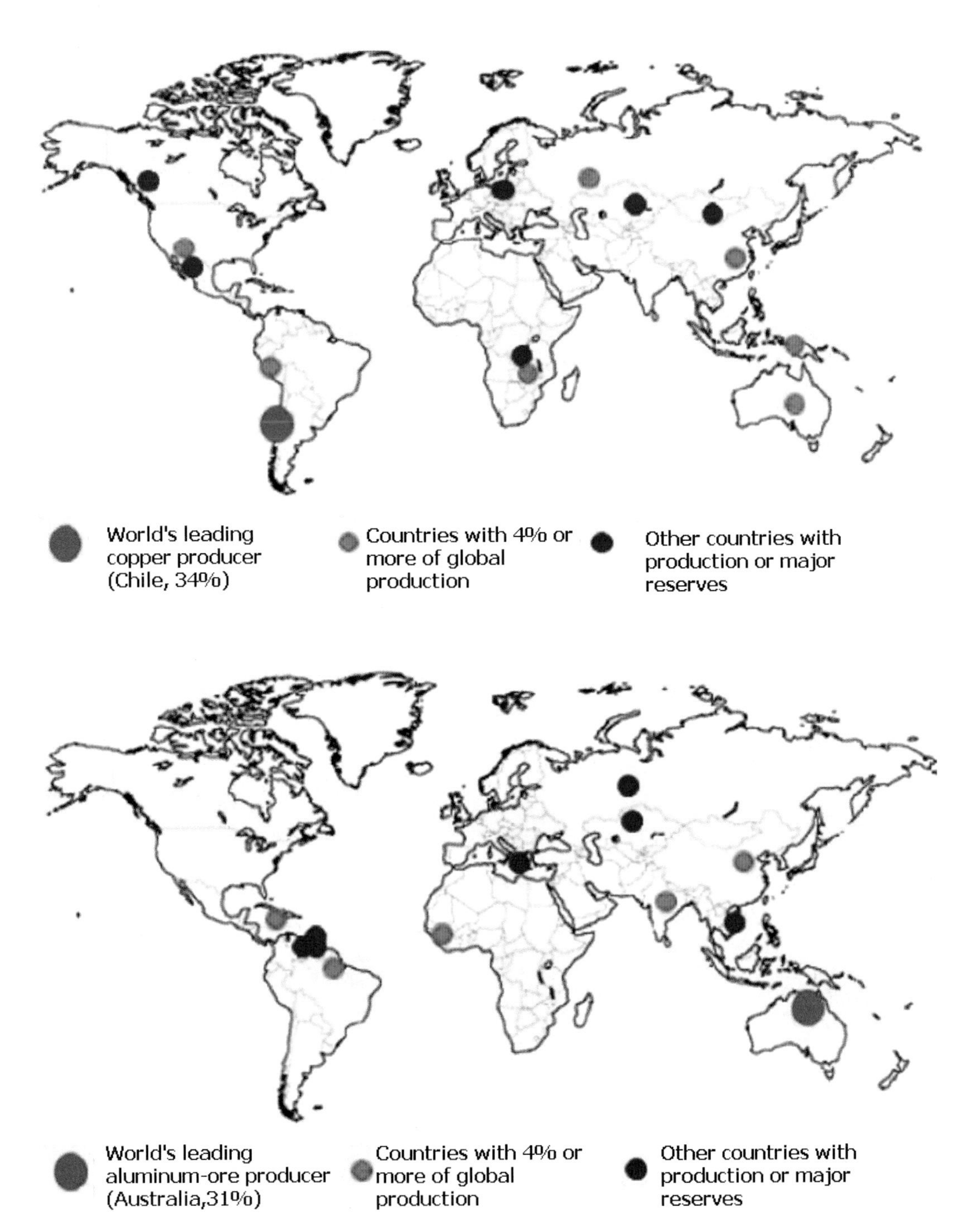

FIGURE 9.1 Location of current copper and aluminum-ore production. Energy critical elements required for several emerging solar technologies are obtained as by-products of refinement.

to exhibit new optical properties that the original materials themselves would not naturally possess. Structuring materials with features less than or close to one wavelength of light can lead to these novel properties, with the optical behavior coming more from the nanopatterning or nanostructuring than from the specific underlying materials. Such sub-wavelength structuring can be used with metals, semiconductors, or dielectrics, including combinations of these. The resulting effective materials or complex structures with sub-wavelength patterning allow control of such properties as spectral and polarization dependence in transmission, refraction, reflection, absorption, and emission of light. If a sub-wavelength structure is periodic in two or three dimensions, it is called a photonic crystal (PhC)[7,8,9] because the structure shows effects on the motions of an optical wave similar to the effects of a semiconductor crystal on an electron wave.[10,11] Since its inception, rapid progress has been made on this topic. Complex PhC structures with varied dimensions have been reported; these structures allow devices to be made that can control properties such as spectral and polarization dependence in transmission, refraction, reflection, absorption, and emission of light.[12,13,14] Interesting metal patterns have been added to such periodic structures to produce negative refractive index and superlens effects;[15,16] these patterns can focus light below the conventional diffraction limit. PhC devices are also being actively deployed in biosensors, semiconductor lasers, and hollow-core fibers.[17,18,19,20]

A new class of dielectric sub-wavelength grating exhibits very different prop-

[7] Yablonovitch, E. 1987. Inhibited spontaneous emission in solid-state physics and electronics. *Physical Review Letters* 58:2059-2062.

[8] John, S. 1987. Strong localization of photons in certain disordered dielectric superlattices. *Physical Review Letters* 58:2486-2489.

[9] Joannopoulos, J.D., P.R. Villeneuve, and S. Fan. 1997. Photonic crystals: Putting a new twist on light. *Nature* 386:143-149.

[10] Yablonovitch, E. 1987. Inhibited spontaneous emission in solid-state physics and electronics.

[11] John, S. 1987. Strong localization of photons in certain disordered dielectric superlattices.

[12] Joannopoulos et al. 1997. Photonic crystals: Putting a new twist on light.

[13] Busch, K., G. von Freymann, S. Linden, S.F. Mingaleev, L. Tkeshelashvili, and M. Wegener. 2007. Periodic nanostructures for photonics. *Physical Review Letters* 444:101-202.

[14] Painter, O., R.K. Lee, A. Scherer, A. Yariv, J.D. O'Brien, P.D. Dapkus, and I. Kim. 1999. Two-dimensional photonic band-gap defect mode laser. *Science* 284(5421):1819-1821.

[15] Pendry, J.B. 2000. Negative refraction makes a perfect lens. *Physical Review Letters* 85:3966-3969.

[16] Zhang, X., and Z. Liu. 2008. Superlenses to overcome the diffraction limit. *Nature Materials* 7:435-441.

[17] Skivesen, N., A. Têtu, M. Kristensen, J. Kjems, L.H. Frandsen, and P.I. Borel. 2007. Photonic-crystal waveguide biosensor. *Optics Express* 15:3169-3176.

[18] Noda, S., M. Yokoyama, M. Imada, A. Chutinan, and M. Mochizuki. 2001. Polarization mode control of two-dimensional photonic crystal laser by unit cell structure design. *Science* 293:1123-1125.

[19] Knight, J.C. 2003. Photonic crystal fibres. *Nature* 424:847-851.

[20] Russell, P. 2003. Photonic crystal fibres. *Science* 299:358-362.

erties from PhCs or traditional gratings. This grating leverages a high contrast in refractive indexes for the grating medium and its surroundings and sub-wavelength period to lead to extraordinary properties; hence the name high-contrast grating.[21,22] They can be easily designed to exhibit super-broadband, high-reflectivity mirrors for light incident in a surface-normal direction and at a glancing angle, ultra-high-Q (over 10^6) resonators with surface-normal output, planar high-focusing power reflectors and lenses (numerical aperture, over 0.9), ultra-low-loss hollow-core waveguides, slow-light waveguides,[23] and high-efficiency vertical to in-plane waveguide couplers.[24]

The United States has several research groups working on these types of materials that can potentially lead to solar cells that more efficiently absorb and trap sunlight and convert it to electrical energy. Researchers have designed materials that can bend visible light at unusual but precise angles, regardless of its polarization,[25] as a step toward perfectly transparent solar cell coatings that would direct all the Sun's rays into the active area to improve solar power output. In addition, researchers are working on novel antireflective solar cell coatings in the hope of getting more light into the cells.[26,27] Finally, metastructured materials can be used to separate the solar spectrum efficiently to optimize solar energy conversion.[28]

The versatility of metamaterials and photonic nanostructures in tailoring optical properties could lead to radical improvements and completely new types of devices. Metamaterials have been proposed for superlens applications[29] where they can to some extent go beyond the conventional diffraction limit of focusing in some special situations. Moreover, PhC devices are being deployed in biosensors,

[21] Vlasov, Y.A., M. O'Boyle, H.F. Hamann, and S.J. McNab. 2005. Active control of slow light on a chip with photonic crystal waveguides. *Nature* 438:65-69.

[22] Mateus, C.F.R., M.C.Y. Huang, Y. Deng, A.R. Neureuther, and C.J. Chang-Hasnain. 2004. Ultrabroadband mirror using low-index cladded subwavelength grating. *IEEE Photonics Technology Letters* 16:518-520.

[23] Chang-Hasnain, C.J. 2011. High-contrast gratings as a new platform for integrated optoelectronics. *Semiconductor Science and Technology* 26:014043.

[24] Stöferle, T., N. Moll, T. Wahlbrink, J. Bolten, T. Mollenhauer, U. Scherf, and R. Mahrt. 2010. An ultracompact silicon/polymer laser with an absorption-insensitive nanophotonic resonator. *Nano Letters* 10:3675-3678.

[25] Burgos, S.P., R. de Waele, A. Polman, and H.A. Atwater. 2010. A single-layer wide-angle negative-index metamaterial at visible frequencies. *Nature Materials* 9:407-412.

[26] Xi, J.Q., M.F. Schubert, J.K. Kim, E.F. Schubert, M. Chen, S.Y. Lin, W. Liu, and J.A. Smart. 2007. Optical thin-film materials with low refractive index for broadband elimination of Fresnel reflection. *Nature Photonics* 1:176-179.

[27] Diedenhofen, S.L., G. Vecchi, R.E. Algra, A. Hartsuiker, O.L. Muskens, G. Immink, E.P.A.M. Bakkers, W.L. Vos, and J.G. Rivas. 2009. Broad-band and omnidirectional antireflection coatings based on semiconductor nanorods. *Advanced Materials* 21:973-978.

[28] Burgos et al. 2010. A single-layer wide-angle negative-index metamaterial at visible frequencies.

[29] Zhang and Liu. 2008. Superlenses to overcome the diffraction limit.

semiconductor lasers, and hollow-core fibers.[30,31,32,33] Higher-dimensional metamaterials or custom nanophotonic structures provide extra design "knobs" to turn to modify optical characteristics and design devices with no precedent in conventional optics. Whereas one-dimensional photonic nanostructures have long been used in commercial products in the form of Bragg reflectors, dielectric interference filters, or antireflection coatings, higher-dimensional photonic nanostructures are not easy to fabricate and require more research on process technology.

TECHNOLOGY CHALLENGES OF NANOSTRUCTURED MATERIALS

Nanostructured materials became a subject of great interest because of the promise to "tailor" materials' innate physical properties when they are made small enough for the wave functions of electrons, phonons, or photons to be significantly confined by the structured boundaries. Control of matter on nanometer scales would allow manipulation of absorption, emission, transmission, refraction, transport, and energy conversion and storage in innovative ways that could have profound implications for many applications.

The applications of nanostructures for optoelectronic devices started when one-dimensional confinement structures, quantum wells, were demonstrated in 1974.[34] Double heterostructured material with a thickness under 20 nm was shown to provide quantum confinement of electrons to alter the material's effective bandgap energy. Today, most semiconductor diode lasers and integrated optoelectronics use quantum wells as a means to control wavelength, reduce threshold, and provide modulation. In the near future, it will be possible to extend this quantum confinement to two-dimensional (quantum wires) and three-dimensional (quantum dots, QDs) to achieve temperature-independent, ultra-low-threshold laser diodes and ultra-broadband semiconductor optical amplifiers.

In addition to changing the effective bandgap due to the quantization effect, nanostructures may enable monolithic integration of lattice-mismatched materi-

---

[30] Block, I.D., L.L. Chan, and B.T. Cunningham. 2006. Photonic crystal optical biosensor incorporating structured low-index porous dielectric. *Sensors and Actuators B: Chemical* 120:187-193.

[31] Park, H.G., S.H. Kim, S.H. Kwon, Y.G. Ju, J.K. Yang, J.H. Baek, S.B. Kim, and Y.H. Lee. 2004. Electrically driven single-cell photonic crystal laser. *Science* 305:1444-1447.

[32] Allan, D.C., N.F. Borrelli, J.C. Fajardo, D.W. Hawtof, and J.A. West. 2001. Photonic Crystal Fiber. U.S. Patent 6 243 522.

[33] Temelkuran, B., S.D. Hart, G. Benoit, J.D. Joannopoulos, and Y. Fink. 2002. Wavelength-scalable hollow optical fibres with large photonic bandgaps for CO_2 laser transmission. *Nature* 420:650-653.

[34] Dingle, R., W. Wiegmann, and C.H. Henry. 1974. Quantum states of confined carriers in very thin Al{x}Ga(1-x}As-GaAs-Al{x}Ga(1-x}As heterostructures. *Physical Review Letters* 33:827-829.

als on a single substrate by drastic increases in their critical thicknesses.[35,36,37] For materials grown on lattice-mismatched substrates, there are thickness limits below which single-crystal material can be grown on substrates with mismatch. The larger the mismatch is, the lower the critical thickness. Typically, this layer is very thin; for 2 percent lattice mismatch, the thickness is about 8 nm, at which the quantization energy can be easily observed. This is therefore called a strained quantum well. For typical III-V semiconductors, such as gallium arsenide (GaAs), indium phosphide (InP), or indium arsenide (InAs), on silicon (Si), the lattice mismatches are 4 percent, 8 percent, and 12 percent, respectively. Such mismatches are so large that the critical thicknesses are too small for device applications. This is not an issue for nanowires or nanopillars. For about 80-nm-diameter GaAs nanowires on Si, the height can be infinite without mismatches. Moreover, the core/shell layers grown later can accommodate much thicker lattice-mismatched materials. Various devices have already been demonstrated on such III-V nanopillars grown on silicon, including room-temperature operation of LEDs, avalanche photodiodes, and optically pumped lasers.[38,39] Such materials are grown at low enough temperature to be compatible with wafers with fabricated complementary metal-oxide-semiconductor (CMOS) circuits. That makes such nanostructured growth promising for integration of various III-V materials device structures on silicon.

Besides III-V on silicon, the recent advancement of germanium (Ge) or germanium tin (GeSn) grown on Si has enabled optoelectronics devices operating in the telecommunication region, 1.3 to about 1.6 µm,[40,41,42,43] to be fabricated in a

[35] Chuang, L.C., M. Moewe, C. Chase, N. Kobayashi, S. Crankshaw, and C. Chang-Hasnain. 2007. Critical diameter for III-V nanowires grown on lattice-mismatched substrates. *Applied Physics Letters* 90:043115.

[36] Glas, F. 2006. Critical dimensions for the plastic relaxation of strained axial heterostructures in free-standing nanowires. *Physical Review B* 74:121302.

[37] Chuang, L.C., M. Moewe, K.W. Ng, T. Tran, S. Crankshaw, R. Chen, W.S. Ko, and C. Chang-Hasnain. 2011. GaAs nanoneedles grown on sapphire. *Applied Physics Letters* 98:123101.

[38] Chuang, L.C., F.G. Sedgwick, R. Chen, W.S. Ko, M. Moewe, K.W. Ng, T.D. Tran, and C. Chang-Hasnain. 2011. GaAs-based nanoneedle light emitting diode and avalanche photodiode monolithically integrated on a silicon substrate. *Nano Letters* 11(2):385-390.

[39] Chen, R., T.-T.D. Tran, K.W. Ng, W.S. Ko, L.C. Chuang, F.G. Sedgwick, and C. Chang-Hasnain. 2011. Nanolasers grown on silicon. *Nature Photonics* 5:170-175.

[40] Fidaner, O., A.K. Okyay, J.E. Roth, R.K. Schaevitz, Y. Kuo, K.C. Saraswat, J.S. Harris, and D.A.B. Miller. 2007. Ge-SiGe quantum-well waveguide photodetectors on silicon for the near-infrared. *IEEE Photonics Technology Letters* 19:1631-1633.

[41] Tang, L., S.E. Kocabas, S. Latif, A.K. Okyay, D. Ly-Gagnon, K.C. Saraswat, and D.A.B. Miller. 2008. Nanometre-scale germanium photodetector enhanced by a near-infrared dipole antenna. *Nature Photonics* 2:226-229.

[42] Tsybeskov, L., and D.J. Lockwood. 2009. Silicon-germanium nanostructures for light emitters and on-chip optical interconnects. *Proceedings of the IEEE* 97:1284-1303.

[43] Michel, J., J. Liu, and L.C. Kimerling. 2010. High-performance Ge-on-Si photodetectors. *Nature Photonics* 4:527-534.

way that is compatible with CMOS electronic device fabrication, opening many new possibilities for the intimate integration of optoelectronics and electronics in manufacturable platforms, Ge-nanostructured emitters[44] with high luminescence quantum efficiency, and Ge photodetectors with large 3-dB bandwidth and high responsivity have been demonstrated all on an Si-compatible platform. Strong optical modulation mechanisms previously observed practically only in III-V materials have been demonstrated in Ge quantum well layers grown on silicon.[45] Such Ge-based materials and structures on silicon are thus very strong candidates for dense integration of optoelectronics and electronics in such applications as optical data interconnections.

Another special feature of nanostructured materials is their large surface-to-volume ratio. For optoelectronic devices, that is detrimental inasmuch as the surface recombination represents dark current[46] and non-radiative recombination. However, for sensors and applications requiring large surface-to-volume ratio, it is very desirable.[47]

The full potential for nanostructured materials is still limited to some extent by non-uniformities and large numbers of defects in currently available materials. In addition, the smaller structures have an associated larger quantization energy; hence the greater the impact of non-uniformity of size.

There are two main approaches to synthesizing nanostructured materials: bottom-up and top-down. The former refers to various self-assembly methods, including molecular-beam epitaxy, metal-organic chemical vapor deposition (CVD), catalyst CVD, electrodeposition, pulsed-laser synthesis, and solution-based synthesis. Small and nearly uniform (for example, a few percent to 10 percent non-uniformity) particle-like structures can be created by using chemical solutions.[48] Such QDs have seen wide applications in biosensing and bioimaging applications, where the uniformity requirement is not very stringent.[49,50] They face major challenges if one is to make electrical contacts to them or form p-n junctions

---

[44] Tang et al. 2008. Nanometre-scale germanium photodetector enhanced by a near-infrared dipole antenna.

[45] Kuo, Y.-H., Y.-K. Lee, Y. Ge, S. Ren, J.E. Roth, T.I. Kamins, D.A.B. Miller, and J.S. Harris. 2005. Strong quantum-confined Stark effect in germanium quantum-well structures on silicon. *Nature* 437:1334-1336.

[46] Dark current is current present even in the absence of a direct source.

[47] Medintz, I.L., A.R. Clapp, H. Mattoussi, E.R. Goldman, B. Fisher, and J.M. Mauro. 2003. Self-assembled nanoscale biosensors based on quantum dot FRET donors. *Nature* 2:630-638.

[48] Alivisatos, P. 2000. Colloidal quantum dots: From scaling laws to biological applications. *Pure and Applied Chemistry* 72:3-9.

[49] Michalet, X., F.F. Pinaud, J.M. Tsay, S. Doose, J.J. Li, G. Sundaresan, A.M. Wu, S.S. Gambhir, and S. Weiss. 2005. Quantum dots for live cells, in vivo imaging, and diagnostics. *Science* 307:538-544.

[50] Medintz, I.L., H.T. Uyeda, E.R. Goldman, and H. Mattoussi. 2005. Quantum dot bioconjugates for imaging, labelling and sensing. *Nature Materials* 4:435-446.

with them, however. Hence, they have limited device or integrated optics applications. Self-assembled InAs QDs can be grown on GaAs substrates through the Stranski-Krastanov growth mode.[51,52] This type of QD has about 5 to 10 percent non-uniformity. However, recent advances have been made in diode lasers and semiconductor optical amplifiers[53,54] to exploit QDs. These are very promising for deployment in the near future.

The top-down approach involves various lithography techniques to define the structures, using e-beam lithography, nanoimprint,[55,56] or dip pen technologies[57] and subsequent etching or growth of materials. These methods can lead to higher uniformity, but they still suffer from surface defects.[58,59]

OPTICAL MATERIALS IN THE LIFE SCIENCES AND SYNTHETIC BIOLOGY

Synthetic biology is a new field of biological research and technology development that combines science and engineering with the goal of designing and constructing novel and useful biological systems not found in nature. Synthetic biology has provided the means of both genetically engineering specific optical properties

[51] Stranski, Ivan N., and Lubomir Krastanov. 1938. Abhandlungen der Mathematisch-Naturwissenschaftlichen Klasse IIb. *Akademie der Wissenschaften Wien* 146:797-810.

[52] Eaglesham, D.J., and M. Cerullo. 1990. Dislocation-free Stranski-Krastanow growth of Ge on Si(100). *Physical Review Letters* 64:1943-1946.

[53] Zhukov, A.E., A.R. Kovsh, V.M. Ustinov, Y.M. Shernyakov, S.S. Mikhrin, N.A. Maleev, E.Y. Kondrat'eva, D.A. Livshits, M.V. Maximov, B.V. Volovik, D.A. Bedarev, Y.G. Musikhin, N.N. Ledentsov, P.S. Kop'ev, Z.I. Alferov, and D. Bimberg. 1999. Continuous-wave operation of long-wavelength quantum-dot diode laser on a GaAs substrate. *IEEE Photonics Technology Letters* 11:1345-1347.

[54] Akiyama, T., M. Ekawa, M. Sugawara, K. Kawaguchi, Hisao Sudo, A. Kuramata, H. Ebe, and Y. Arakawa. 2005. An ultrawide-band semiconductor optical amplifier having an extremely high penalty-free output power of 23 dBm achieved with quantum dots. *IEEE Photonics Technology Letters* 17:1614-1616.

[55] Vieu, C., F. Carcenac, A. Pepin, Y. Chen, M. Mejias, A. Lebib, L. Manin-Ferlazzo, L. Couraud, and H. Launois. 2000. Electron beam lithography: Resolution limits and applications. *Applied Surface Science* 164:111-117.

[56] Colburn, M., S.C. Johnson, M.D. Stewart, S. Damle, T.C. Bailey, B. Choi, M. Wedlake, T.B. Michaelson, S.V. Sreenivasan, J.G. Ekerdt, and C.G. Willson. 1999. Step and flash imprint lithography: A new approach to high-resolution patterning. *Proceedings of the SPIE* 3676:379.

[57] Lee, K.B., S.J. Park, C.A. Mirkin, J.C. Smith, and M. Mrksich. 2002. Protein nanoarrays generated by dip-pen nanolithography. *Science* 295:1702-1705.

[58] Cao, X.A., H. Cho, S.J. Pearton, G.T. Dang, A.P. Zhang, F. Ren, R.J. Shul, L. Zhang, R. Hickman, and J.M. Van Hove. 1999. Depth and thermal stability of dry etch damage in GaN Schottky diodes. *Applied Physics Letters* 75:232-234.

[59] Tanaka, S., Y. Kawaguchi, N. Sawaki, M. Hibino, and K. Hiramatsu. 2000. Defect structure in selective area growth GaN pyramid on (111)Si substrate. *Applied Physics Letters* 76:2701.

into living organisms and manufacturing optically active materials, both of which are now routinely used in life science research.

One of the major advances of the last two decades in the life sciences (the subject of a Nobel Prize in 2008) was the development of genetic engineering techniques that allow the programming of individual cells to produce protein molecules that fluoresce in response to specific stimuli. The approach is used for optical detection of the turning on and turning off of specific genes in cells in response to drugs or environmental conditions. Similarly, neurons engineered to produce fluorescent dyes based on proteins that report the active and inactive states of neurons in living animals are being used to map out the neuronal wiring of living brains and to track the flow of information through neural circuits in live animals.

New, highly efficient inorganic dyes have been developed that can be chemically linked to nucleic acids and provide an optical readout of nucleic acid sequences in high-throughput DNA sequencing instruments. The optical materials are a critical element in the technology that will ultimately enable the $1,000 genome, which will help to make possible a new era of personalized medicine.[60]

New nanostructured materials have also demonstrated new methods for labeling cells and intracellular organelles with biocompatible optical materials by using nano-scale semiconductor structures (QDs) and nanometer gold spheres and rods. The new materials provide several advantages, including greatly improved resistance to photo-bleaching, tunable and very narrow spectral features, and the ability to functionalize the nanoparticle surface with antibodies to allow it to bind specifically to a wide variety of biological surfaces.

FINDINGS

This chapter illustrates the strategic role that materials and nanostructuring can play in the development of new optical systems. Below are the findings of the committee regarding strategic materials for optics.

Finding: There is much promise in tailoring existing materials in novel ways to produce innovative results. The new metamaterials and photonic nanostructures enable original optical properties that can be developed for innovative functions that could not be exhibited in traditional materials. Realizing the full potential of nanostructured materials is still hampered by non-uniformities and many defects. The smaller the structure, the more issues with non-uniformity.

Finding: Gallium, germanium, indium, selenium, silver, and tellurium are all critical elements for development of thin-film photovoltaics (TFPV).

[60] For more information, see discussions in Chapter 6 and Appendix C of this report.

Finding: Most current white-light LEDs require rare materials to fluoresce in the proper wavelengths. Many of the efficient phosphors contain rare earth elements. The two key rare earth materials that provide color LED lighting are europium and terbium.

RECOMMENDATIONS

The committee presents the following recommendations with respect to strategic materials for optics.

Recommendation: The U.S. R&D community should increase its leadership role in the development of nanostructured materials with designable and tailorable optical material properties, as well as process control for uniformity of production of these materials.

Recommendation: The United States should develop a plan to ensure the availability of critical energy-related materials, including solar cells for energy generation and fluorescent materials to support future LED development.

10

Displays

INTRODUCTION

Displays are a critical enabling technology for the information age. Over the last two decades, liquid-crystal displays (LCDs) have become the dominant type of display, displacing the cathode-ray tube (CRT). During this period, LCD technology improved dramatically in several aspects, including resolution, quality, reliability, size, cost, and capability. The 1998 National Research Council report *Harnessing Light: Optical Science and Engineering for the 21st Century*[1] mentioned 10.4-inch LCD displays and postulated the potential for 40-inch plasma displays. Today, LCD displays are ubiquitous. Small ones are used for mobile appliances (cell phones) and e-readers, and large ones for desktop computers, table-based displays, TVs, and wall-mounted displays. However, there is always a drive toward bigger displays.[2]

LCDs will probably dominate displays for the next decade. There are technical challenges in LCD displays that can be turned into opportunities. For a detailed description of the various display technologies, see Appendix C in this report.

Although they are less developed than LCDs, the use of organic light-emitting diodes (OLEDs) is a growing trend in displays. The fact that the principal pixel component in an OLED display is an LED eliminates the need for external back-

[1] National Research Council. 1998. *Harnessing Light: Optical Science and Engineering for the 21st Century.* Washington, D.C.: National Academy Press.

[2] Ultimately, if large generic displays were available at low cost, they could replace larger and larger information devices until movie theater systems and billboards were included.

lighting technology. However, several challenges need to be solved before OLED technology becomes pervasive for large displays, as discussed later in this chapter.

One of the most important trends in recent years is the rapid spread of touch displays. Touch technology already has matured to a point where its performance, reliability, and robustness are outstanding. Touch technology was invented in the United States, and many improvements are still coming from U.S. academic and industrial research and development (R&D) laboratories.[3] Given that significant innovations and improvements are expected in the next decade, this field has potential for continued U.S. leadership.

One of the technologies changing how people study, spend leisure time, and get information involves e-readers and computer tablets. Within the next generation, it is possible that people will be dependent on e-readers and tablets for many of their educational and general reading needs. With respect to e-readers, a revolution is taking place, from a simple electrophoretic ink (E-ink) product, to a higher-resolution E-ink, to color-passive LCD, to full-blown LCD-based e-readers. As the displays are improved, there are opportunities to develop low-cost, durable, and green materials and manufacturing processes.

An interesting technology that might have a major impact in the future but will require major investment and innovation is that of flexible displays. Flexible material (such as glass or a polymer) could be used for newspapers, magazines, and work papers. The idea is to simulate the paper media that many people are still using, but with a flexible organic or inorganic material that is essentially a computer system with a wireless transceiver. Such flexible displays could communicate with servers to download information and display it as though it were printed on paper. The flexible materials should be lightweight, sturdy, robust, and reusable and should provide exceptional text, picture, and video quality whether indoors or in bright sunlight (similar to printed-paper quality). To create flexible displays, however, it is necessary to develop flexible substrate materials and processes that exhibit chemical resistance, thermal stability, and endurance of high temperatures and pressures during fabrication.

Most of the current state-of-the-art display technologies require backlighting subsystems to provide the necessary brightness. LCDs do not produce light themselves, and so they need illumination to produce a visible image. In general, backlights are needed to illuminate the display panel from the side or the back.

---

[3] A Scopus search on "touch display" gives a total of 2,706 publications, of which the following institutions are the top 10 publishers: University of Tokyo (47), Massachusetts Institute of Technology (39), University of British Columbia (29), University of Calgary (23), University of Toronto (22), KAIST (formerly Korea Advanced Institute of Science and Technology) (22), University of Oxford Medical Sciences Division (20), Stanford University (20), Microsoft Research (20), and Carnegie Mellon University (19). SOURCE: Scopus database. Available at http://www.scopus.com/home.url. Accessed November 28, 2011.

Several backlighting technologies exist, and they depend critically on advances in optical and photonic technologies. Solid-state lighting (LEDs) is now replacing most other lighting approaches for backlighting.[4,5]

Another technology with potential for big future impact and leadership by the United States is three-dimensional holographic displays.[6] Uses today are mainly for military applications, but the technology could become widely available if the technology advances and cost declines.

THE NEAR FUTURE

The next few years should see a continuation of the trend of ever-larger displays at lower prices. In addition, higher-quality LED illumination with higher refresh rates (up to 240 hertz [Hz]) will be developed. In the United States, the display market has transitioned to flat panels; LCD displays now account for a high percentage of new sales. Plasma captures a smaller share of the TV market and is expected to decline as a fraction of the total TVs produced.[7] There is considerable interest in OLED TVs, but they are at least several years away from capturing a significant share of the market. However, when the remaining technical challenges are solved, OLED displays should become a significant competitor to LCD displays.[8] Another big trend is "three-dimensional-ready" televisions, although most of them are limited by the need for special glasses for viewing. Innovations are needed for three-dimensional displays that do not require special glasses. Additional trends for display products include thinner and lower-power-consumption displays. Several countries are beginning to impose restrictions on the total power consumption of display units, forcing manufacturers to come up with better designs, especially for lighting, which consumes approximately 25 to 40 percent of the total display unit power.[9] On the notebook computer side, where low power consumption is a major consideration, LED lighting is the dominant technology. For desktops, fluorescent lighting is still being used because thickness and power consumption

---

[4] Anandan, M. 2008. Progress of LED backlights for LCDs. *Journal of the Society for Information Display.* Volume 16(2):287-310.

[5] Flanders, V. 2012. *Liquid Crystal Display and Technology.* New Delhi: World Technologies.

[6] Zebra Imaging. 2012. Available at http://www.zebraimaging.com. Accessed July 31, 2012.

[7] In 2010, LCDs increased by 31 percent to 187.9 million units, while plasma grew by 29 percent to 18.24 million units. Data from Displaybank's Month-to-Month FPD TV Shipment Information report; see Slipokoyenko Y., ed. 2012. *Plasma vs LCD vs LED.* Available at http://www.hdtvreviewspro.com/plasma-vs-lcd-vs-led/. Accessed August 2, 2012.

[8] Semenza, P. 2010. Display week 2010 review and market outlook: Can OLED displays make the move from the mobile phone to the TV? *Information Display* 26(8):14-17.

[9] Lubarsky, G. 2011. "A Holistic Approach to Reducing Backlight Power Consumption." *How 2 Power Today.* Available at http://www.how2power.com/newsletters/1110/articles/H2PowerToday1110_design_TexasInstruments.pdf. Accessed November 28, 2011.

are not critical. Higher refresh rates and three-dimensional displays are beginning to penetrate the desktop market.

Although traditionally associated mostly with "science fiction" movies, the ability to create and view three-dimensional optical holographic images has potential applications in which full three-dimensional viewing is of great benefit, such as in defense, telemedicine, high-skill-level training, and collaborative rapid prototyping. A U.S. company, Zebra Imaging,[10] is taking a leadership role in trying to commercialize this technology, and its ZScape holographic motion displays were named by *Time* magazine as one of the top 50 inventions of 2011.[11] With this technology, teams of people can collaborate in real time by viewing lifelike, interactive images from any angle and with fine detail.

OVERVIEW OF DISPLAYS

This section briefly describes the different display technologies. For a more in-depth description, see Appendix C.

Liquid-Crystal Displays

Polarized light and its manipulation are fundamental to the operation of an LCD. Such manipulation allows the creation of light valves (devices for varying the amount of light from an illumination source to a desired target), with millions of miniature valves in a single high-definition display.[12] To achieve color, each pixel is subdivided into subpixels, with red, green, and blue (RGB) filters. Transistors on a glass substrate control the voltage and thus the amount of light passed through each subpixel. The early motivation for color LCDs was use in notebook computing. Improvements in LCDs resulted in wider applications and led to the displacement of competing technologies, including CRTs in traditional television sets.

There has been at least some market interest in extending LCDs to three-dimensional displays, exploiting polarization in a different way.[13] In one technique, a polarizer on the glasses worn by the viewer matches the orientation of the polarized output of the display. Another possible approach for creating three-

[10] Zebra Imaging. 2012. "Motion Displays." Available at www.zebraimaging.com/products/motion-displays. Accessed July 30, 2012.

[11] Grossman, L., M. Thompson, J. Kluger, A. Park, B. Walsh, C. Suddath, E. Dodds, K. Webley, N. Rawlings, F. Sun, C. Brock-Abraham, and N. Carbone. 2011. Top 50 inventions. November 28. *Time* magazine. Available at http://www.time.com/time/magazine/article/0,9171,2099708,00.html. Accessed November 15, 2012.

[12] For information on the control of polarization in LCD to define color, see Appendix C.

[13] For more information, see Appendix C.

dimensional displays is the use of lenticular arrays[14,15] to create a fixed number of viewing zones (usually about seven to nine) in which the left and right images are coherent although with reduced resolution (see Figure 10.1). At present, the acceptance of the latter approach has been limited by restriction of the positions from which the display can be observed; image quality suffers significantly if an improper position is used.

Another trend for LCDs is in providing greater color fidelity than is currently possible. That is related to another recent trend: the replacement of fluorescent tubes with LEDs in the display backlight. The most common backlight now consists of white LEDs, which themselves consist of blue emitters passing through a color-converting phosphor, typically cerium-doped yttrium aluminum garnet (Ce:YAG).

A problem with such white LEDs in an LCD backlight application is that the color filters are not perfect. The light from the green and the red filters overlaps with yellow wavelengths passed by both filters. That results in colors that are less true, so, for example, a completely green image appears more yellow-green than true green.

It has been suggested that the color-filter deficiencies might be addressed by substituting LEDs that emit only in relatively narrow wavelength bands of RGB. Although that might be done with RGB rather than white LEDs, another possibility is that quantum dots (QDs) would substitute for phosphor (see Figure 10.2). QDs would have the advantage of providing much more precise and narrow-wavelength color conversion so that emission of the white LED with QD color conversion would not need to emit at all in the yellow band where the green and red filters overlap. The result would be more pure colors and a broader color gamut.

Touch Displays

Of the several kinds of touch technologies, three are most prevalent: resistive, in which a touching object presses down and causes two separated layers to connect to each other at that point; optical, in which a touching object appears as a shadow, and cameras can locate and measure the size of the object; and capacitive, in which an electrically conducting finger touching a screen results in a distortion of the screen's electrostatic field and a change in capacitance. Resistive touch technology was the first to be used, most commonly in bank automated teller machines and cash registers. It has low cost but low resolution. Optical touch technology was invented and commercialized (for example, the HP 150 PC in 1984) with much

[14] Johnson, R.B., and G.A. Jacobsen. 2005. Advances in lenticular lens arrays for visual display. *Proceedings of SPIE—The International Society for Optical Engineering*. Volume 5874.

[15] Kim, H., and J. Hahn. 2010. Optimal design of lens array for wide-viewing angle multi-view lenticular 3D displays. *Proceedings of the International Meeting on Information Display and International Display Manufacturing Conference and Asia Display*, October 11-15, 2010, Seoul, Republic of Korea.

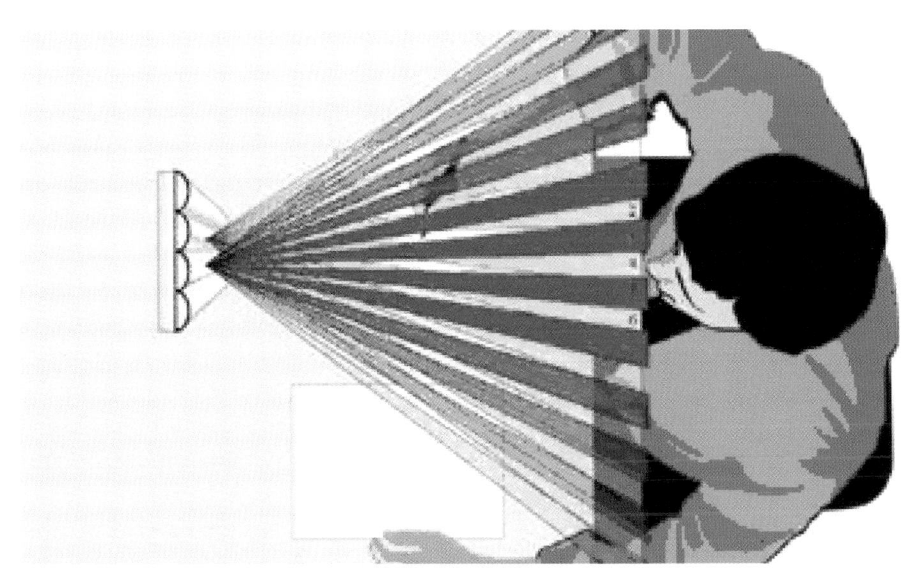

FIGURE 10.1 Multiple autostereoscopic viewing zones with lenticular array. SOURCE: International Organisation for Standardisation (ISO) and Organisation Internationale de Normalisation (IEC). 2008. "Introduction to 3D Video." Available at http://mpeg.chiariglione.org/technologies/general/mp-3dv/index.htm. Reprinted with permission.

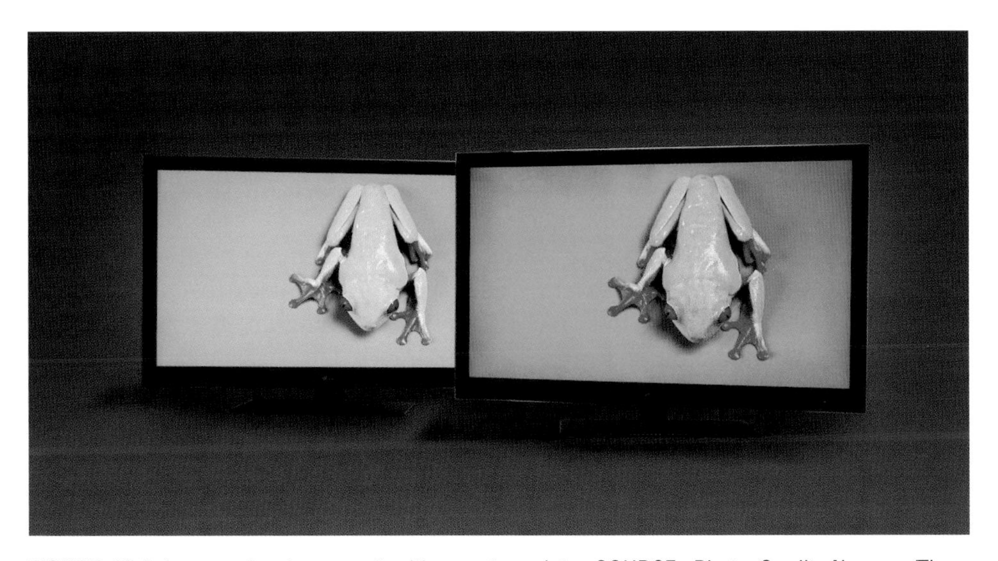

FIGURE 10.2 Improved color gamut with quantum dots. SOURCE: Photo Credit: Nanosys/Thom Sanborn. Reprinted with permission.

better resolution and performance, but it was not easily scaled to larger displays. Not until capacitive touch technology was commercialized on a large scale (for example, the Apple iPhone) did consumers begin to expect mobile appliances and laptops to incorporate touch. Regardless of the sensing mechanism (resistive, capacitive, or optical), touch technologies have two important aspects that rely on optics: the display function (brightness, contrast, resolution, refresh rate, and so on) and the construction of the touch sensors.

Capacitive coupling in touch displays exploits the fact that the signal driven on one conductor will couple to an adjacent conductor. Before 2007, this phenomenon was not widely perceived as a good means of implementing touch, although it had been used on all the notebook touchpads since the mid-1990s and on the iPod since 2001. However, as a result of Apple's introducing its first iPhone, which incorporated projected capacitive multi-touch, considerable new interest in this technology has been generated.

In projected capacitive touch, the display area is divided into a grid of unit cells of roughly 5 mm × 5 mm each, the size commonly selected for the typical size of a fingertip.[16] The most common implementation of projected capacitive touch involves a touch sensor array bonded directly behind a cover glass. In the display aperture area, the conductors are transparent and typically made of indium tin oxide (ITO).[17,18] The fundamental challenge in projected capacitive touch technology is the poor conductivity of the transparent conductors. The technology requires that the touching instrument (digit) be a conducting one. Thus, a gloved hand or nonmetallic pen would not be detected. In addition, the system is typically optimized for the size of a finger, and so using a stylus, even if it is conductive, would result in errors due to poor interpolation. One potential substitute for capacitive touch is optical, but thus far the application of optical sensors has been limited mostly to very large displays. In wall-size optical touch displays, infrared (IR) LEDs and imagers are typically positioned in the upper corners of the array, and retroreflectors are along the sides and bottom. The reflectors return the IR to the imagers unless interrupted by the shadow of a touch.

A plastic front polarizer, which would be prone to damage from touch unless protected by a cover glass, is currently used for LCDs of all sizes. One potential trend could be to eliminate this polarizer by substituting an internal polarizer. Although this would be difficult to achieve, an internal polarizer would sit on the liquid-crystal side of the front color filter glass. With that configuration, the face of

---

[16] For details about touch displays, their composition, and challenges faced in their development, see Appendix C.

[17] Chopra, K.L., S. Major, and D.K. Pandya. 1983. Transparent conductors—A status review. *Thin Solid Films* 102(1):1-46.

[18] Granqvist, Claes G. 2007. Transparent conductors as solar energy materials: A panoramic review. *Solar Energy Materials and Solar Cells* 91(17):1529-1598.

the display would be the color filter glass rather than the polarizer plastic. Achieving such a substitution would motivate efforts to eliminate the stand-alone cover glass and integrate the protection of a cover into the color filter glass itself.

OLED Displays

OLEDs make use of the electrophosphorescence exhibited by some organic molecules to create light directly at the diode itself when desired; the color of the diode can be adjusted by the choice of organic molecules. Therefore, an array of OLEDs can directly generate the displayed image rather than light a modulator, such as an LCD, to create the displayed image. Because the image is directly generated, light is not wasted by attenuating it with the modulator that creates the image, and these displays can be thinner because generating the image is a single-step process. OLED-based displays can therefore be brighter than LCDs and consume less power.[19,20] Two principal ways are used to supply the electrical bias to the individual OLEDs in a display: passive- and active-matrix designs.[21] In a passive-matrix scheme, the individual anodes and cathodes are connected by perpendicular conducting strips, and external circuitry applies voltages to the rows and columns that are necessary to determine which pixels will be turned on. At present, the external circuitry required for a passive-matrix design results in higher power consumption than is needed with an active-matrix display. Active-matrix displays incorporate a thin-film transistor (TFT) array as part of the device to supply the power to the individual OLEDs. This biasing scheme uses less power than is needed for the passive-matrix design and has a higher refresh rate, and so it is suitable for large displays.[22]

Because non-rigid substrates can be used, OLED displays can be more robust than displays that incorporate glass. That makes OLEDs potentially useful in applications in which they will be subjected to rough handling, such as in cell phones or other consumer electronics. Furthermore, in the case of completely flexible substrates, the displays could be used in applications in which no other display currently can be contemplated, such as being integral with an article of clothing.

OLEDs offer several advantages over both LCDs and LEDs for small and large

---

[19] Kelley, T.W., P.F. Baude, C. Gerlach, D.E. Ender, D. Muyres, M.A. Haase, D.E. Vogel, and S.D. Theisset. 2004. Recent progress in organic electronics: Materials, devices, and processes. *Chemistry of Materials* 16(23):4413-4422.

[20] Iwamoto, M., Y. Kwon, and T. Lee. 2011. *Nanoscale Interface for Organic Electronics*. Singapore. Hackensack, N.J.: World Scientific.

[21] Ju, S., J. Li, J. Liu, P. Chen, Y. Ha, F. Ishikawa, H. Chang, C. Zhou, A. Facchetti, D.B. Janes, and T.J. Marks. 2008. Transparent active matrix organic light-emitting diode displays driven by nanowire transistor circuitry. *Nano Letters* 8(4):997-1004.

[22] For more details about OLED displays, see Appendix C.

displays: they can have substrates that are flexible and more impact-resistant than glass or rigid plastic; they have higher brightness than LEDs; they do not require backlighting, as LCDs do, which is a significant power drain that limits battery life; they offer ease of deposition on very large substrates; and they have a larger field of view and a wider operating temperature range than LCDs. However, present OLEDs also have significant disadvantages, including shorter lifetimes[23,24] and lower efficiency.

Flexible Displays

Future flexible displays could be extremely thin, light, and inexpensive. Moreover, flexible displays could enable the display market to expand to new applications, including e-paper and large signage.

In recent years, engineers worldwide have been developing the organic materials and manufacturing processes to make flexible displays a reality. Such new materials could be OLEDs, liquid crystals, and electrophoretic particles. Moreover, plastic substrates and printable semiconductors are now available to help to create flexible back planes. Other key technologies are needed for future flexible LCDs or OLEDs, such as the use of stable and heat-resistant organic materials and low-temperature printing.

In the next decade, flexible substrates (such as thin-film polymers) could have a large impact, enabling the possibility of printing reports on e-paper[25,26] that is both thin and reusable. Cell phones could have large displays that could be unfurled. Displays might even be placed on non-flat clothing surfaces. Commercial laboratories are working on flexible displays[27] that can be 15-inch displays and larger.

---

[23] Wellmann, P., M. Hofmann, O. Zeika, A. Werner, J. Birnstock, R. Meerheim, G. He, K. Walzer, M. Pfeiffer, and K. Leo. 2005. High-efficiency p-i-n organic light-emitting diodes with long lifetime. *Journal of the Society for Information Display* 13(5):393-397.

[24] Luiz, Pereira. 2010. *Organic Light Emitting Diodes: The Use of Rare Earth and Transition Metals.* Singapore: Pan Stanford.

[25] Burns, S.E., K. Reynolds, W. Reeves, M. Banach, T. Brown, K. Chalmers, N. Cousins, M. Etchells, C. Hayton, K. Jacobs, A. Menon, S. Siddique, P. Too, C. Ramsdale, J. Watts, P. Cain, T. Von Werne, J. Mills, C. Curling, H. Sirringhaus, K. Amundson, and M.D. McCreary. 2005. A scalable manufacturing process for flexible active-matrix e-paper displays. *Journal of the Society for Information Display* 13(7):583-586.

[26] Blankenbach K., L-C Chien, S.-D. Lee, and M.H. Wu. 2011. "Advances in Display Technologies, e-papers and Flexible Displays." *Proceedings of SPIE—The International Society for Optical Engineering.* January 26-27. San Francisco, Calif. Bellingham, Wash.: SPIE.

[27] Choi, M.-C., Y. Kim, and C.-S. Ha. 2008. Polymers for flexible displays: From material selection to device applications. *Progress in Polymer Science* 33(6):581-630.

Projection Displays

Although projectors are widely used in schools and for presentations to large audiences, price has been the major factor in their adoption rather than the size, the power consumption, or, to some extent, the brightness of these projectors. In contrast, size, power consumption, and brightness are the dominant technical specifications that inhibit the incorporation of projectors in handheld devices of various types. The market for such battery-powered "personal projectors" is expected to be much larger than that for "traditional" mains-powered projectors.[28]

Currently, three major technologies are used to project images that are created or transmitted electronically: digital light processors (DLPs), which use an array of micromirrors to create an image by the reflection of light illuminating the mirrors; liquid crystal on silicon (LCoS), which creates an image by modulating the intensity of the light by means of liquid crystals; and beam steering of one or more lasers, which is used to "write" an image directly on a screen. All three are candidates for future development for incorporation in handheld devices. Such devices have strict requirements on energy consumption, as well as size.

Each of those technologies has advantages and disadvantages, and there is no clear overall winner yet. DLP projectors incorporating RGB LEDs have high image quality and low power consumption, but they have lower resolution than LCoS projectors. Laser scanning also has low power consumption, but laser speckle noise affects image quality. In all cases, the challenge is to provide an image that is large enough to be advantageous over a handheld device's built-in screen and that is bright enough to be seen under less than ideal lighting conditions and is of a quality that end users consider acceptable and to provide this capability without unacceptable battery drain.

Three-Dimensional Holographic Displays

A display can sometimes create the illusion of being three-dimensional when a stereoscopic technology is used to produce images. The human visual system, however, experiences a two-dimensional plane, and it is not true three-dimensional images that are being displayed. In contrast, modern holographic displays are able to produce true three-dimensional images (or holograms) that do not require any

[28] The technology for storing images has completely outpaced the technology of the projectors needed to display them. A carousel holding 140 35-mm slides is comparable in volume to that of the projector. Similarly, 140 plastic foils are comparable in volume to that of a portable overhead projector. In contrast, today it is possible to carry more than 10,000 high-resolution images on a tiny flash drive. This capability is driving the desire to create portable projectors, comparable in size to a cell phone, that are capable of displaying these electronically stored images.

special eyewear (see the two images in Figure 10.3.[29] The first interactive three-dimensional holographic system was demonstrated in 1990 at the Massachusetts Institute of Technology Media Lab.[30] Note that a simple hologram can be created by superimposing two coherent plane waves; the resulting interference pattern produces a straight-line fringe across the recording medium.

In an electroholographic display, the three-dimensional object is converted into a fringe pattern. To do that, a computer graphic stage is used to perform lighting, shading, occlusion, and rendering to two-dimensional images. For some cases, such as magnetic resonance imaging (MRI) data, this stage is trivial inasmuch as the data may already be stored as three-dimensional voxels, and each voxel might already have the color and shading information. Later, the fringe generation stage computes a large two-dimensional holographic fringe based on the result of the computer graphic stage. It is worth mentioning that significant computing power is necessary to produce realistic computer-generated holograms (real-time three-dimensional displaying and telepresence), and recent progress in digital signal processing has enabled the present advances.

After generation of fringe patterns, spatial-light modulators (SLMs) are used to convert the three-dimensional information in electronic bits into optical photons with a computed holographic fringe pattern. It should be noted that either amplitude modulation or phase modulation can be used in three-dimensional holographic imaging.[31] LCDs tend to use phase modulators because of their optical efficiency, and they can have many gray-scale levels that are useful for generating complex images. Acousto-optic modulator SLMs have also been used in holovideo because of their fast response, and this technique was used in the first demonstration of three-dimensional holographic displays (see Figure 10.4).[32]

The applications of three-dimensional displays are potentially far-reaching and include medical training, military simulation, situational awareness, and entertainment.

---

[29] Blanche P.-A., A. Bablumian, R. Voorakaranam, C. Christenson, W. Lin, T. Gu, D. Flores, P. Wang, W.-Y. Hsieh, M. Kathaperumal, B. Rachwal, O. Siddiqui, J. Thomas, R.A. Norwood, M. Yamamoto, and N. Peyghambarian. 2010. Holographic three-dimensional telepresence using large-area photorefractive polymer. *Nature* 468:80-83.

[30] Reichelt, S., R. Haussler, N. Leister, G. Futterer, and A. Schwerdtner. 2008. "Large Holographic 3D Displays for Tomorrow's TV and Monitors—Solutions, Challenges, and Prospects." *IEEE Lasers and Electro-Optics Society Annual Meeting*, November 9-13, 2008, Newport Beach, Calif.

[31] Lucente, M. 2003. "Interactive Holographic Displays: The First 10 Years." In *Holography: The First 50 Years*, Fournier, J.M., ed. Berlin: Springer-Verlag.

[32] Lucente, M. 2003. "Interactive Holographic Displays: The First 10 Years."

FIGURE 10.3 Pictures of colored holograms. (*Left*) Hologram of two model cars recorded. (*Right*) Hologram of a vase and flowers. SOURCE: Reprinted, with permission, from Blanche P.-A., A. Bablumian, R. Voorakaranam, C. Christenson, W. Lin, T. Gu, D. Flores, P. Wang, W.-Y. Hsieh, M. Kathaperumal, B. Rachwal, O. Siddiqui, J. Thomas, R.A. Norwood, M. Yamamoto, and N. Peyghambarian. 2010. Holographic three-dimensional telepresence using large-area photorefractive polymer. *Nature* 468:80-83.

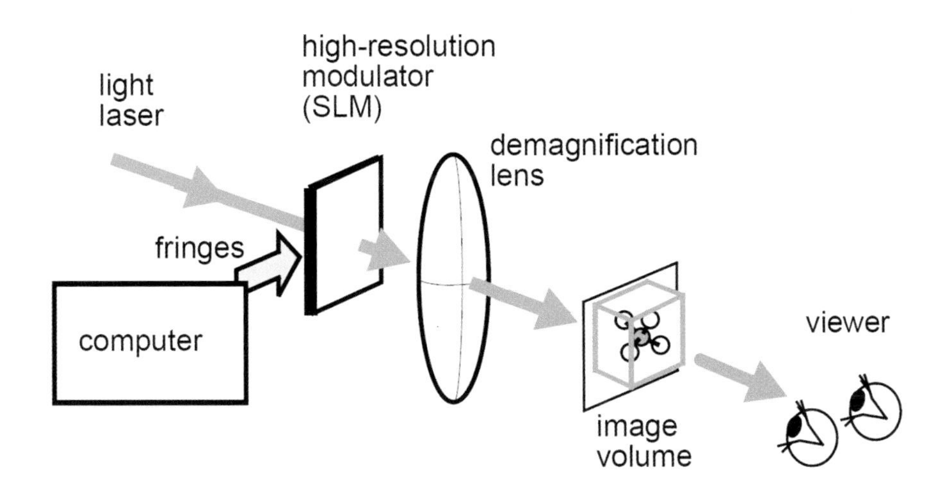

FIGURE 10.4 Holographic optical modulation using a typical high-resolution spatial-light modulator (SLM). At present, a minimum of 2 million modulation elements is required to produce even a small three-dimensional image the size of a thumb. SOURCE: Lucente, M. 2003. "Interactive Holographic Displays: The First 10 Years." In *Holography: The First 50 Years*, Fournier, J.M., ed. Berlin: Springer-Verlag. Reprinted with permission.

DISPLAY PRODUCT MANUFACTURING

As discussed in Chapter 7, the bulk of LCD panel production occurs outside the United States; the leading countries are Korea, Taiwan, Japan, and China.[33] In 1998, when *Harnessing Light* was published, most production had already moved outside the United States. At that time, Japan was dominant, and Korea was just starting to play a big role. The same holds true in the production of touch sensors; Taiwan holds a slightly stronger position than the others. The segment of the touch-display technologies in which the United States leads is the touch controller circuits, in whose production the major firms are Atmel,[34] Broadcom,[35] and Synaptics.[36] In the United States, Corning is producing large sheets of the high-quality glass needed for the display panels (color filter and thin-film transistors) in factories in Kentucky.

The committee believes that the role best played by U.S. companies in the future of displays is in the research and development of new technologies. The committee would like to see a future in which new manufacturing remains in this country.

FINDINGS

Key Finding: Although traditional LCD manufacturing has migrated out of the United States, there are opportunities in future display technologies that can give U.S. industry competitive leadership advantages in R&D and manufacturing. They include energy-efficient backlighting, OLED displays and flexible displays, and real-time three-dimensional-holographic displays.

Finding: An overarching issue for U.S. leadership and revolutionary advances in displays involves new materials that are low cost, durable, green, and easy to process. For example, the dominant LCD technology consumes a significant amount of energy; backlighting subsystems consume about 25 to 40 percent of the total display power.

RECOMMENDATIONS

Solid-state lighting (with LEDs) is discussed in Chapter 5. Not only is great progress being made, but it is a field that the United States should emphasize.

[33] National Research Council. 2008. *Innovation in Global Industries: U.S. Firms Competing in a New World (Collected Studies)*, J.T. Macher and D.C. Mowery, eds. Washington, D.C.: The National Academies Press.

[34] Atmel Corporation. 2012. See http://www.atmel.com/Default.aspx/. Accessed July 31, 2012.

[35] Broadcom Corporation. 2012. See http://www.broadcom.com/. Accessed July 31, 2012.

[36] Synaptics, Inc. 2012. See http://www.synaptics.com/. Accessed July 31, 2012.

Recommendation: U.S. private display companies should invest in and take advantage of the development of solid-state lighting for more efficient and uniform illumination of displays.

Recommendation: U.S. private companies and the Department of Defense should ensure a leadership role by funding R&D related to new materials for flexible, low-power, holographic and three-dimensional display technologies.

Appendixes

A

Statement of Task, with Introductory Information

The statements below introduced and presented the statement of task for the Committee on Harnessing Light: Capitalizing on Optical Science Trends and Challenges for Future Research.

BACKGROUND

The National Research Council will convene a committee to: (1) Review updates in the state of the science that have taken place since publication of the National Research Council report *Harnessing Light* in 1998.[1] (2) Identify the technological opportunities that have arisen from recent advances in optical science and engineering. (3) Assess the current state of optical science and engineering in the United States and abroad. (4) Prioritize a set of research grand-challenge questions to fill identified technological gaps. (5) Recommend actions for the development and maintenance of global leadership in the photonics-driven industry.[2]

INTRODUCTION

The discipline of optical science and engineering (OSE) has been foundational for many of the scientific and technical advances of the past 100 years and

---

[1] National Research Council. 1998. *Harnessing Light: Optical Science and Engineering for the 21st Century.* Washington, D.C.: National Academy Press.

[2] See the full formal statement below, in the section "Statement of Task."

is one of the key transformative disciplines driving innovation in the 21st-century economy. In 1998, the National Research Council through the Board on Physics and Astronomy and the National Materials Advisory Board issued a landmark report, *Harnessing Light: Optical Science and Engineering for the 21st Century*, that (1) captured the importance of optics and photonics, (2) described the major challenges facing the field, (3) made strategic recommendations for national policy, and (4) impacted the discussions of institutions around the world. As the first comprehensive report in the field of optics, *Harnessing Light* clearly demonstrated that optics is an enabling technology to many educational, governmental, health care, industrial, and military organizations. While the original *Harnessing Light* report has been extremely useful to both U.S. and foreign, academic, industrial, and governmental organizations, in the past 10 years enormous progress has been made in photonics sciences and technologies. Irrespective of the economic conditions, OSE is headed toward another strong growth period, driven by developments in advanced materials, solid state lighting, solar technologies, sensors, lasers, imaging, fiber-optic communications, digital photography, diagnostic medicine, computing/processing, and consumer displays/TVs. The impacted markets encompass critical issues that affect society, ranging from energy, data storage, and health care to manufacturing, communications, and security. A revisiting of the technology and policy issues would be quite timely. The new report would address the role that photonics plays in national competitiveness and innovation. For example, the infusion of photonics-related technologies into mass-market applications will require careful thought in order to economically and appropriately harness the necessary multifunctional manufacturing capabilities. The new study could identify national strengths and weaknesses in relation to current and future needs including economic impact, workforce needs, and future research directions. Given that the 2007 National Research Council report *Controlling the Quantum World: The Science of Atoms, Molecules, and Photons*[3] includes much of the area of basic optical science, this study would consider the technology areas where optics is an enabler that can dramatically impact the economy of the country.

STATEMENT OF TASK

A committee of the National Academies will be convened to:

1. Review updates in the state of the science that have taken place since publication of the National Research Council report *Harnessing Light*;

[3] National Research Council. 2007. *Controlling the Quantum World: The Science of Atoms, Molecules, and Photons*, Washington, D.C.: The National Academies Press.

2. Identify the technological opportunities that have arisen from recent advances in and potential applications of optical science and engineering;
3. Assess the current state of optical science and engineering in the United States and abroad, including trends in private and public research, market needs, examples of translating progress in photonics innovation into competitiveness advantage (including activities by small businesses), workforce needs, manufacturing infrastructure, and the impact of photonics on the national economy;
4. Prioritize a set of research grand-challenge questions to fill identified technological gaps in pursuit of national needs and national competitiveness;
5. Recommend actions for the development and maintenance of global leadership in the photonics-driven industry—including both near-term and long-range goals, likely participants, and responsible agents of change.

In carrying out this charge, the committee will consider the materials necessary for the technological development of optics.

THE COMMITTEE

This project will be executed by an ad hoc National Research Council committee appointed to carry out this study and produce the report, under the oversight of the Board on Manufacturing and Engineering Design. In recognition of the increasing importance of materials science to innovations in engineering and manufacturing, the National Materials and Manufacturing Board was formed. This Board combines the charges of two preexisting boards: the National Materials Advisory Board and the Board on Manufacturing and Engineering Design. The committee will consist of about 16 experts chosen to cover the breadth of fields that are relevant to the charge. The committee membership will ideally include experts and researchers from universities, national laboratories, and industrial research centers. The committee will meet in person a total of four times over about 12 months.

B

Acronyms and Abbreviations

AAAS	American Association for the Advancement of Science
ABL	Airborne Laser Laboratory
AC	alternating current
ACF	anisotropic conducive film
AFOSR	Air Force Office of Scientific Research
AIDS	acquired immune deficiency syndrome
AIP	American Institute of Physics
ALTB	Airborne Laser Test Bed
ALTSim	alternative liquid fuels simulation model
AMD	age-related macular degeneration
AMO	atomic, molecular, optics
AO	adaptive optics
APD	avalanche photodiode
APPLE	Adaptive Photonic Phase Locked Elements (program)
APRS	advanced plasma reactive sputtering
APS	American Physical Society
ARGUS-IS	Autonomous Real-Time Ground Ubiquitous Surveillance-Imaging System
ARO	Army Research Office
ARPA	Advanced Research Projects Agency
ASML	Advanced Semiconductor Materials Lithography (Dutch company)
ATL	advanced tactical laser
ATP	Advanced Technology Program

BER	bit error rate
BRCA1	breast cancer susceptibility gene 1
BRCA2	breast cancer susceptibility gene 2
CAD	computer-aided design
CCD	charge-coupled device
CCT	correlated color temperature
Ce:YAG	cerium-doped yttrium aluminum garnet
CIA	Central Intelligence Agency
CIGS	copper indium gallium selenide
CLEO	Conference on Lasers and Electro-Optics
CMOS	complementary metal oxide semiconductor
CMU	Carnegie Mellon University
CNC	computer numerically controlled
CO	carbon monoxide
CO_2	carbon dioxide
COIL	chemical oxygen iodine laser
CPV	concentrating photovoltaic
CRI	Color Rendering Index
CSP	concentrating solar power
CT	computed tomography
CTE	coefficient of thermal expansion
CVD	chemical vapor deposition
DARPA	Defense Advanced Research Projects Agency
DAS	distributed acoustic sensing
DC	direct current
DHS	Department of Homeland Security
DLP	digital light processor
DNA	deoxyribonucleic acid
DOD	Department of Defense
DOE	Department of Energy
DPP	discharge-produced plasma
DPSK	differential phase-shift keying
DQPSK	differential quadrature phase-shift keying
DR	diabetic retinopathy
DSO	Defense Sciences Office
DSP	digital signal processing
ECE	energy critical element
ECOC	European Conference on Optical Communication

EDFA erbium-doped fiber amplifier
EE electrical engineering
EECS electrical engineering and computer science
EERE Energy Efficiency and Renewable Energy
EM electromagnetic
EO electrooptical
EPMD Electronics, Photonics, and Magnetic Devices (Division, NSF)
ER emergency room
EU European Union
EUV extreme ultraviolet

FACS flow cytometry
FDI foreign direct investment
flops floating point operations per second
FOV field of view
FPA focal plane array
FPD flat panel display
FTTH fiber to the home
FY fiscal year

GaAs gallium arsenide
GFP green fluorescent protein
GFR glomerular filtration rate
GHz gigahertz
GPS Global Positioning System
GPT general-purpose technology

HALOE halogen occultation experiment
HAMR heat-assisted magnetic recording
HDTV high-definition television
HEL JTO High Energy Laser-Joint Technology Office
HELLADS High Energy Liquid Laser Area Defense System
HGP Human Genome Project
HIV human immunodeficiency virus
Ho:YAG holmium-doped yttrium aluminum garnet
HT high-throughput

IAD ion-assisted deposition
IBS ion beam sputtering
IC integrated circuit
IEA International Energy Agency

IEEE	Institute of Electrical and Electronics Engineers
IGZO	indium gallium zinc oxide
InAs	indium arsenide
InP	indium phosphide
IP	intellectual property
IPO	initial public offering
IQE	internal quantum efficiency
IR	infrared
IRR	internal rate of return
IRS	Internal Revenue Service
ISAT	Information Science and Technology Board (DARPA)
ISIS	Integrated Sensor Is Structure (program)
ISR	intelligence, surveillance, and reconnaissance
IT	information technology
ITAR	International Traffic in Arms Regulations
ITO	indium tin oxide
ITRS	International Technology Roadmap for Semiconductors
IWGN	Interagency Working Group of Nanotechnology
JBSDS	Joint Biological Stand-off Detection System
JHPSSL	Joint High Power Solid State Laser
JTEC	Japanese Technology Evaluation Center
JTO	Joint Technology Office
LACOSTE	Large Area Coverage Optical Search-while-Track and Engage
LADAR	laser radar
LAN	local area network
LASIK	laser-assisted in situ keratomileusis
LCD	liquid-crystal display
LCM	laser capture micro dissection
LCOE	levelized cost of energy
LCoS	liquid crystal on silicon
LED	light-emitting diode
LELE	litho-etch-litho-etch
LENS™	Laser Engineered Net Shaping
LEOS	Lasers and Electro-Optics Society
LFLE	litho-freezing-litho-etch
LIA	Laser Institute of America
LIDAR	light detection and ranging
LIGO	Laser Interferometer Gravitational-Wave Observatory
LISA	Laser Interferometer Space Antenna

LLE	litho-litho-etch
LOCI	Ladar and Optical Communications Institute
LP	linear polarization
LPP	laser-produced plasma
MANPADS	man portable air defense systems
MDR	multi-drug resistant
MEMS	microelectromechanical system
MIMO	multiple-input, multiple-output
MIPS	millions of instructions per second
MIT	Massachusetts Institute of Technology
MLD	Maritime Laser Demonstrator
MOCVD	metal-organic chemical vapor deposition
MONET	multiwavelength optical networking
MOPA	master oscillator power amplifier
MOS	metal oxide semiconductor
MRF	magnetorheological finishing
MRI	magnetic resonance imaging
mRNA	messenger ribonucleic acid
MTHEL	Mobile Tactical High Energy Laser
MTO	Microsystems Technology Office
MWIR	mid-wavelength infrared
NA	numerical aperture
NAE	National Academy of Engineering
NAICS	North American Industry Classification System
NAS	National Academy of Sciences
NASA	National Aeronautics and Space Administration
NATO	North Atlantic Treaty Organization
NCI	National Cancer Institute
Nd:YAG	neodymium-doped yttrium aluminum garnet
NEI	National Eye Institute
NHLBI	National Heart, Lung, and Blood Institute
NIH	National Institutes of Health
NIST	National Institute of Standards and Technology
NMMB	National Materials and Manufacturing Board
NNAP	National Nanotechnology Advisory Panel
NNI	National Nanotechnology Initiative
NRC	National Research Council
NRDA	Nanotechnology Research and Development Act
NREL	National Renewable Energy Laboratory

NSE	nanoscale science and engineering
NSET	Nanoscale Science, Engineering, and Technology
NSF	National Science Foundation
NVCA	National Venture Capital Association
OADM	optical add/drop multiplexer
OAI	off-axis illumination
OCT	optical coherence tomography
OFC	Optical Fiber Communications Conference
OFDM	orthogonal frequency division multiplexing
OIDA	Optoelectronics Industry Development Association
OITDA	Optoelectronics Industry and Technology Development Association
OLED	organic light-emitting diode
ONR	Office of Naval Research
OOK	on-off keying
OPC	optical proximity correction
OSA	Optical Society of America
OSE	optical science and engineering
OSTP	Office of Science and Technology Policy
PC	personal computer
PCAST	President's Council of Advisors on Science and Technology
PCF	photonic-crystal fiber
PCR	polymerase chain reaction
PDM	polarization division multiplexed
PhC	photonic crystal
PIC	photonic integrated circuit
PISA	Program for International Student Assessment
PON	passive optical network
PRK	photorefractive keratectomy
PSA	prostate-specific antigen
PSK	phase-shift keying
PSM	phase-shifting masks
PV	photovoltaic
PVOCT	phase variance optical coherence tomography
QAM	quadrature-amplitude-modulation
QCMC	quantum communication, measurement, and computing
QCSE	Quantum-Confined Stark Effect
QD	quantum dot

QELS	quantum electronics and laser science
QKD	quantum key distribution
QPSK	quadrature-phase-shift-keying
QW	quantum well
R&D	research and development
RCA	Radio Corporation of America
RET	resolution enhancement technology
RF	radio frequency
RGB	red green blue
RIN	relative intensity noise
RMS	root mean square
RNA	ribonucleic acid
ROADM	reconfigurable optical add/drop multiplexer
RT-PCR	real-time polymerase chain reaction
S&T	science and technology
SAIL	Synthetic Aperture Imaging Ladar
SAM	Solar Advisor Model
SAR	synthetic aperture radar
SBIR	Small Business Innovation Research
SBS	synthesis-based sequencing
SDM	space division multiplexing
SEMATECH	Semiconductor Manufacturing Technology
SEMI	Semiconductor Equipment and Materials International
SIC	Standard Industrial Classification
SIM	sensing, imaging, and metrology
SLS	selective laser sintering
SOP	state of polarization
SPDT	single-point diamond turning
SPIE	International Society for Optics and Photonics
SRC	Semiconductor Research Corporation
SS	solid state
STEM	science, technology, engineering, and mathematics
SWaP	size, weight, and power
TFPV	thin-film photovoltaics
TFT	thin-film transistor
THEL	Tactical High Energy Laser
TIP	Technology Innovation Program

UAV	unmanned aerial vehicle
UNCTAD	United Nations Conference on Trade and Development
UNESCO	United Nations Educational, Scientific, and Cultural Organization
UNIDO	United Nations Industrial Development Organization
USPTO	U.S. Patent and Trademark Office
UV	ultraviolet
VCSEL	vertical cavity surface emitting laser
WDM	wavelength division multiplexing
YAG	yttrium aluminum garnet

C

Additional Technology Examples

ENABLING TECHNOLOGIES

This appendix gives examples of additional technologies in the area of optics and photonics as they relate to many of the fields described in the chapters of this report. The Committee on Harnessing Light: Capitalizing on Optical Science Trends and Challenges for Future Research believes that this compilation puts additional emphasis on how optics and photonics truly are enabling technologies, and at the same time it provides the reader with further examples that highlight the many complex ways that optics and photonics support the foundation of many common areas not always directly associated with the fields of optics and photonics.

DEFENSE AND NATIONAL SECURITY

This section discusses the changes in many of the areas that were addressed in "Optics in National Defense," Chapter 4 of the National Research Council's (NRC's) 1998 *Harnessing Light: Optical Science and Engineering in the 21st Century*.[1] The subsections below provide an update for the areas of surveillance, night vision, laser systems operating in the atmosphere and in space, fiber-optic systems, and special techniques (e.g., chemical and biological species detection, laser gyros, and optical signal processing).

[1] National Research Council. 1998. *Harnessing Light: Optical Science and Engineering for the 21st Century*. Washington, D.C.: National Academy Press.

Surveillance

Surveillance still plays a critical role in the detection and assessment of hostile threats to the United States. High-resolution imaging satellites have been deployed for more than 50 years to provide critical data for U.S. defense experts over denied airspace. The progress in optical sensors over the past decade has created an exponential growth in intelligence, surveillance, and reconnaissance (ISR) data from both passive and active sensors. This progress includes not just an increase in area coverage rate, but also an increase in sensor capabilities and performance. Material advances have made collection at new wavelengths feasible, and improved components provide new data signatures including vibrometry, polarimetry, hyper-spectral signatures, and three-dimensional data that mitigate camouflage for targets of interest.

A key advance since the NRC's 1998 study is the dramatic increase in the application of active optical sensors for surveillance. The primary impetus for this increase has been the advances made in laser technology, including advances in robustness, efficiency, and optical power (see Figure C.1) for many wavelengths. In order for optical sensors to be widely fielded, they must also meet eye-safety requirements, which have driven advances in sources and amplifiers for 1.5 and 2 μm wavelength lasers.

In order to maximize atmospheric transmission, there has been a push for longer wavelength amplifiers in regimes with efficient detectors. Recent advances in thulium (Tm)-doped fiber amplifiers enable laser sensor operation in the 1.9 to 2.1 μm wavelength range. Average power levels are approaching the kilowatt level, and pulsed amplifiers with peak powers approaching 100 kW have been demonstrated.[2,3,4,5] Several vendors are offering lasers with output powers up to 150 W. There are also several vendors offering narrow line-width, rapidly tunable 2.1 μm laser sources. The atmospheric transmission at this wavelength combined with the availability of commercial amplifiers, sources, and detectors makes 2.1 μm an attractive wavelength for long-range laser sensing. The efficiency of the current

---

[2] Cristensen, S., G. Frith, and B. Samson. 2008. "Developments in Thulium-Doped Fiber Lasers Offer Higher Powers." *SPIE Newsroom*. DOI: 10.1117/2.1200807.1152. Available at http://spie.org/x26003.xml. Accessed July 31, 2012.

[3] Moulton, P.F., G.A. Rines, E.V. Slobodtchikov, K.F. Wall, G. Firth, B. Samson, and A.L.G. Carter. 2009. Tm-doped fiber lasers: Fundamentals and power scaling. *IEEE Journal of Selected Topics in Quantum Electronics* 15(1):85-92.

[4] Sudesh, V., T.S. McComb, R.A. Sims, L. Shah, M. Richardson, and J. Stryjewsky. 2009. Latest developments in high power, tunable, CW, narrow line thulium fiber laser for deployment to the ISTEF. *Proceedings of SPIE* 7325:73250B.

[5] McComb, T.S., R.A. Sims, C.C.C. Willis, P. Kadwani, V. Sudesh, L. Shah, and M. Richardson. 2010. High-power widely tunable thulium fiber lasers. *Applied Optics* 49(32):6236.

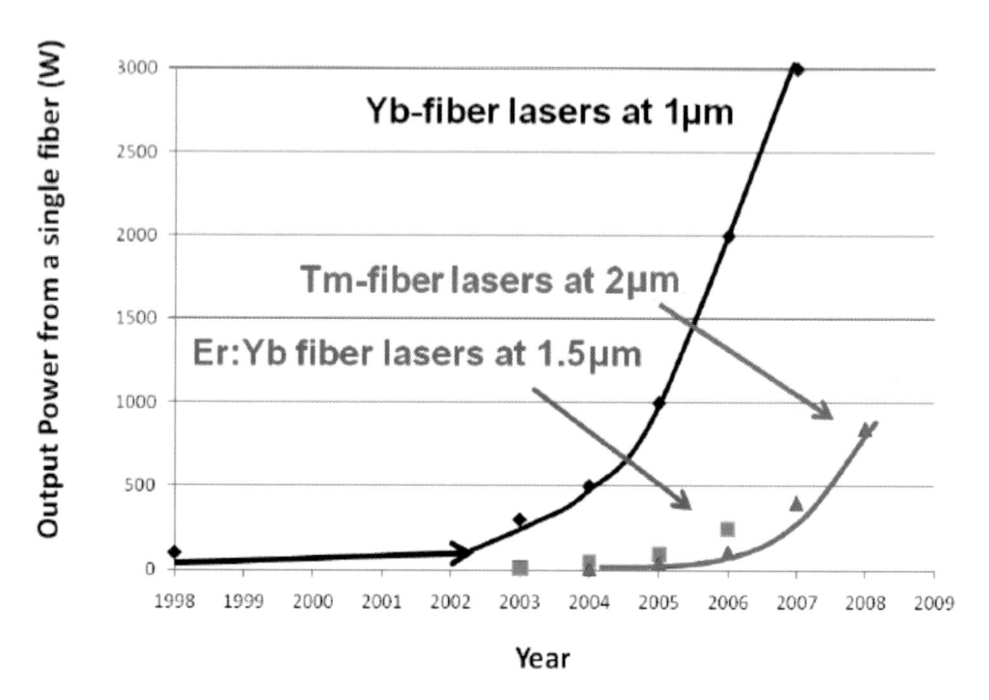

FIGURE C.1 Progress in output power for 1 μm, 1.5 μm, and 2 μm optical amplifiers. This plot tracks the historical progress in output power from a single fiber amplifier. Ytterbium (Yb)-fiber amplifiers (1 μm) have a significant lead over other technologies primarily due to their low quantum defect and the abundance of high-brightness pump sources at 915 to 975 nm. However, thulium (Tm)-fiber amplifiers (1.9 to 2.1 μm) have been demonstrated with average output powers approaching 1 kW and pulsed operation approaching 100 kW without sacrificing beam quality. There are several commercial units available with an average power ~150 W. Their efficiency is set by the efficiency of the pump source, which should continue to improve over time as demand for Tm-lasers increases. SOURCE: Cristensen, S., G. Frith, and B. Samson. 2008. "Developments in Thulium-Doped Fiber Lasers Offer Higher Powers." *SPIE Newsroom.* DOI: 10.1117/2.1200807.1152. Available at http://spie.org/x26003.xml. Reprinted with permission.

commercial amplifiers is approximately 6.25 percent[6] limited by the efficiency of the pump source. However, there have been many investments made in these areas, which should improve the efficiency over time.

[6] See, for example, IPG Photonics, "2 Micron CW Fiber Lasers." Available at http://www.ipgphotonics.com/products_2microns.htm. Accessed July 31, 2012.

Night Vision

Night-vision capabilities continue to be an important tactical tool for the warfighter. In fact, the proliferation of equipment over the past few decades has led to a significant amount of surplus equipment available at very low cost, which has eroded the tactical advantage for the United States that existed for some time. During the First Gulf War, the United States "owned the night," with U.S. night-vision systems significantly outperforming Iraqi night-vision sensors.[7] However, the current commercially available night-vision sensors are nearly equivalent to the best U.S. night-vision systems. Since the NRC's 1998 *Harnessing Light*[8] study, there have been substantial improvements in sensitivity, performance for uncooled systems, and expanded wavelength applicability, which have enabled practical thermal imaging systems for size, weight, and power (SWaP)-constrained platforms.

Laser Rangefinders, Designators, Jammers, and Communicators

The significant increase in laser diode efficiency coupled with the decrease in cost has enabled recent advances in laser designators. However, similar advances in night vision and imaging detector arrays have limited the use of laser designators and led to ground force casualties in recent engagements. Therefore, there is a greater push for SWaP improvements to enable designators on small unmanned platforms (e.g., micro-unmanned aerial vehicles [UAVs]), which will also carry over to active sensors and optical communication systems. Early laser designator systems used neodymium-doped yttrium aluminum garnet (Nd=YAG) lasers at 1 μm. However, improvements in laser materials and efficiency have enabled a wider range of wavelengths to be implemented.

A large investment in laser communications had been made prior to publication of the NRC's 1998[9] report and has continued since that time. Optical communication in fibers has been steadily advancing in the past decade. The high carrier frequency of light, combined with the low attenuation in fiber, makes it attractive for telecommunications applications. For free-space applications, the short wavelength improves directivity by minimizing diffraction when compared to radio-frequency (RF) communications. This is one of the key motivators for moving to optical communications, which minimize the probability of interception, jamming, and detection while dramatically minimizing the power needed for a given communication bandwidth, since most of the energy can be focused on the receiver.

---

[7] National Research Council. 2010. *Seeing Photons: Progress and Limits of Visible and Infrared Sensor Arrays*. Washington, D.C.: The National Academies Press.

[8] National Research Council. 1998. *Harnessing Light*.

[9] National Research Council. 1998. *Harnessing Light*.

Laser Weapons

The Missile Defense Agency demonstrated the potential use of directed energy to defend against ballistic missiles when the Airborne Laser Test Bed (ALTB) successfully destroyed a boosting ballistic missile on February 11, 2010.[10] The experiment, conducted at Point Mugu Naval Air Warfare Center-Weapons Division Sea Range off the coast of central California, served as a proof-of-concept demonstration for the directed-energy technology. The ALTB is a pathfinder for the nation's directed-energy program and its potential application for missile defense technology. For the demonstrations, a short-range threat-representative ballistic missile was launched from an at-sea mobile launch platform. Within seconds, the ALTB used onboard sensors to detect the boosting missile and used a low-energy laser to track the target. A second low-energy laser was fired to measure and compensate for atmospheric disturbances. Finally, the ALTB fired its megawatt-class high energy laser, heating the boosting ballistic missile to critical structural failure. The entire engagement occurred within 2 minutes of the target missile launch, while its rocket motors were still thrusting. This was the first directed-energy lethal interception demonstration against a liquid-fuel boosting ballistic missile target from an airborne platform. This revolutionary use of directed energy is very attractive for missile defense, with the potential to attack multiple targets at the speed of light, at a range of hundreds of kilometers, and at a low cost per interception attempt compared to the cost with current technologies.

Fiber-Optic Systems

Fiber-optic systems have continued to evolve to achieve higher performance with lower power in a smaller volume. Fiber-optic systems (e.g., gyros, communication links) have several attractive attributes including low loss, high transmission rates, and freedom from electromagnetic interference. Therefore, they have continued to be adopted into military platforms as they are upgraded.

Special Techniques

The special techniques (i.e., chemical and biological species detection, laser gyros, and optical signal processing) evaluated in the NRC's 1998 report[11] have evolved in different ways. Optical signal processing has also advanced, but not at the pace forecasted at that time. Importantly, recent advances in optical integrated

[10] Missile Defense Agency, U.S. Department of Defense. 2010. "Airborne Laser Test Bed Successful in Lethal Intercept Experiment." *MDANews Release.* Available at http://www.mda.mil/news/10news0002.html. Accessed August 2, 2012.

[11] National Research Council. 1998. *Harnessing Light.*

circuits should enable significant advances in optical signal processing over the next decade.

Chemical and Biological Species Detection

Weapons of mass destruction, including nuclear, biological, and chemical weapons, continue to be a high-priority threat. Long-range chemical and biological detection has advanced considerably since the 1998 report.[12] One example is the Joint Biological Stand-off Detection System (JBSDS), a light detection and ranging (LIDAR)-based system that is designed to detect aerosol clouds out to 5 km in a 180 arc and to discriminate clouds with biological content from clouds without biological material at distances of 1 to 3 km or more. This system will provide advance warning of the presence of potential biological weapon aerosol cloud hazards so that a commander can implement individual and collective protective measures for assigned forces.

Laser Gyros for Navigation

Laser gyros were already very mature at the time of the 1998 report.[13] They are critical in maintaining precision navigation when the Global Positioning System (GPS) is unavailable due to platform constraints or jamming. A new advance since the 1998 report is in the area of star-trackers, which can augment inertial navigation systems to improve long-term stability.

Optical Signal Processing

Optical processing has not changed very much since the NRC's 1998 study.[14] It continues to be very promising, since some mathematical functions can be performed very rapidly using optical analog techniques. One example is optical correlations that rely on Fourier transforms. Optical correlators compare two-dimensional image data at very high speeds. They were invented in the mid-1960s and have traditionally been used in high-cost military applications such as the analysis of satellite photographs. With recent advances in liquid-crystal technology, optical correlators have become more commercially viable—at a fraction of the high costs previously associated with such high-performance systems. Image data that are entered into the optical system are compared during the correlation process in terms of two criteria, similarity and relative position. Typically, the comparison

[12] National Research Council. 1998. *Harnessing Light.*
[13] National Research Council. 1998. *Harnessing Light.*
[14] National Research Council. 1998. *Harnessing Light.*

is done between a reference image (e.g., from a database) and an input image (e.g., from an external camera or sensor).

ENERGY

Cost of Solar Technology

In a 2008 study, leading experts in the solar field working on various technologies were asked to estimate the probability of any technology reaching two different "dollars per watt" benchmarks.[15] The two price benchmarks were $1.20/W installed cost, the point at which a technology can be considered commercially viable, and $0.30/W installed cost, the point at which a technology would likely emerge as dominant in supplying utility-scale energy. The experts were asked to gauge the probability that any solar technology would meet these price points by the years 2030 and 2050. The results of this study are shown in Figure C.2, in which a larger circle corresponds to a larger number of experts giving this probability.

Hybrid Solar and Wind Power Systems

Another approach to harvesting the Sun's energy is a solar updraft tower, which uses the greenhouse effect to create a hybrid of solar and wind power. This technology uses a large base area sealed by a transparent material. The air heats to approximately 70°C due to the greenhouse effect. This air is then forced out of the high central tower, referred to as a "solar chimney," as shown in Figure C.3. This is expected to produce wind, which will then be used to power turbines. Two 200-MW plants have been proposed for installation in western Arizona.[16]

Supporting Technologies for Solar Power

The solar power industry has several supporting technologies that are crucial to further development but not directly involved in converting the Sun's energy into electric power. Technologies such as mounts for solar modules and electronics are also crucial to the commercialization of solar power technologies.

Work has been done to model the competitiveness of a given solar technology quantitatively given the variability of many uncertain factors. Several modeling programs are being developed, one of which has become widely available: the Solar

---

[15] Curtright, A., M. Granger Morgan, and D. Keith. 2008. Expert assessments of future photovoltaic technologies. *Environmental Science and Technology* 42(24):9031-9038.

[16] Southern California Public Power Authority. 2008. "La Paz Solar Tower Project." Available at http://www.scppa.org/pages/projects/lapaz_solartower.html. Accessed July 30, 2012.

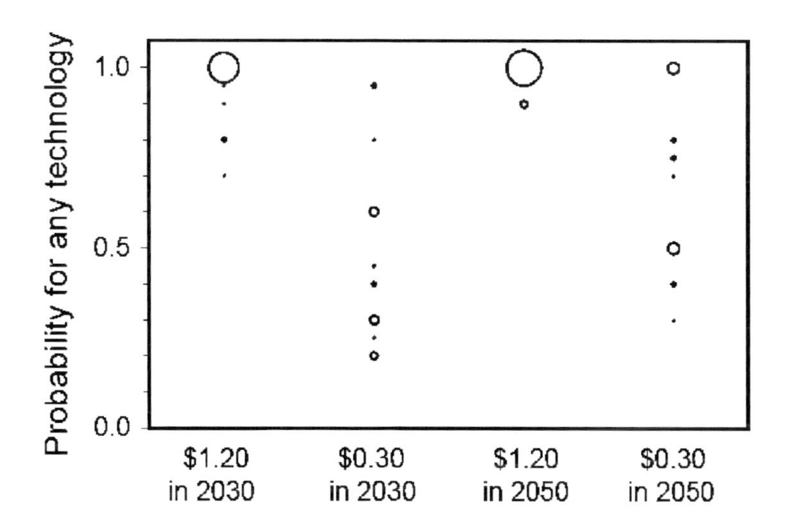

FIGURE C.2 Estimates of the probability of any solar technology reaching two dollars-per-watt price points—$1.20/W and $0.30/W—by 2030 and 2050, according to experts in the solar field. Larger circles correspond to a larger number of experts giving this probability. SOURCE: Estimates based on Curtright, A., M. Granger Morgan, and D. Keith. 2008. Expert assessments of future photovoltaic technologies. *Environmental Science and Technology* 42(24):9031-9038. Reprinted with permission.

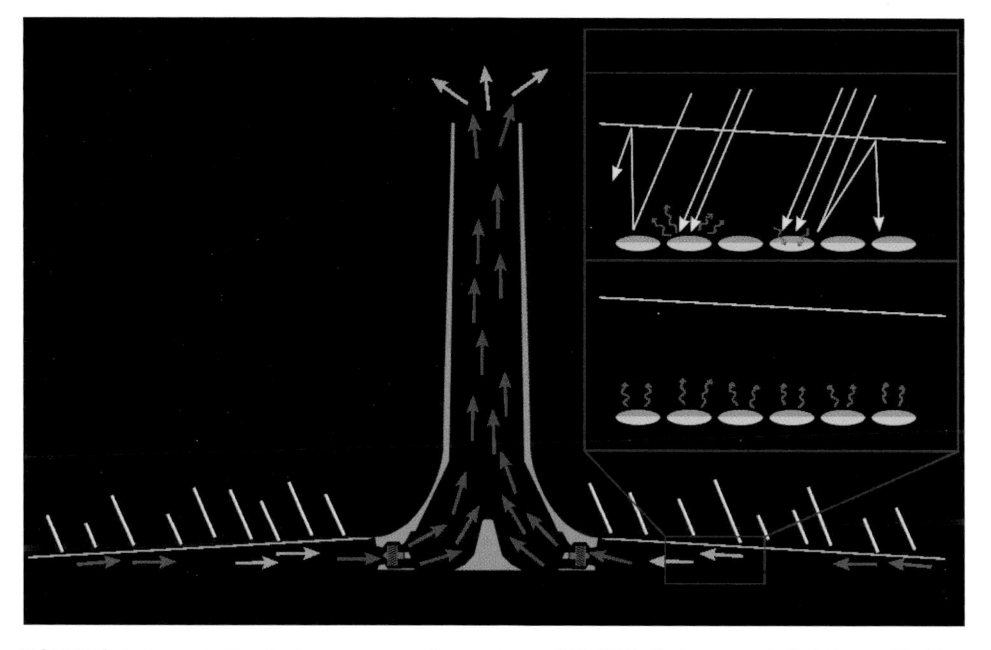

FIGURE C.3 An updraft solar tower power plant scheme. SOURCE: Redrawn and slightly modified by Cryonic07. Original jpg-drawing made by fr:Utilisateur:Kilohn limahn.

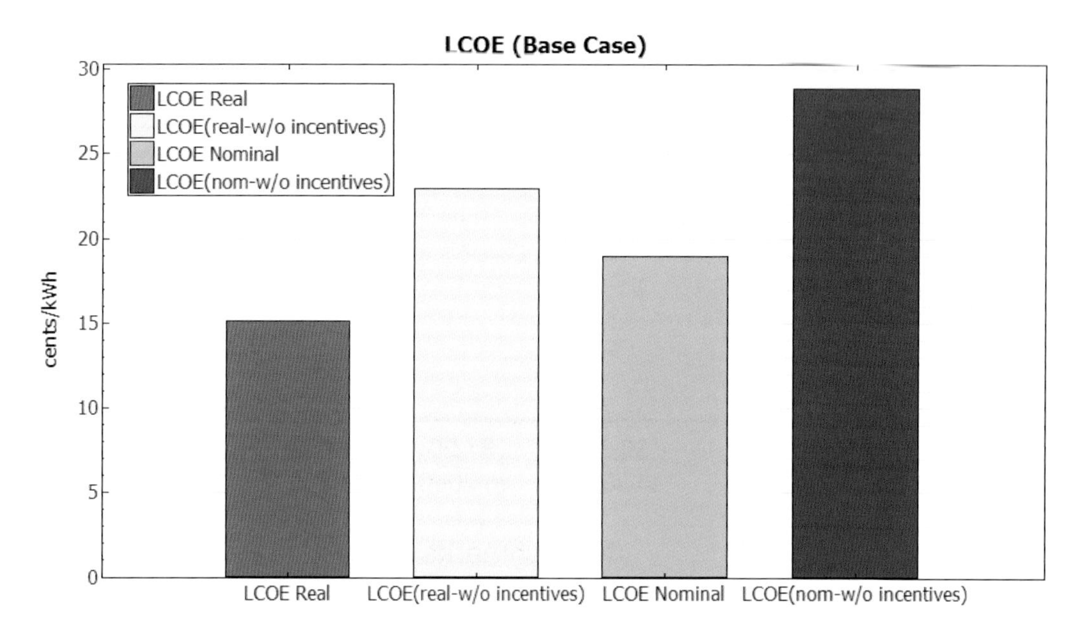

FIGURE C.4 A sample output graph produced from the Solar Advisor Model. SOURCE: Mooney, D., M. Mehos, N. Blair, C. Christensen, S. Janzou, and P. Gilman. 2006. "Solar Advisor Model (SAM) Overview." P. 23 in *16th Workshop on Crystalline Silicon Solar Cells and Modules: Materials and Processes; Extended Abstracts and Papers.* Sopori, B.L., ed. Proceedings of the NREL/BK-520-40423 workshop held August 6-9, 2006, in Denver, Colorado. Golden, Colo.: National Renewable Energy Laboratory. Reprinted with permission.

Advisor Model (SAM), produced by the National Renewable Energy Laboratory (NREL) and Sandia National Laboratories.[17] This system considers a wide range of module performance options, financing options, and government subsidies and will calculate the levelized cost of energy (LCOE) for an expected situation. A sample output of SAM is shown in Figure C.4, although the program produces an enormous amount of information and only a small part of the output is represented in the figure.

Models that can predict the performances of solar power relative to alternate sources of energy have been developed. Among these is the program ALTSim, developed at Hobart and William Smith Colleges to determine the viability of

[17] Mooney, D., M. Mehos, N. Blair, C. Christensen, S. Janzou, and P. Gilman. 2006. "Solar Advisor Model (SAM) Overview." P. 23 in *16th Workshop on Crystalline Silicon Solar Cells and Modules: Materials and Processes; Extended Abstracts and Papers.* Sopori, B.L., ed. Proceedings of the NREL/BK-520-40423 workshop held August 6-9, 2006, in Denver, Colo. Golden, Colo.: National Renewable Energy Laboratory.

biofuels competing with fuels such as oil and coal.[18] Elaboration and integration of economic modeling programs such as these would enable more quantitative comparisons among technologies and allow more informed investment decisions.

A large portion of the module cost for solar applications, particularly those that require concentration, comes from the tracker that the system must be mounted on. These systems must be able to keep accurate alignment with the Sun, moving a large panel area. Maintaining the proper alignment is made more difficult by the requirement that the system stay aligned despite the significant wind loading generated by the solar panel carried. The tracker can comprise approximately half the cost of a current concentrating system. Reducing this cost while not compromising the performance or lifetime of a tracker would make these systems much more commercially viable.

Although the majority of the cost for a solar plant is the solar collection module, the electronics required to interface with the power grid and power storage still form a substantial portion of the cost. Solar panels produce direct current (DC) power, which must be transformed into alternating current (AC) to be sent into the power grid, requiring the use of an inverter. In some cases solar power can be matched to the load—for example, in the southwestern United States, where air conditioning drives peak load. Solar power up to some level can be used to handle this peak and is thus matched to the load. If solar power is also to be used at night, however, photovoltaic (PV) devices require a battery, or some other method of storage to store the power produced to be used when the system is not producing power. Charging a battery requires a charge controller, further adding to the system costs. Improving the performance or reducing the cost of any of these devices will make solar power more cost-effective. An approximate cost breakdown of these components is presented in Table C.1.[19]

The electronics cost is substantially reduced for concentrating solar power (CSP) systems, as the generator can produce AC power directly to the grid and the battery is unnecessary. These systems do have a large additional cost of thermal storage systems for on-demand generation, which costs approximately 30 percent as much as the solar module.[20] Technical or manufacturing advances in this field will drive the cost of CSP plants down.

---

[18] Drennen, T., R. Williams, and A. Baker. 2009. *Alternative Liquid Fuels Simulation Model (AltSim)*. Sandia Report SAND2009-7602. Albuquerque, N. Mex.: Sandia National Laboratories.

[19] Solarbuzz. 2011. "Solar Buzz Retail Pricing." SB_Retail_Pricing_111013.xls. Available at http://www.solarbuzz.com/facts-and-figures/retail-price-environment/module-prices. Accessed June 22, 2011.

[20] Greenpeace International. 2009. "Global Concentrating Solar Power Outlook 09." Available at http://www.greenpeace.org/international/en/publications/reports/concentrating-solar-power-2009/. Accessed July 31, 2012.

TABLE C.1 Approximate Cost Breakdown Between Module Cost and Various Complementary Electronics Required

Module	Unit	September 2010	September 2011
Module	US $/Wp (?125 W)	$3.61	$2.65
	Euro €/Wp (?125 W)	€3.23	€2.43
Inverter	US $/continuous Watt	$0.715	$0.714
	Euro €/continuous Watt	€0.558	€0.500
Battery	US $/output Watt hour	$0.207	$0.213
	Euro €/output Watt hour	€0.161	€0.149
Charge Controller	US $/Amp	$5.87	$5.93
	Euro €/Amp	€4.58	€4.15
Solar Systems	Residential ¢/kWh	34.28	29.53
	Commercial ¢/kWh	24.32	19.97
	Industrial ¢/kWh	18.95	15.56

SOURCE: Frost and Sullivan analysis.

HEALTH AND MEDICINE

The roles played by imaging, optics, and photonics in modern medicine are mentioned in Chapter 6. Some of the details of the technologies used are examined below.

Optics and Photonics in the Emergency Room

In the modern emergency room, the technologies mention take advantage of photons with energies chosen so that they can penetrate deep into the human body. The performance of present-day CT instruments has improved dramatically over the past decades owing to advances in x-ray sources and the introduction of sophisticated multi-element, high-efficiency x-ray detectors. These instruments provide images with submillimeter resolution over large volumes of the body in mere seconds. These almost instantaneous three-dimensional images provide visual evidence of life-threatening disorders, saving precious minutes in life-and-death situations in the emergency room.

Optics and Photonics in Diagnostics

The high-speed blood work mentioned in Chapter 6 can also help determine the status of the patient's immune system. When the AIDS epidemic was first detected in the early 1980s, the cause of the disease was unknown. It took several years for the human immunodeficiency virus (HIV) to be identified as the infectious agent and several more years for the affected immune system cell types to be determined. Optical methods based on laser spectroscopy of stained immune system cells provided the key technology that enabled the identification of the specific immune system cells impacted by the virus, out of the dozens of circulating cell types in the immune system. Currently, optical methods based on flow cytometry are still the standard of care for monitoring the immune system status of HIV patients and determining the efficacy of specific antiviral agents used to control the viral load in the patient's bloodstream.

Pathologists also utilize optics and photonics. Today's microscopes are interfaced to computers to speed diagnosis, assist the pathologist in recognizing diseased tissue, and provide a permanent electronic record of the images of tissue biopsies. Computerized microscopes combined with genetically engineered fluorescent antibody stains highlight the tumor region, identifying and quantifying the specific mutations and molecular changes in the tissue that are contributing to tumor growth. Knowing the molecular makeup of a tumor helps guide the physician and patient in choosing the most appropriate drug therapies. Although the microscope was invented more than 300 years ago, it still plays a major role as a primary diagnostic tool in the clinical laboratory.

The Human Genome Project Outcomes

One of the major motivations for sequencing of the human genome was to find the particular genes that determine the likelihood that individuals will develop specific diseases. Optical instrumentation was essential for the successful completion of the Human Genome Project (HGP). Using the HGP data, researchers have demonstrated conclusively that mutations in DNA repair genes (BRCA1 and BRCA2) dramatically increase the risk of breast cancer in both men and women. Individuals with these mutations can choose mastectomy or other preventive measures to reduce greatly their risk of developing tumors. Mutations in other genes (P450) are used to predict a patient's ability to tolerate chemotherapy, allowing the oncologist to determine the optimal dosage more precisely. These key portions of a patient's genome are measured today in optical instruments that can quantify millions of sequences in a single test using optical lithography technology, which was developed originally to manufacture integrated circuits and was modified to work with DNA molecules. The activity levels of specific genes can provide insight into the causes of tumor growth and thus allow oncologists to prescribe the most effective drugs. The activity or expression levels of specific genes can be measured from tumor biopsies using extremely sensitive fluorescence techniques based on amplifying and measuring the activity levels of genes using a technique called real-time polymerase chain reaction (RT-PCR). This technique is so sensitive and specific that it can detect in a sample the presence of a single molecule with a unique genetic sequence, even in the presence of billions of DNA molecules with slightly different sequences. Doctors can evaluate data indicating gene expression levels in a tumor biopsy and use these results to predict the aggressiveness of certain types of cancer and the response of specific tumors to hormonal therapy, thus helping to determine the optimal drug options for each patient.

Biomedical Optics in Everyday Life

Glucose Monitors

Optical methods are also used to monitor chronic conditions, such as the monitoring of the glucose level in diabetic patients by means of compact, cell-phone-sized, battery-operated readers that analyze the concentration in the blood serum. These devices use optically active reagents that react with glucose in a small blood sample placed on a very inexpensive disposable paper strip and change the optical properties of the strip depending on the amount of glucose in the sample. These measurements can be done in a few seconds using a device that easily fits in a shirt pocket or purse, providing people who have diabetes with portable methods for measuring and maintaining safe levels of blood sugars.

Cosmetic Biomedical Optics

Far-infrared lasers are selectively absorbed at the surface of the skin, providing a sterile method for removing or modifying large areas of skin either to serve cosmetic purposes or to aid in the recovery of burn victims. Subsequent healing and regrowth can provide a smooth, more physically appealing surface. Hair follicles also selectively absorb certain infrared (IR) wavelengths emitted by solid-state lasers. Heating follicles above 50°C destroys the cells that cause hair to grow, thus eliminating unwanted hair without surgery.

Both nearsighted and farsighted patients can be treated with precision shaping of the cornea with lasers, which can also correct for astigmatism and other eye aberrations, eliminating the need for glasses in many patients. Contact lenses and standard eyewear provide both disposable and permanent options for correcting vision for those who prefer to avoid surgery.

Proteomic Analysis Through Protein and Tissue Arrays

High-throughput protein-detection instruments that trap individual protein molecules on microscopic beads have recently been developed. These protein molecules are in turn detected by genetically engineered enzymes that bind to the proteins and efficiently generate a fluorescent signal. This signal is highly concentrated near the microscopic bead and can be detected using high-sensitivity, low-noise, charge-coupled-device (CCD) imaging devices. Very low concentrations of protein molecules can be detected with such high efficiency that individual protein molecules can be counted digitally, providing very high accuracy and precision. This technology increases the sensitivity of the detection of proteins by almost three orders of magnitude. Such improvements in sensitivity provide the possibility of detecting early signs of cancer recurrence. Medical studies are currently underway to use these new approaches to look for low levels of prostate-specific antigen (PSA) in prostate cancer patients who have had their prostates removed. The presence of PSA is evidence that prostate cells maybe proliferating in other organs, which indicates a recurrence of the cancer through metastasis. Another potential application of protein-detection technology would allow the early detection in peripheral blood of very low levels of proteins indicative of Alzheimer's disease. The current protocol involves proteomic analysis of spinal fluid requiring much more invasive lumbar puncture procedures.

Ophthalmology

Excessive blood vessel growth is one symptom of diabetic retinopathy (DR), one of the most common causes of late-onset vision loss. A common method for diagnosing retinal disease such as DR involves photographing the blood vessels in the retina. Current methods use an injectable fluorescent dye (sodium fluorescein) and excite the dye in the retina using flash photography or a scanning laser system that raster scans over the patient's retina. Adverse reactions to the intravenous dye occur in a significant fraction of elderly patients and particularly in those with hypertension. Intravenous administration of fluorescein can cause nausea, vomiting, hives, acute hypotension, and anaphylactic shock. A version of optical coherence tomography (OCT), called phase variance optical coherence tomography (PVOCT), appears to have the potential to eliminate the need for the administration of an intravenous fluorescent dye, thus making this procedure much safer. PVOCT measures the changes in the light reflected from the retina due to the motion of the blood flowing through the retina. These signals can be detected without using fluorescent dyes, and the resulting images have resolution and contrast comparable to the standard dye-based procedures. An additional advantage of the PVOCT protocol is that three-dimensional information about the vasculature is obtained, providing further valuable information that can aid in diagnosis of retinal disease.[21]

Head trauma can result in partially detached retinas, which can be "welded" back in place using pulsed lasers. Occluded lenses caused by cataracts can be sec-

[21] Kim, D.Y., J. Fingler, J.S. Werner, D.M. Schwartz, S.E. Fraser, and R.J. Zawadzki. 2011. In vivo volumetric imaging of human retinal circulation with phase-variance optical coherence tomography. *Biomedical Optics Express* 2(6):1504-1513.

tioned and more easily removed using pulsed lasers. Using lasers with different output colors, doctors can repair almost every major element in the eye: cornea, lens, and retina.

Millions of individuals have had their vision permanently corrected using the photorefractive keratectomy (PRK) procedure that uses excimer lasers to modify the lens in the eye precisely. New methods involving ultrafast lasers allow the removal of an ultrathin corneal flap prior to PRK surgery (laser-assisted in situ keratomileusis, or LASIK). This flap is placed over the machined cornea and helps protect the corneal surface and promotes faster healing. LASIK eliminates the use of a scalpel to cut this sensitive tissue and provides a much more precise and thinner flap covering the surgical incision. Moreover, in the LASIK procedure the precise laser incision allows a tab to be retained, which assists in repositioning the flap over the incision. This process reduces the risk of infection by eliminating physically touching the eye with a scalpel and provides a much more precise method for determining the diameter and symmetry of the incision; see Figure C.5.

One of the most common causes of visual impairment is the development of cataracts or cloudiness in the eye lens, which leads to reduced visual acuity and difficulty in seeing at night. The standard treatment procedure involves the surgical removal of the clouded lens and its replacement with a clear plastic lens. Common cataract procedures involve cutting the lens capsule surrounding the lens and then segmenting the occluded lens prior to manual removal by the surgeon. Current methods require making an incision in the eye and inserting scissors to cut the lens capsule surrounding the clouded lens manually. This method can often be imprecise and result in uneven or torn lens capsules, which compromise the placement of the replacement lens. This process can now be performed with phenomenal precision using laser surgery guided by OCT data.[22]

Advances in Endoscopic Surgery

Prosthetic devices can restore hearing in many older adults who have experienced degenerative hearing loss and in children born with hearing deficits. Inserting these devices often requires delicate surgery in the close environment of the ear canal, surrounded by very delicate tissue structures. Far-infrared lasers can be used to resculpt the inner ear to allow effective incorporation of a prosthetic device. The tissues in the ear canal strongly absorb mid-infrared (mid-IR) laser light, which can be used to cut and oblate tissue to allow insertion of the prosthetic device. Until recently lasers were not easily employed in this application, since directing the beam into the ear canal was complicated by the complex physiology and small confines of the ear canal. Building on advances in fiber-optic technology stimulated by the

[22] See Chapter 6 in the main text of this report for a description.

(a)

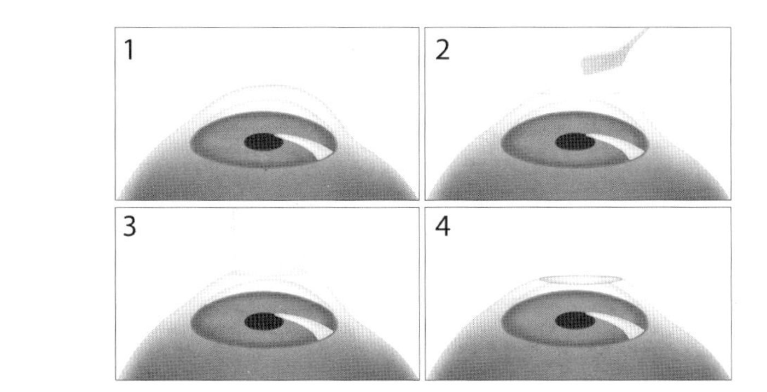

Photorefractive Keratectomy (PRK)

(b)

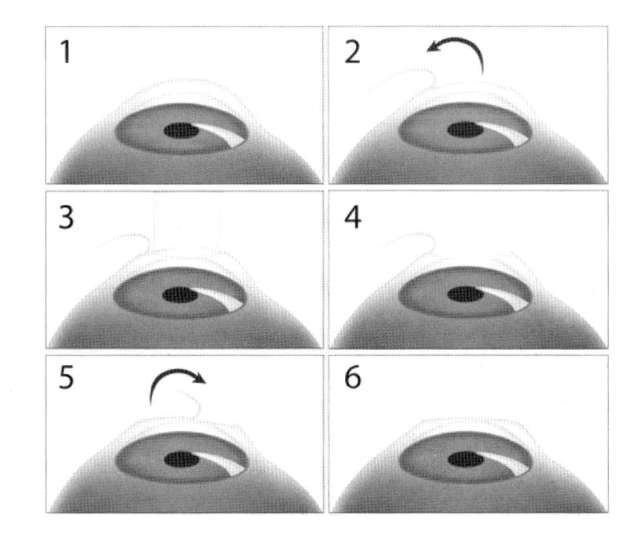

Lasik Eye Surgery

FIGURE C.5 (a) Steps in the photorefractive keratectomy (PRK) procedure. (b) Steps in the laser-assisted in situ keratomileusis (LASIK) procedure.

telecommunications industry, several new types of fiber-optic cables have been developed that allow the effective delivery of mid- and far-IR lasers, wavelengths that are highly absorbed by soft tissue, into previously inaccessible portions of the human anatomy like the inner ear. Thus mid-IR lasers can now be used in a variety of surgical procedures that were not previously possible. Novel fiber optic designs use nanostructured geometries to confine the far-IR light of a carbon dioxide (CO_2) laser into a flexible fiber cable, as shown in Figure C.6. These geometries

FIGURE C.6 Omniguide fiber for otology. SOURCE: Courtesy of OmniGuide, Inc.® Cambridge, Massachusetts. Reprinted with permission.

utilize submicron layers of optical materials surrounding a hollow core, which allows the far-IR light from the CO_2 laser to be delivered to delicate bone and tissue structures of the inner ear. This new fiber-coupled source has found great utility in otology, providing effective methods for the surgical implantation of prosthetic devices restoring hearing to adults and children who have hearing impairments. These advances have relied heavily on fundamental research in nanophotonics and photonic crystal devices, technology originally developed primarily for the telecommunications industry.

Similarly, kidney stones can be very effectively fragmented using fiber-coupled, near-IR pulsed lasers. High-energy, short pulses from these lasers are absorbed by the kidney stone, creating a thermal shock wave, disrupting the stone. New, low-water-content fibers allow a holmium-doped (Ho):YAG laser to be focused into an optical fiber, which is incorporated into a flexible endoscope. This endoscope can be threaded through the urethra and guided to the kidney stone. This laser effectively fragments kidney stones and provides a low-morbidity, less expensive, and highly effective alternative to previous treatments using high-intensity sound shock waves to disrupt the stones.

Advances in Oxygen Saturation Measurements

When firefighters come upon an unconscious person, they must often assess whether the victim has been subjected to high levels of carbon monoxide (CO). A rapid assessment of the levels of CO in the blood is critical so that intervention can take place before brain damage occurs. Exposure to CO and the effect of a number of common drugs on the blood's ability to transport oxygen throughout the body alter the spectroscopic properties or color of the blood, providing key indications of the cause and expediting the appropriate intervention. Exposure to CO reduces the oxygen-carrying capacity of blood and causes the IR transmission of the blood to increase, whereas exposure to certain drugs decreases the absorption in the orange region of the visible spectrum. These optical signatures, as indicated in Figure C.7, can be used to diagnose quickly and noninvasively the cause of hypoxia and can help determine the most effective treatment.

Rapidly growing tumors require high blood flow to supply sufficient nutrients and oxygen to support tumor growth. Using near-IR wavelengths of light, which can penetrate through most normal tissues but are absorbed preferentially by highly oxygenated blood, tumors can be detected by monitoring the increase in oxygenated blood flow to regions of tissue deep beneath the surface of the skin.

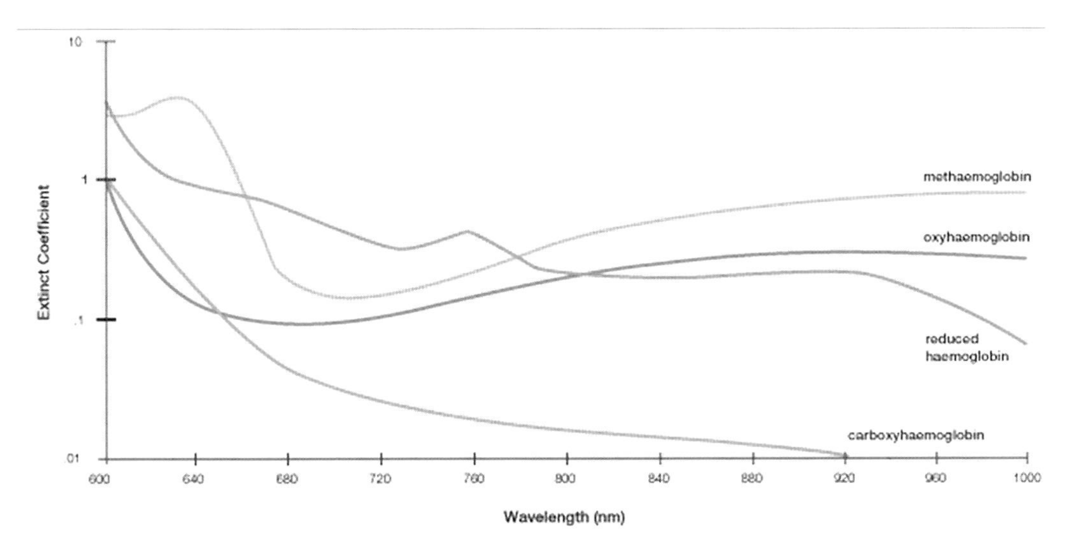

FIGURE C.7 Blood changes color depending on the hemoglobin status. The absorption profiles of four different hemoglobin statuses are shown. Simple disposable probes that can be mounted on a patient's finger can precisely measure the absorption level of the blood at different wavelengths and help determine the extent and cause of hypoxia. SOURCE: Masimo. 2012. "Rainbow SET Pulse CO-Oximetry." Available at http://www.masimo.co.uk/Rainbow/about.htm. Reprinted with permission.

This increased blood flow can be measured and tumor images can be generated by positioning multiple sources and detectors around the affected tissue and observing "shadows" of the tumor due to this increased blood flow required by the tumor.

Biomedical Optics in Research

The polymerase chain reaction (PCR) was discovered in the early 1980s. The PCR provides one of the most sensitive and specific methods for measuring nucleic acid molecules in vitro. Using high-efficiency fluorescent organic dyes and low-noise detectors, single DNA molecules with specific sequences can be detected even when contained in a sample with a large concentration of background DNA molecules. This high sensitivity and specificity have allowed the development of techniques that separate the sample into multiple individual wells, where the number of wells is large enough that only a single molecule is likely to be in any individual well. The absolute number of molecules in the original sample can be determined simply by counting the number of wells in which single molecules are detected. This approach to quantitative PCR, called digital PCR, allows a significant increase in the accuracy with which low levels of specific nucleic acid molecules can be detected. Digital PCR may also provide a method for developing precise standard reference materials and detection protocols for ultra-low trace concentrations of nucleic acids.

Cancer biopsies often contain heterogeneous mixtures of various types of normal and tumor cells. It is thought that even within a tumor, specific tumor cells (cancer stem cells) have an enhanced ability to reproduce and to survive chemotherapy treatment, leading to the post-treatment regrowth of the tumor. Cancer stem cells are rare and are located close to both normal tissue and non-stem tumor cells. Studying these rare cells requires isolating and removing them from the surrounding tissue sample. Automated laser-based methods for performing these micro-dissections have been developed, providing a fast and precise method for excising single cells from the complex tissue environment found in most biopsies. These methods combine ultraviolet (UV) lasers used for isolating the single cells, with IR lasers which capture the cells, and allow an efficient method for extraction of the sample from the biopsy. Automated laser capture micro-dissection (LCM) has been combined with microfluidic technology, allowing the macromolecules to be extracted efficiently from the micro-dissected samples and reagents that provide a means to copy and thus amplify specific target molecules of messenger ribonucleic acid (mRNA) or DNA within the microscopic samples to allow precise quantification. The abilities to isolate, extract, and analyze single cells on the basis of optical imaging and laser technologies have created the new field of microgenomics, which has found applications in many fields of research including microbiology, neuroscience, developmental biology, and forensics.

ADVANCED MANUFACTURING

CNC Grinding and Polishing

Computer numerically controlled (CNC) grinding and polishing have brought a deterministic approach to centuries-old processes. Applying technology from the machine tool industry, CNC grinding and polishing have made significant advancements in the ability to produce precision aspheric components. Aspheric geometry is much more difficult to generate and polish than are spheres, however. Five-axis CNC grinding and polishing equipment, as shown in Figure C.8, makes it possible to produce these challenging geometries. Computer controllers dynamically adjust cutting paths for tool wear and can be programmed for edging, beveling, sagging, concave, and convex surface grinds.

Polymer Molding

Molded plastic lenses have become commonplace in consumer and commercial products. Mobile phones, DVD players, digital cameras, and conferencing systems have incorporated polymer lenses. Hydraulic injection molding presses and injection compression molding presses are now employed by most optical

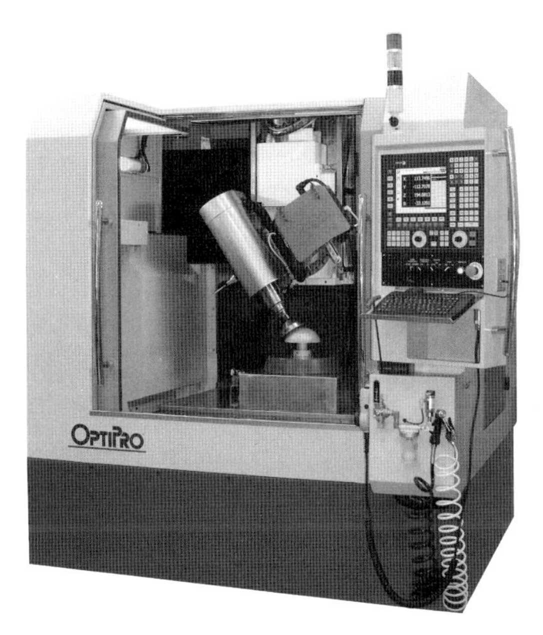

FIGURE C.8 (*Left*) A commercially available 5-axis computer numerically controlled grinder. (*Right*) A closeup image of the grinding head. SOURCE: Courtesy of Optipro.

component molding shops. Presses in the 20- to 50-ton range are frequently used and have excellent process control. Single- and multiple-cavity (2 and 4 cavities) molding tools with replaceable inserts are standard in the industry. For components associated with higher-volume products which have a long product life, higher tool cavitation can often be justified. In these cases, molds of 16, 18, or 20 cavities may be employed. Automation to pick and degate parts is employed for high-volume products in an effort to reduce costs.

In the past few years, several new materials have been proven to be moldable. Table C.2 indicates some of the polymers commonly used today. The capability of the molding process has also been advanced. Depending on material, lens type, and size, newer polymer molding processes routinely hold tight tolerances of 1 fringe accuracy during testing.

TABLE C.2 Table of Commonly Used Polymers

	Unit	Acrylic	Acrylic Copolymer	Polystyrene	Polyetherimide
Trade name		Plexiglas	UVT	Styron	Ultem
Refractive Index					
n_f (486.1 nm)		1.497		1.604	1.689
n_d (589 nm)		1.419	1.49	1.59	1.682
n_c (656.3 nm)		1.486		1.585	1.653
Abbe value V_d		57.2	50-53	30.8	18.94
Transmission	1%	92-95	92-95	87-92	82
Maximum continuous service temp.	°F °C	161 72	190 88	180 82	338 170
Water absorption	3%	0.3	0.25	0.2	0.25
Haze	%	1-2	2	2-3	
$dN/dN \times 10^{-5}$	/°C	-8.5	−10 to −12	−12	
Color/tint		Water clear	Water clear	Water clear	Amber
Benefit 1		High transmission and purity	High transmission and purity	High index	Impact resistant
Benefit 2		Scratch resistance Chemical resistance	Excellent UV properties	Clarity	Thermal and chemical resistance
Benefit 3		High Abbe value Low dispersion High melt flow	82% transmission at 924-301 nm, 1 mm CT		High index

SOURCE: Courtesy of Syntec.

Glass Molding

Glass molding technology, first developed in the early 1970s, has become increasingly available in the past decade. Prior to the turn of the century, most glass molding technology was proprietary and only available to the corporations that developed it. During the past decade, domestic and offshore manufacturers have marketed systems capable of producing precision-molded glass lenses. Figure C.9 shows a molding machine developed and produced in the United States and commercially available around the world, and Figure C.10 shows molding inserts and molded asphere lenses.

Not all glass is moldable. Several factors are important in the determination of moldability. Constituents in the glass and the glass transition temperature are factors to be considered. Molding trials are conducted to ensure moldability. Figure C.11 shows glass types that are moldable by one U.S. manufacturer with its

Polycarbonate	Methylpentene	ABS	Cyclic Olefin Polymer	Nylon	NAS	SAN
Lexan	TPX	Acrylon	Zeonex	Polyamide	Methyl	Styrene Acrylonitrile
1.593	1.473		1.537		1.575	1.578
1.586	1.467	1.538	1.53	1.535	1.533-1.567	1.567-1.571
1.58	1.464		1.527		1.558	1.563
34	51.9		55.8		35	37.8
85-91	90	79-90.6	90-92	88	90	88
255			253	179.6	199.4	174-190
124			123	82	93	79-88
0.15			<0.01	3.3	0.15	0.2-0.35
1-3	5	12	1-2	7	3	3
−11.8 to −14.3			−8		−14	−11
Water clear	Slight yellow	Durable	Water clear		Water clear	Water clear
Impact strength	Chemical resistance		High transmission and purity	Chemical resistance	Good index range	Stable
Temperature resistance			Low moisture absorption			
			Good thermal stability			

FIGURE C.9 Commercially available glass molding machine. SOURCE: Courtesy of Moore Nanotech, LLC.

FIGURE C.10 Generic molding inserts and molded asphere lenses. SOURCE: Courtesy of Rochester Precision Optics.

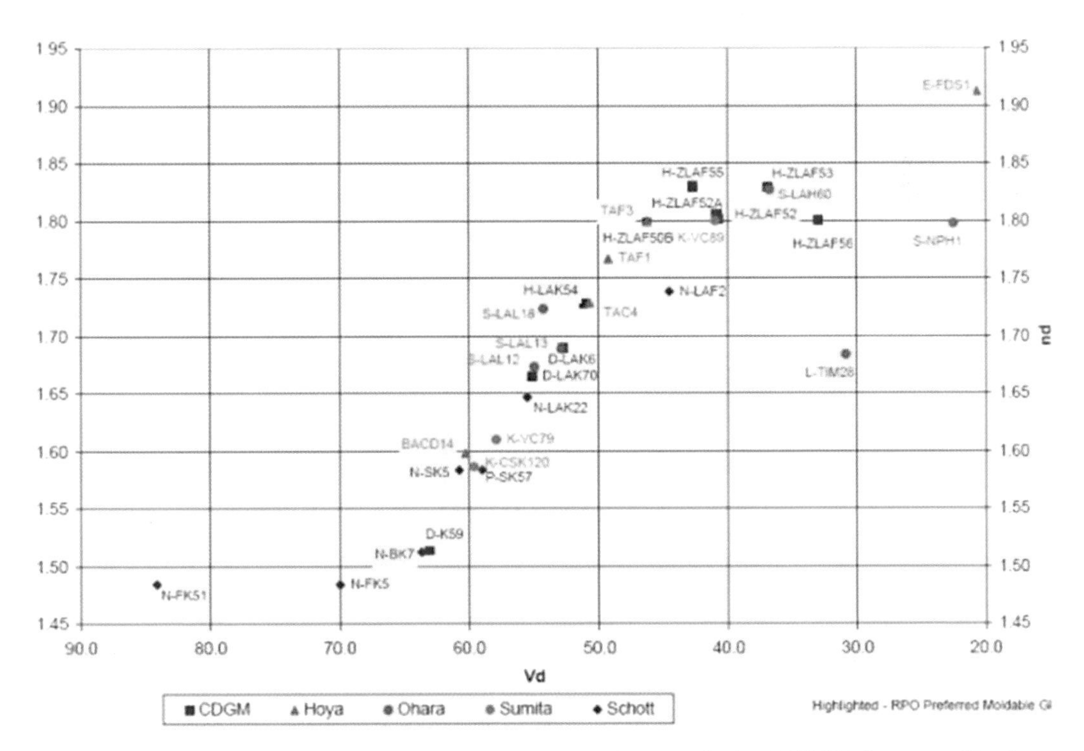

FIGURE C.11 Moldable glasses types available from Rochester Precision Optics. SOURCE: Courtesy of Rochester Precision Optics.

proprietary process. Figure C.11 indicates that lenses can be produced from glass with indices of refraction that range from approximately 1.48 to above 1.9, with Abbe numbers from 84.47 to 20.88, respectively.

The quality of lenses produced from the glass molding process is highly dependent on the ability to produce molding tools. Molding inserts are single-point diamond turned or ground and post-polished. The resulting inserts and tools are very capable with surface roughness as low as 5 A root mean square (RMS) and a surface accuracy of 1/10 λ. High-precision tools are capable of producing lenses with tolerances for power of 3 fringes and irregularity of ½ fringe.

Magnetorheological Finishing

The use of magnetorheological finishing (MRF) has increased in the fabrication of optical components. MRF requires the starting surfaces to be polished and is a deterministic polishing process that uses interferometry to provide feedback to the polishing tools. Figure C.12 shows a schematic of the polishing process. Ferrous-laden fluid passes through an electromagnetic field where its viscosity is stiffened, allowing the creation of a precise and repeatable polishing tool. When the fluid rotates out of the magnetic field, it is collected and recirculated. Constant monitoring of the polishing process parameters such as pressure and flow rate and

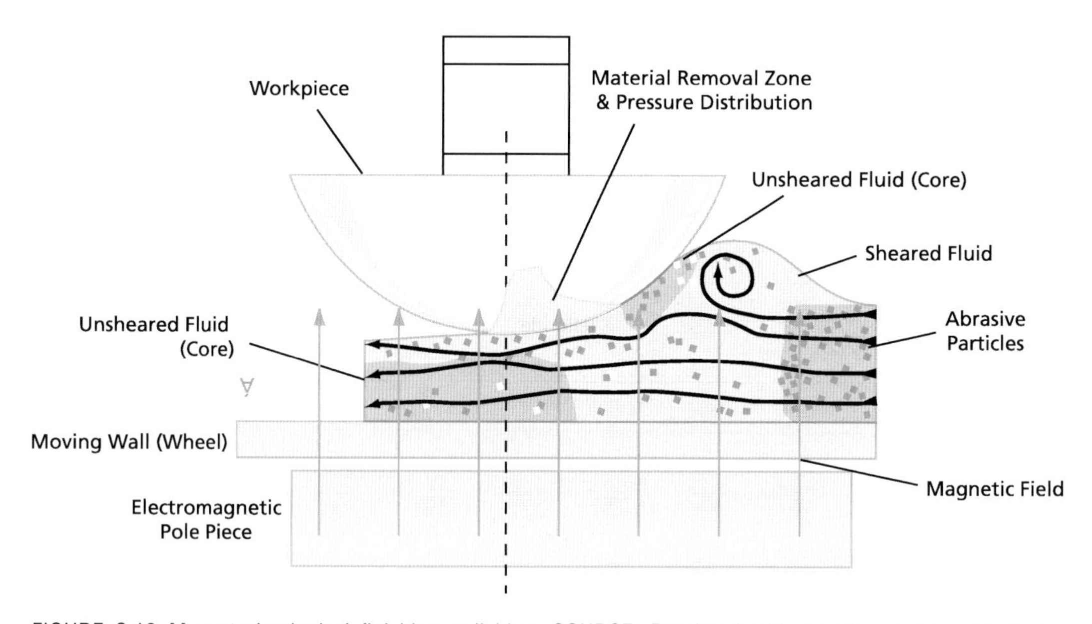

FIGURE C.12 Magnetorheological finishing polishing. SOURCE: Reprinted with permission from *Photonics Tech Briefs*.

the addition of process fluids provides for a predictable removal rate. Very accurate surfaces, particularly aspheric surfaces, can be produced using the MRF process. Surface accuracies of better than 1/20 λ peak to valley are standard for this process.

Single-Point Diamond Turning

Single-point diamond turning (SPDT) as a fabrication process for optical components has grown in popularity since the mid-1970s. SPDT is now routinely used to produce finished optical elements as well as mold inserts for polymer lenses and glass molding. The machining process uses single-crystal diamond cutting tools combined with nanometer-precision positioning to generate spherical surface geometries as well as more complex geometries such as toroids, aspheres, and diffractives.

The materials best suited to SPDT include metals, crystals, and polymers. Box C.1 provides a list of optical materials that have been demonstrated to be machinable. It is important to note that Box C.1 is not an exhaustive list of the materials that are machinable in the SPDT process, but rather a general listing. It is known that ferrous metals generate excessive tool wear; non-ferrous metals are thus the preferred metals. Metals such as aluminum 6061 and electroless nickel can be machined to produce a very high quality optical surface. As demand has increased for optical components that transmit in the infrared, the capability to machine IR materials has improved. Single-crystal materials such as those listed in Box C.1 can be reliably machined to

BOX C.1
Materials Suitable for Single-Point Diamond Turning

Metals	Nonmetals	Polymers
Aluminum	Calcium Fluoride	Cyclic Olefin
Brass	Magnesium Fluoride	Polymethylmethacrylate
Copper	Cadmium Telluride	Polycarbonates
Beryllium Copper	Zinc Selenide	Polyimide
Bronze	Zinc Sulphide	Polystyrene
Gold	Gallium Arsenide	
Silver	Sodium Chloride	
Lead	Calcium Chloride	
Platinum	Germanium	
Tin	Strontium Fluoride	
Zinc	Sodium Fluoride	
Electroless Nickel	Potassium di-hydrogen phosphate	
	Potassium titanyl phosphate	
	Silicon	

produce the required components. Polycrystalline materials are more difficult to machine and generally considered not diamond-machinable. Polymers have become popular in optical systems. SPDT provides an excellent solution for the manufacture of polymer prototypes, low and mid production, and inserts for polymer molds.

Optical Coating

Optical thin-film coatings technology has been advanced in response to requirements in multiple and diverse markets including telecommunications, health and medical, biometrics and defense markets. Evaporation deposition processes in which materials are deposited by means of a transformation from solid to vapor back to solid have been the most widely used processes in the optical industry. Although the coatings satisfy requirements, they are often porous and sensitive to humidity and thermal conditions. The ion-assisted deposition (IAD) process, as highlighted in Figure C.13, more tightly "packs" coating layers, yielding a more robust coating. These evaporation processes, both assisted and unassisted, are widely

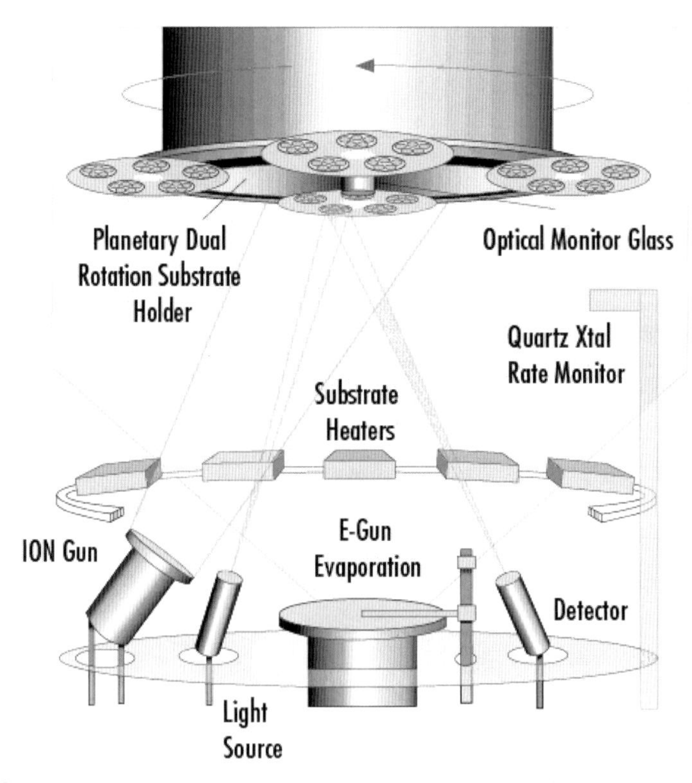

FIGURE C.13 Ion-assisted deposition with sources, heaters, and substrates labeled as well as detectors for monitoring the process. SOURCE: Courtesy of Edmunds Optics, Inc.

used for antireflection coatings, mirrors, filters, and beam splitters, all typically with less than 100 layers.

Responding to industry requirements for higher-precision coatings, ion beam sputtering (IBS) was developed. This process uses an ion gun to excite ions such that they collide with the source, resulting in sputtering of material from the source to the part being coated. The process provides very high quality coatings, although the cost of the equipment and its maintenance is high relative to IAD. IBS is often chosen for high- and ultrahigh-precision coatings.

In the past few years a technology newer than IBS, advanced plasma reactive sputtering (APRS), has been developed and is especially effective for complex coatings of more than 200 layers per run totaling more than 20 μm. Figure C.14 shows the APRS system from Leybold Optics. The APRS system uses two dual-magnetron sources, which operate at mid-frequency. Material oxidation occurs when the part being coated passes through an oxygen plasma. The deposition rate of approximately 0.5 nm/sec is similar to the deposition rates experienced with evaporation; however, the resultant coatings are denser and more stable.

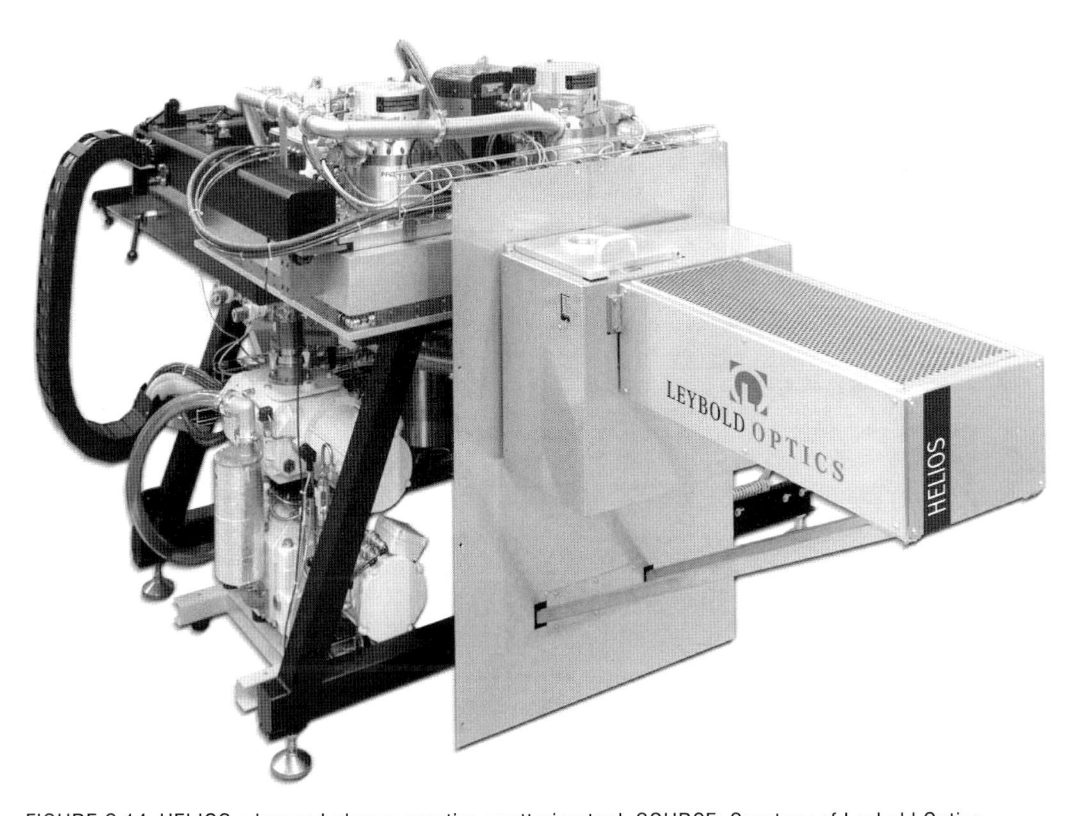

FIGURE C.14 HELIOS advanced plasma reactive sputtering tool. SOURCE: Courtesy of Leybold Optics.

Metrology

Metrology is an important enabling technology in the optics industry. It has long been said, "If you can't measure it, you can't make it." Over the past decade there have been advances in interferometry improving the ability to measure increasingly more challenging optics, particularly aspheres. Extremely tight tolerances for some of these aspheric geometries are the challenge for available metrology. Testing variability for ultrahigh-precision optics, as in optics for lithographic applications, often needs to be 3 to 5 times smaller than the tolerance of the optic being measured. Recent advancements, for example in stitching interferometry, shown in Figure C.15, have provided the industry with the capability of making

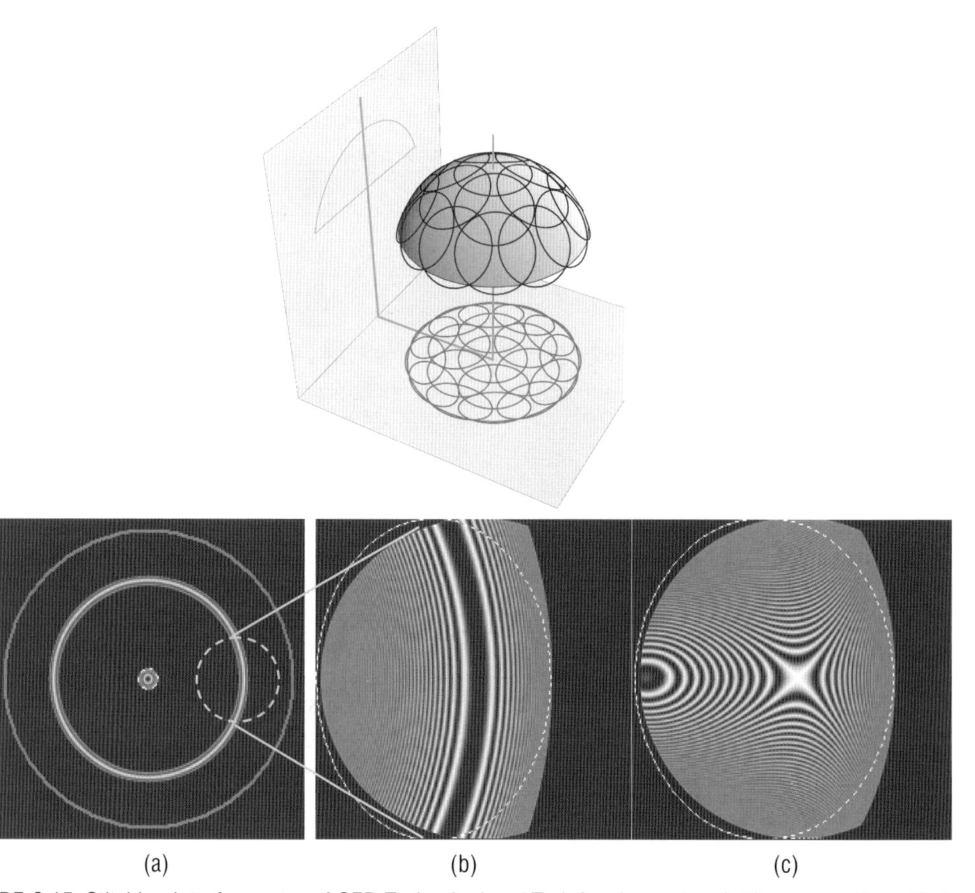

FIGURE C.15 Stitching interferometer of QED Technologies. (*Top*) A sub-aperture lattice covers the entirety of a high-numerical-aperture part. (*Bottom*) Simulated fringe patterns for an approximately 50 micon-depatture asphere: (a) the lack of data shows because the fringes are not resolved; (b) magnification allows for some resolution; (c) correct choice of local best-fit sphere increases fringe resolution even further. SOURCE: Images provided courtesy of QED Technologies International, Inc.

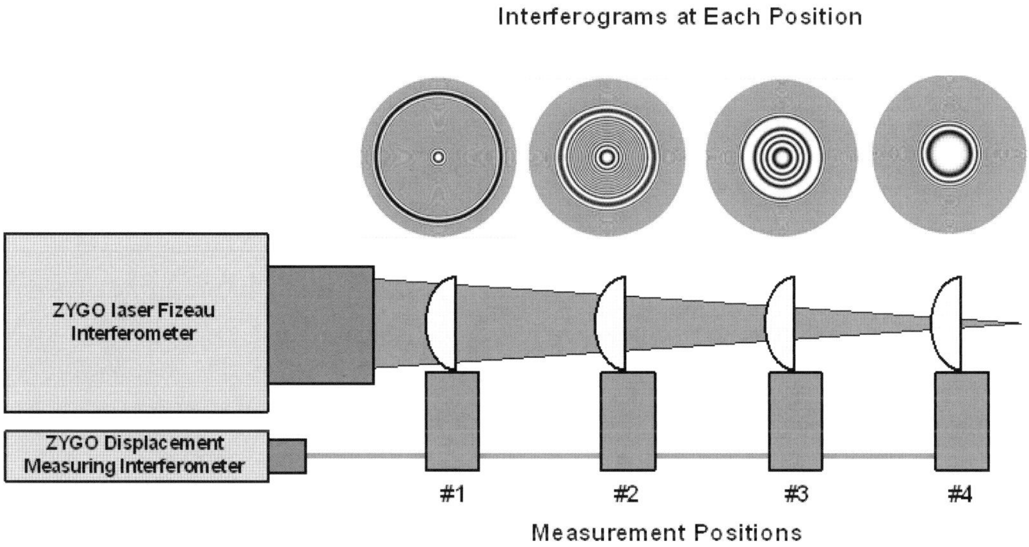

FIGURE C.16 The Verifire Asphere system of Zygo combines interferometer technology. SOURCE: Courtesy of Zygo Corporation. Reprinted with permission.

full-aperture measurements on optics exceeding 200 mm in diameter including aspheres with up to 650 microns of aspheric departure. Another advancement, shown in Figure C.16, combines laser Fizeau interferometry and displacement-measuring interferometry to reduce measurement uncertainty.

Gray-Scale Lithography for Diffractive- and Micro-Optic Components

Gray-scale lithography has become an important method for the fabrication of diffractive- and micro-optical components for optical systems applications spanning the range from the deep UV (193 nm) to the infrared (10.6 μm). The process involves the exposure of photoresist coated on a substrate. The photoresist is exposed using a focused laser beam, which is scanned across the surface of the photoresist using air-bearing translation stages. As the laser beam is scanned, the intensity is modulated, so that when the photoresist is developed, the desired surface-relief pattern is obtained. In the early work in the 1980s, surface-relief profiles were typically in the range of 1 to 5 μm; now one can obtain virtually any continuous, surface-relief profile with depths up to 100 to 120 μm. The developed photoresist master can then be replicated using UV cast and cure materials, or it can be used to create a nickel electroform, which in turn can be used in high-volume manufacturing processes, such as polymer-injection molding and roll-to-roll film manufacturing. Examples of surface-relief optical components, produced using this gray-scale lithography manufacturing method, are shown in Figure C.17.

Numerous and important applications exist for these lithographically generated, surface-relief optical components: for example, efficient extraction and distribution of light from LED sources for general lighting, laser-beam shaping for sensors systems, front- and rear-projection screens, and imaging and display systems. Or, one can also place the substrate with the patterned photoresist into a reactive-ion etcher chamber in which ions are used to bombard the surface and transfer (or etch) the surface-relief pattern directly into the substrate material. Common substrate materials for reactive-ion-etched optical components include fused silica, silicon, and germanium.

FIGURE C.17 Different types of regular microlens arrays. SOURCE: Reprinted, with permission, from Fan, Xiqiu, Honghai Zhang, Sheng Liu, Xiaofeng Hu, and Ke Jia. 2006. NIL—A low-cost and high-throughput MEMS fabrication method compatible with IC manufacturing technology. *Microelectronics Journal* 37(2):121-126.

DISPLAYS

Polarization in Liquid-Crystal Displays

If two pieces of polarized material are placed in series between an observer and a light source and then rotated relative to one another, at one point the transmission of light will be blocked as the polarization of the two pieces of material is perpendicular. If from that point they are rotated another 90°, the polarizers will be aligned and about half the light will pass (with the other half absorbed by the first polarizer). A variable amount of light can be passed by rotating the polarizers in between the extreme perpendicular and aligned positions, as shown in Figure C.18. This form of light modulation automatically loses a factor of two in the best case unless the initial light source is polarized.

A liquid-crystal display (LCD) works roughly the same way; however, rather than rotating the polarizers, the light itself is rotated by a liquid crystal. Light from a source is passed through two fixed polarizers with liquid crystal in between. The polarizers are crossed, but in the absence of electric current, the thickness of the liquid-crystal layer is such that it rotates the polarization by 90°, so light is passed. Applying an electric field alters the alignment of the liquid crystal so that the light is not rotated and light is thus blocked. Varying the strength of the field varies the degree of alignment and thus the amount of light passed, as indicated in Figure C.19. A wide variety of colors can be achieved by varying the amount of light passed through each of the three subpixels.

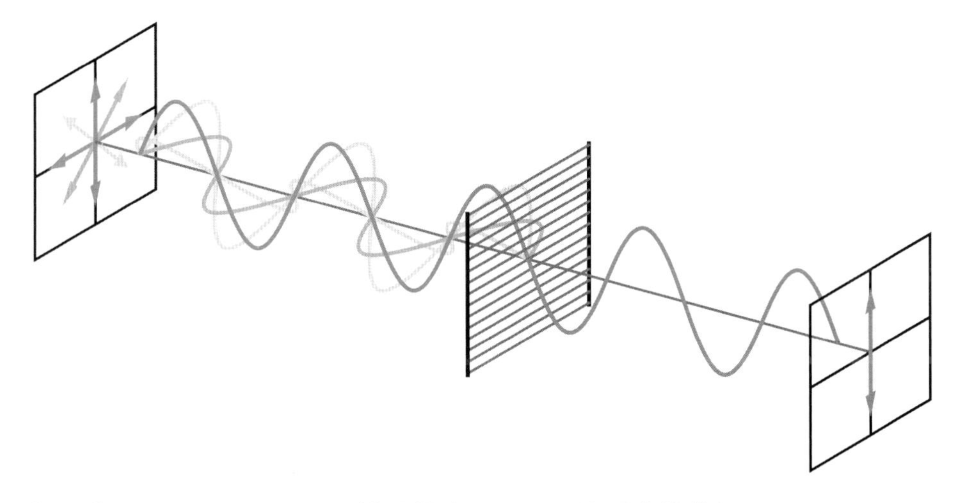

FIGURE C.18 Polarized light waves. SOURCE: Image created by Bob Mellish.

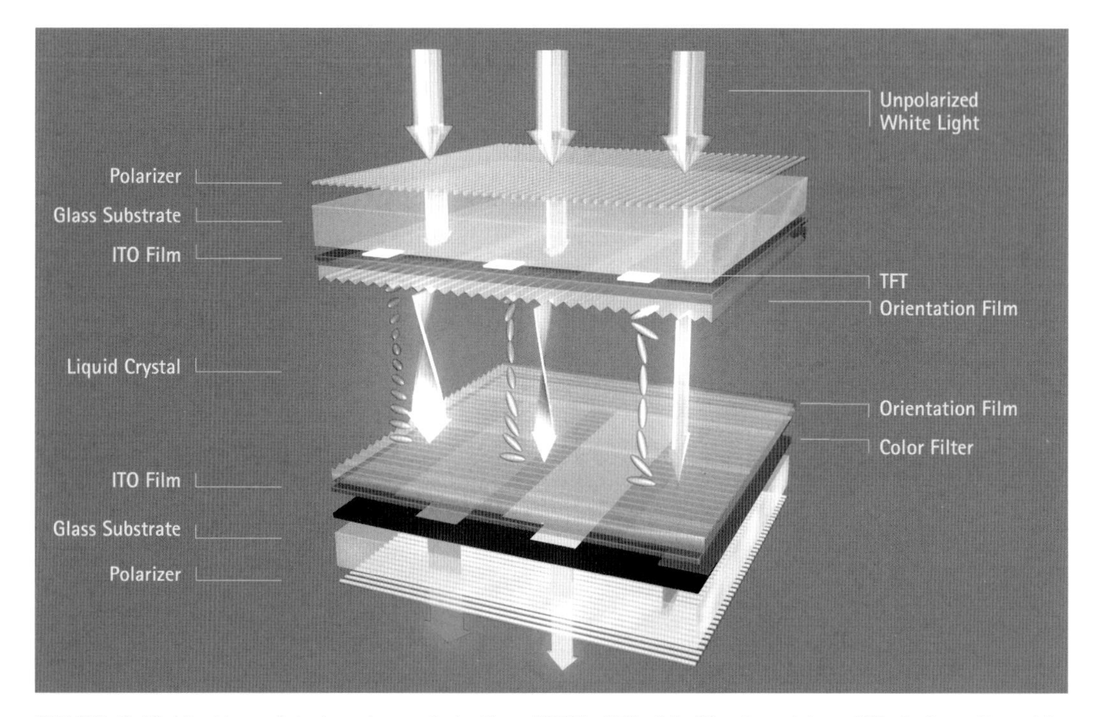

FIGURE C.19 Liquid-crystal changing polarization. NOTE: TFT, thin-film transistor; ITO, indium tin oxide. SOURCE: © Merck KGaA, Darmstadt, Germany. Reprinted with permission.

Three-Dimensional Technology in LCDs

One method used to achieve a three-dimensional effect using an LCD display that was mention in Chapter 10 alternates showing an image for one eye, then a slightly displaced image for the other eye. Wireless connection between a pair of glasses and the display keeps the polarization of glasses in sync, only passing light to the intended eye. The first-generation three-dimensional LCD televisions used such active shutter glasses, seen in Figure C.20.

The shutter glasses were themselves crude LCDs, consisting of a single large pixel per eye. One of the challenges of this approach is brightness. In an LCD, the entire screen is not switched at once, but the screen is painted line by line, typically from top to bottom. As the screen is being refreshed for one eye, the top of the display shows the desired image for that eye, but the bottom of the display still shows the previous image for the other eye. The display is constantly refreshing, so only for an instant during each cycle would the screen show an image for only one eye. To overcome this problem, an image for a given eye was shown twice in succession and the shutter glasses opened during the second painting. Unfortunately,

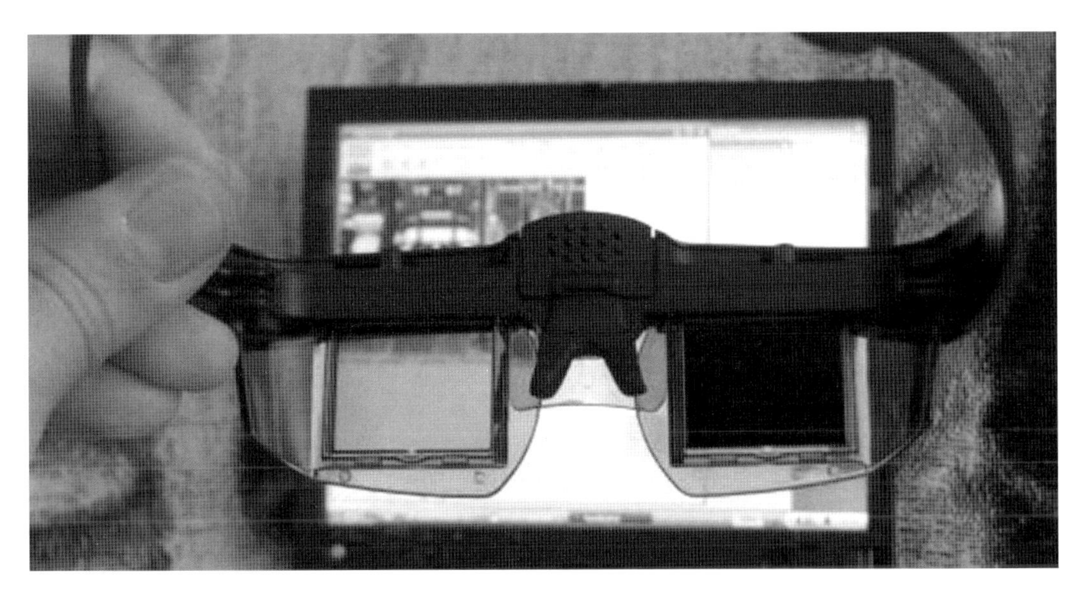

FIGURE C.20 Shutter glasses blocking left image from entering right eye. Image courtesy of E. Svedberg.

this solution meant that the shutter glasses for any one eye were passing light only during one out of four frames. Thus, the apparent brightness of the display was only one-quarter that of the same display when used in a two-dimensional mode without glasses.

More recently an alternative approach to three-dimensional displays has been introduced, placing a patterned optical retarder on the face of the display. The role of this retarder is to continually twist the polarization of every other row, as shown in Figure C.21. To view the display in three dimensions, observers wear passive glasses that have polarizers 90° opposed. The left eye can then only observe the odd rows, say, while the right eye can only observe the even rows. The screen brightness is thus increased because both eyes are constantly receiving signal.

Unfortunately, the trade-off of this alternate scheme is a reduction in vertical resolution. Rather than each eye receiving the 1,080 rows of a standard high-definition set, each receives only 540 rows. This reduction in vertical resolution has motivated some to suggest that future sets be made with the standard 1,920 horizontal pixels but with double the number of vertical pixels, to 2,160, so that each eye then can receive full high definition when viewing in three-dimensional mode. In two-dimensional mode, the even and odd row pairs could mimic one another to result in standard high definition, as input sources with doubled vertical resolution may be rare.

Another possible approach for creating three-dimensional displays puts two

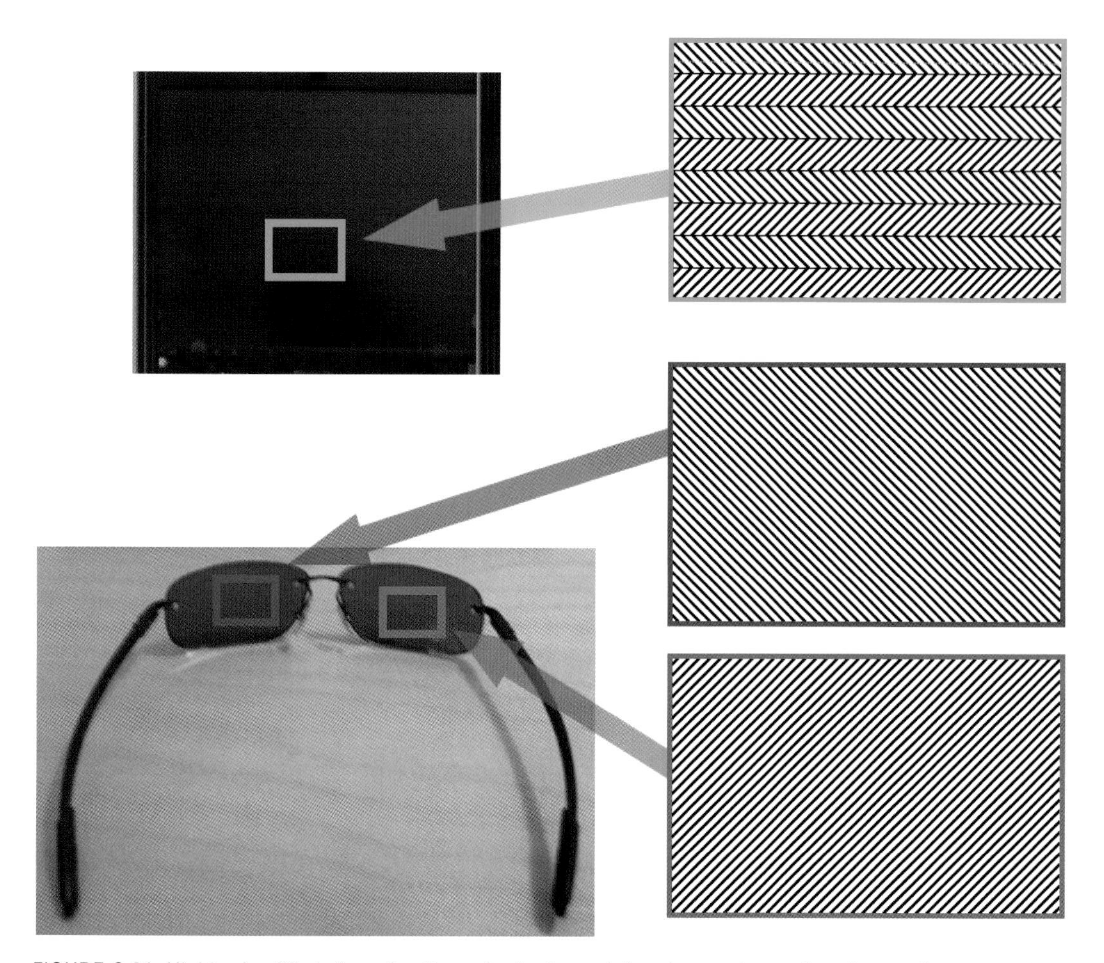

FIGURE C.21 Highly simplified view of patterned retarder and the observer's passive glasses. Image courtesy of E. Svedberg.

liquid-crystal displays in series. In this active retarder scheme, the rear display has the traditional role, while the role of the front display is to rotate the polarization on a frame-by-frame basis. The viewer then wears passive glasses to filter out images not intended for a given eye.

One might expect brightness to be an issue with an active retarder as it is with active shutter glasses. However, the brightness issue can be greatly reduced by synchronizing the two displays and segmenting the backlight. That is, during the refreshing of a given row or set of rows, the backlight can be turned off. As the set of rows is updated for the alternate eye, both in terms of image from the rear display and polarization from the front display, the backlight is turned on for those rows. The dark period is then only during the refreshing of a set of rows rather than affecting the entire display.

Other Three-Dimensional Display Methods

The discussion of three-dimensional displays so far has involved glasses of one sort or another. Unfortunately, the need for glasses, even the relatively more comfortable passive glasses, is thought to be stifling the broad adoption of these displays. So the question arises as to how to create three-dimensional displays without the need for glasses. There are actually two cases to consider: that of the individual observer and that involving a group of observers. Of the two, the case of the individual observer is the easier, provided that the position from which the individual is observing is known. This may be a reasonable assumption when considering a handheld device. In such a situation, there are several means of alternating the delivery of a left image only to the left eye and a right image only to the right eye. For example, the backlight might have an illumination source on the left and right side within the display, with those sources alternating and being steered by some projection film to one eye, then the other. The far more difficult problem is having a three-dimensional display without the need for glasses when multiple viewers can be positioned in a wide variety of locations. As was mentioned in Chapter 10, the most popular approach is the use of viewing zones created by lenticular arrays.

Touch Displays

The signal detection in capacitive touch displays is dependent on the grid of unit cells, defined by a unique combination of row and column electrodes. When a signal is applied to a row electrode, the proximity of column electrodes results in coupling that can be measured. By sweeping through the rows, measurement can be made of the entire screen.

As illustrated in Figure C.22, the signal radiates a small distance through nonconductive materials, such as the cover glass, and one might say that such coupling projects through the cover glass. This coupling is attenuated by a finger touching the cover glass, which provides a path to ground through the body. This reduction in capacitive coupling can be measured, and based on the readings from each unit cell the center of the touch position or positions can be interpolated to higher resolution than the cell spacing. This imaging of the touch positions has enabled the multi-touch capability.

While the conductors inside the display aperture area must be transparent, outside the display aperture, and beneath the black border that commonly surrounds the display, are metal conductors that have lower resistivity than the ITO, providing for reduced signal loss en route to an integrated circuit (not shown)

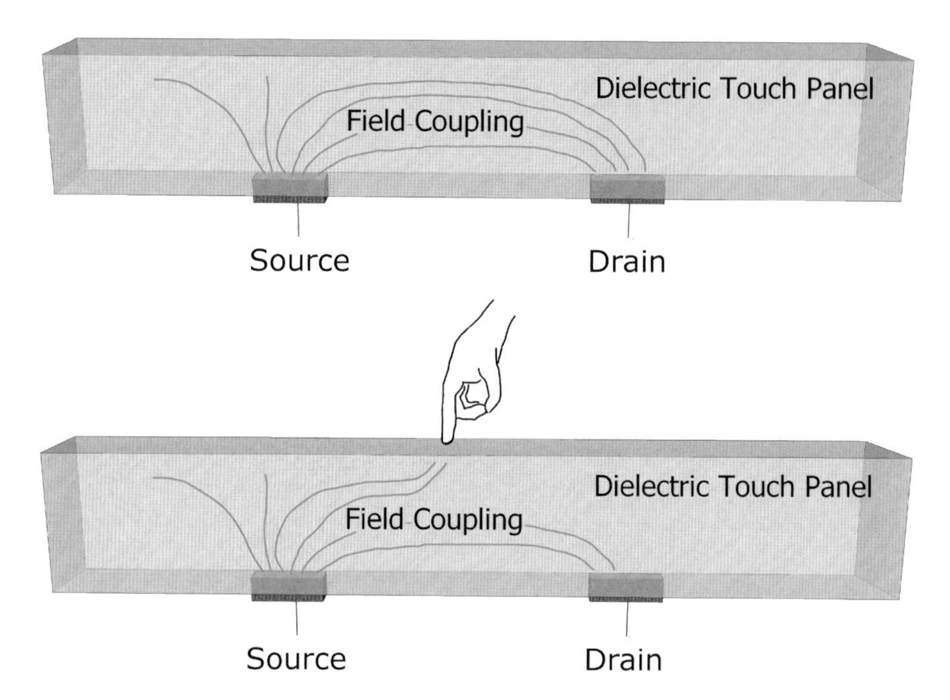

FIGURE C.22 Projected capacitive touch operation. (*Top*) The electric field couples between the source and drain through the dielectric touch panel. (*Bottom*) A grounding by means of touch effects the coupling and can be used to sense position of the touch. (Not to scale.)

mounted to a flex circuit bonded to the sensor glass with anisotropic conductive film, as shown in Figure C.23.

There have been efforts within the industry to eliminate the separate substrate that is dedicated to the touch sensor function. Of the existing substrates considered for integration with this function, the leading candidates are the underside of the cover glass and the face of the color filter glass.

Although the underside of the cover may seem quite attractive, subtle aspects work against this choice. In particular, touch sensors based on dedicated substrates are typically fabricated on large glass sheets, which are then diced into the smaller sheets needed for the display, even though such large-scale lithography runs counter to the manner in which cover glass is made.

The common way for cover glass to be fabricated achieves high retained strength so that it can survive damage inflicted by everyday use. This is done by putting all surfaces under compression. After cutting the cover to shape, it is then dipped into an ion exchange acid bath where smaller sodium ions are exchanged for larger potassium ions, putting all surfaces under compression. Since glass fails

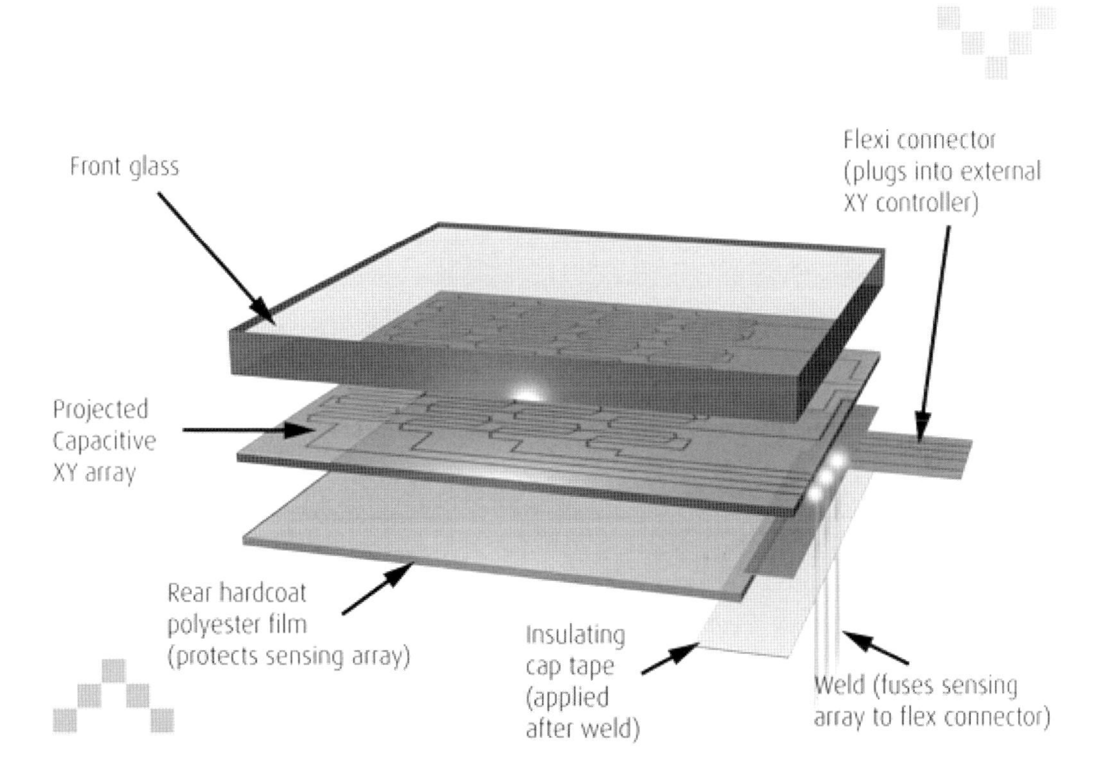

FIGURE C.23 Dedicated touch sensor glass. SOURCE: Courtesy of Zytronic Displays Limited.

under tension, not compression, the compressive stress layer increases damage resistance.

Unfortunately, dipping the entire cover in the acid bath results in all exposed glass surfaces becoming ion exchanged, not just the large front and back faces. If a large sheet of uncut cover glass were to be ion exchanged, after which the transparent conductors were patterned, dicing individual covers from the large sheet would be difficult because the sheet already had been ion exchanged. Even if success in dicing were achieved, this still would be problematic because none of the exposed edges would have been ion exchanged, and thus central tension would be exposed. As integrating touch into the cover is difficult, the face of the color filter glass is an alternative location to consider. However, the fundamental challenge here is that the switching noise from the transistors painting the display image can be coupled into the touch electrodes. Designers of touch ICs have made some progress toward devising schemes to avoid and/or reject such noise.[23]

One of the implications of integrating touch into the face of the LCD's color filter glass is the elimination of air gaps. Although the dedicated touch substrate is commonly bonded directly to the cover glass, the bonded assembly itself is only sometimes bonded to the front polarizer of the display. Such direct bonding results in additional coupling of display noise, which is a particularly challenging problem with larger displays, like those of tablets, since ITO is such a poor conductor. As a result, although direct bonding can improve optical transmission by eliminating Fresnel reflections on either side of the air gap, eliminating the air gap is more challenging with larger displays because of problems with achieving high yields with this process.

These weaknesses and those mentioned in Chapter 10 have motivated inventors to look for other technologies to compete with projected capacitive.

As mentioned in Chapter 10, optical touch displays have limitations. However, the size and conductivity of the touching object in such a system are immaterial, and there is no reduction in optical transparency as there is with the not-completely-transparent conductors and any air gap of a projected capacitive touch system; see Figure C.24. However, optical touch capability has not been achieved on a large scale and—while the multi-touch experience is now commonly expected because of the widespread use of modern handheld devices—has yet to be developed.

Display Frames

A trend in the use of LCDs in arrays is the reduction in the gap, or bezel, between the pixels of adjacent displays for application in video walls. Although still not completely seamless, gaps have been reduced to the single-digit millimeter range, as represented in Figure C.25. This change benefits the image quality, but it has repercussions for touch. In particular, a cover glass can be bonded to a relatively wide bezel, but as that bezel shrinks, it becomes less feasible to use the bezel as a mount. Although very narrow brackets might be used to affix cover glass to large-format displays, in time that might change to direct bonding of cover glass.

This prospect has numerous challenges, such as cutting the glass after ion exchange as mentioned above in the discussion of integrated touch. Thermal expansion would be an additional challenge, as current color filter substrates match the thermal expansion of the glass with the thin-film transistors (TFTs), which are

[23] See, for example, "Development of IPS LCD with Integrated Touch-Panel by Hitachi Displays." 2010. Available at http://japantechniche.com/2010/10/08/development-of-ips-lcd-with-integrated-touch-panel-by-hitachi-displays/. Accessed July 27, 2012.

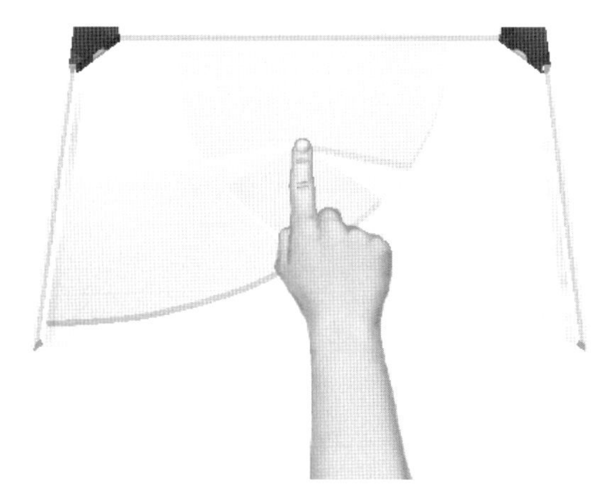

FIGURE C.24 Optical sensors in the upper corners of the array detecting touch. SOURCE: NextWindow. 2012. "Optical Touch Overview." Available at http://www.nextwindow.com/optical/. Reprinted with permission.

FIGURE C.25 Narrow-bezel liquid-crystal display array. SOURCE: Courtesy of NEC Display Solutions of America.

switched to control the light valves that form the image. As these transistors are based on silicon, the glass substrate material is chosen to match the coefficient of thermal expansion of silicon, which is relatively low. Cover glass, by contrast, has relatively high thermal expansion, as this is currently thought necessary to achieve a glass capable of ion exchange. Nevertheless, if an internal polarizer were to be achieved, researchers would be highly motivated to create a glass that combines cover and color filter functionality.

OLED Displays

In order to form a light-emitting device, the light-emitting organic layer of an organic light-emitting diode (OLED) requires several other layers. These include a transparent substrate, which can be either rigid or flexible, depending on the application; a transparent conducting anode; a conducting organic layer; the organic light-emitting layer; and a cathode, which may or may not be transparent, depending on the application.

In operation, an electrical potential is applied across the OLED by connecting a battery or other power source between the cathode and the anode, causing a current to flow. The current flow results in electrons being removed from the molecules in the emissive organic layer, creating holes. When these holes are filled at the interface with the conducting organic layer, the electrons give up their excess energy as photons. The intensity of the emitted light is determined by the total current flow, and the color is determined by the energy level of the hole that is filled by the electrons. This, in turn, is determined by the properties of the organic molecules, allowing OLEDs to be used in color displays.

OLEDs can be made on transparent substrates to form an all-transparent display, or on an opaque or reflective substrate. In the former case, it makes possible what is known as a heads-up display, since only the displayed information interrupts the visual field.

Flexible Displays

"Flexible display technology" is a term used for a desirable technology for the next generation of cell phones, military devices, and reading devices. A device with flexible display technology would enable the user to overcome the fear of breaking, bending, or scratching the device.

One type of flexible display technology uses organic films constructed from OLEDs, which in turn are made from layers of organic material and the conductive materials needed to inject electrons and holes. When a voltage of proper polarity is applied to the conductive layers, electrons from one layer combine with the holes from the other, releasing light. If these OLEDs were constructed from polymers with high flexibility, they could be the basis of lightweight, portable, rollup displays, or displays that could be used on curved surfaces.

Another promising material technology is amorphous oxides. Some amorphous oxides can form thin films that are transparent and electrically conductive, which is why they already serve as the see-through electrode layer in displays and solar cells. It was this combination of qualities that led to the surge in research that began in 1996, when Hideo Hosono and his colleagues at the Tokyo Institute of Technology first noted the merits of amorphous transparent conducting oxides. The biggest problem when amorphous silicon is deposited on flexible plastic is switching and drifts.

Amorphous oxides could do more than simply serve as passive electrodes. They could also replace amorphous silicon as the active semiconducting material in TFTs. The advantages of oxide semiconductors over amorphous silicon are motivating much work in the display industry. Only 2 years after the first oxide-based transistors were reported, Korea's LG Electronics Co. revealed a prototype OLED display that used indium gallium zinc oxide (IGZO) transistors to drive its pixels. Other companies followed quickly, with oxide-based displays of their own. The U.S. $100 billion flat-panel-display industry has been built on amorphous silicon, and the new materials will have to compete with its 30-year head start. However, amorphous silicon is a mature technology, and most limitations arise from fundamental physical and chemical properties requiring breakthroughs.

Amorphous oxide semiconductors will likely challenge amorphous silicon. When this will happen depends mainly on the development time for a large-scale

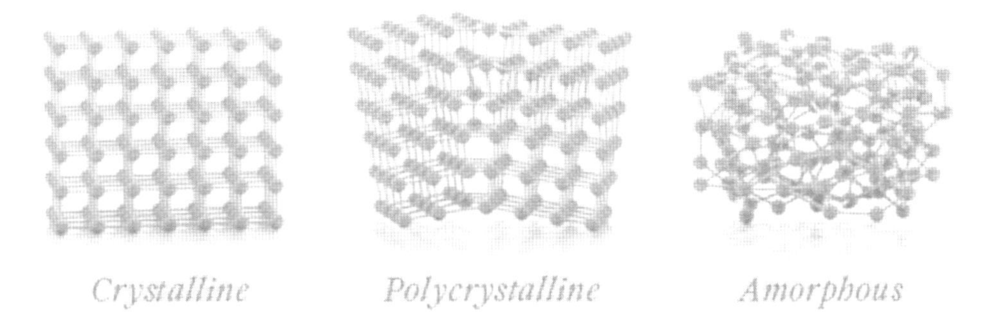

FIGURE C.26 Crystalline, polycrystalline, and amorphous atomic structures. SOURCE: Reprinted, with permission, from Wager, J.F., and Hoffman, R. 2011. Thin, fast, and flexible. *IEEE Spectrum* 48(5). Copyright 2011 by IEEE.

manufacturing capability incorporating these materials. However, the oxide TFT fabrication process is very similar to that used for amorphous (for atomic structures, see Figure C.26) silicon devices; thus the display industry can leverage much of the existing infrastructure and know-how. A key advantage amorphous oxides hold over amorphous silicon is their higher charge-carrier mobility.

D

Biographies of Committee Members

PAUL McMANAMON, *Co-Chair*, is a consultant at Exciting Technology, LLC, and is half-time technical director for the Ladar and Optical Communications Institute (LOCI) at the University of Dayton. Dr. McManamon had been a member of the scientific and technical cadre of senior executives in the Sensors Directorate, Air Force Research Laboratory, Wright-Patterson Air Force Base, Ohio, before becoming chief scientist of the directorate. The Sensors Directorate consisted at that time of approximately 1,100 people. It is responsible for developing new sensor technology for the Air Force. Dr. McManamon was responsible for the directorate's technical portfolio. He served 32 months as acting chief scientist for avionics for the Avionics Directorate in Wright Laboratory. He was the technical lead for more than 500 scientists and engineers, and he was responsible for the technical content of all electro-optical and microwave sensor development, electron device development, and automatic target recognition as well as avionics systems, concepts, and simulation. Prior to serving in that position, he had been the Sensor Directorate's senior scientist for infrared sensors, developing multidiscriminate electro-optical sensors, including multifunction laser radar technology, novel electro-optical countermeasure systems, and optical phased-array beam steering technology. Dr. McManamon is widely recognized in the electro-optical community and was elected as the 2006 president of SPIE, the International Society for Optical Engineering, of which he is also a fellow. He is a member of the executive board and a fellow of the Military Sensing Symposia, a fellow of the Air Force Research Laboratory, a fellow of the Institute of Electrical and Electronics Engineers, and a fellow the Optical Society of America.

ALAN E. WILLNER, *Co-Chair,* is currently the Steven and Kathryn Sample Chaired Professor of Engineering in the Ming Hsieh Department of Electrical Engineering, Viterbi School of Engineering, University of Southern California (USC). Professor Willner received his Ph.D. in electrical engineering from Columbia University, in 1988 and an Honorary Doctorate from Yeshiva University in 2012. He was a post-doctoral member of the technical staff at AT&T Bell Laboratories and a member of the technical staff at Bellcore. At USC, he is the associate director of the Center for Photonics Technology and was an associate director for student affairs for the National Science Foundation Engineering Research Center in Multimedia. Professor Willner is a member of the Defense Sciences Research Council, has served on many scientific advisory boards for small companies, and has advised several venture capital firms. Professor Willner has received the following honors/awards: International Fellow of the U.K. Royal Academy of Engineering, Presidential Faculty Fellows Award from the White House, Packard Foundation Fellowship, Guggenheim Foundation Fellowship, National Science Foundation National Young Investigator Award, Fulbright Foundation Senior Scholar Fellowship, Optical Society of America (OSA) Forman Engineering Excellence Award, Institute of Electrical and Electronics Engineers (IEEE) Photonics Society Engineering Achievement Award, OSA Leadership Award, and 2001 Eddy Paper Award from Pennwell Publications for the Best Contributed Technical Article. He is a fellow of the American Association for the Advancement of Science, IEEE, OSA, and SPIE. He was an invited foreign dignitary representing the sciences for the 2009 Nobel Prize Ceremonies in Stockholm. His professional activities have included serving as president of the IEEE Photonics Society (formerly the Lasers and Electro-Optics Society, or LEOS), co-chair of the Science and Engineering Council of the OSA, vice president for technical affairs of IEEE Photonics Society, Photonics Division chair of OSA, editor-in-chief of the OSA *Optics Letters*, editor-in-chief of *IEEE/OSA Journal of Lightwave Technology*, editor-in-chief of *IEEE Journal of Selected Topics in Quantum Electronics*, chair of the IEEE TAB Ethics and Member Conduct Committee, and general and program co-chair of the Conference on Lasers and Electro-Optics. Professor Willner has 975 publications, including 1 book, 24 U.S. patents, 18 keynote or plenary talks, and 16 book chapters. His research is in optical communications, optical signal processing and networks, fiber optics, and optical technologies.

ROD C. ALFERNESS (NAE) is the dean of engineering at the University of California at Santa Barbara. Prior to this position, he was the chief scientist at Bell Laboratories, Alcatel-Lucent. His previous positions were with the Bell Laboratories as the research senior vice president and the optical networking research senior vice president. Dr. Alferness also was the chief technical officer and advanced technology and architecture vice president of the Optical Networking Group, Lucent Technologies. Prior to that role, he was head of the Photonics Networks Research

Department of Lucent Bell Laboratories, Holmdel, New Jersey. Dr. Alferness joined Bell Labs in 1976 after receiving a Ph.D. in physics from the University of Michigan, where his thesis research, under the supervision of Professor Emmett Leith, concerned optical propagation in volume holograms. His early research at Bell Labs included the demonstration of novel waveguide electro-optic devices and circuits—including switch/modulators, polarization controllers, tunable filters—and their applications in high-capacity light wave transmission and switching systems. This research led to the early development of titanium-diffused lithium niobate waveguide modulators that are now deployed as the high-speed signal-encoding engine in fiber-optic transmission systems around the world. Dr. Alferness has also made contributions in photonic integrated circuits in indium phosphide, including widely tunable lasers, as well as in photonic switching systems and reconfigurable wavelength division multiplexed (WDM) optical networks. In the mid-1990s, he was an originator of and the Bell Labs program manager for the Defense Advanced Research Projects Agency-funded MONET project, which demonstrated the feasibility of wavelength-routed optical networks that are now being implemented for both backbone and metro networks. Dr. Alferness has authored more than 100 papers, 5 book chapters, and 35 patents. He is a member of the National Academy of Engineering and a fellow of the Optical Society of America (OSA) and the Institute of Electrical and Electronics Engineers (IEEE) Lasers and Electro-Optics Society (LEOS). Dr. Alferness received the 2005 IEEE Photonics Award. He has served as an elected member of the LEOS Administrative Committee and was the president of IEEE LEOS in 1997. He was general co-chair of the 1994 Optical Fiber Communications Conference. Dr. Alferness has served as associate editor for *Optics Letters* and for *Photonics Technology Letters.* He has served on many IEEE and OSA committees, including fellows and awards committees. He also currently serves on the European Conference on Optical Communication Executive Management Committee. He served as the editor-in-chief of the IEEE and OSA-sponsored *Journal of Lightwave Technology* from 1995 to 2000. He served as an elected member of the OSA board of directors from 2001 to 2003.

THOMAS M. BAER is the executive director of the Stanford Photonics Research Center and a member of the Applied Physics Department at Stanford University. His current research is focused on developing imaging and analysis technology for exploring the molecular basis of developmental biology and neuroscience. From 1996 to 2005, Dr. Baer was the chief executive officer and chair of Arcturus Bioscience, Inc., a biotechnology company located in Mountain View, California, that he established in 1996. Arcturus Bioscience pioneered the area of microgenomics. Prior to establishing Arcturus, Dr. Baer was the vice president of research at Biometric Imaging. From 1981 to 1992, he was at Spectra-Physics, Inc., where he held positions as vice president of research and Spectra-Physics Fellow. While Dr. Baer was at Spectra-Physics, his research focused on ultrafast lasers, optical pulse compression, diode-pumped solid-state lasers, and nonlinear optics. He has made major contributions in the areas of biotechnology, quantum electronics, and laser applications and is listed as an inventor on 60 patents and is a co-author on many peer-reviewed publications in a number of scientific fields. His commercial products have received many industry awards for design innovation. Co-founder of four companies in Silicon Valley, he was named entrepreneur of the year for emerging companies in Silicon Valley in 2000 by the *Silicon Valley Business Journal*. He graduated with a B.A. degree in physics (magna cum laude) from Lawrence University and received his M.S. and Ph.D. degrees in atomic physics from the University of Chicago. He is also an alumnus of the Harvard Business School, and in 1994 he received the Distinguished Alumni Award from Lawrence University. He has been elected fellow in two international scientific societies—the American Association for the Advancement of Science and the Optical Society of America (OSA)—and served as the president of the OSA in 2009.

JOSEPH BUCK is the vice president of program development at Boulder Nonlinear Systems, Inc. He is currently focused on integrating nonmechanical beam control capabilities into optical communications and ladar remote sensing and imaging systems. His work spans Department of Defense, academic, and commercial applications and markets. He has led cross-disciplinary teams to develop multifunction ladar sensors for three-dimensional imaging, vibrometry, polarimetry, and optical aperture synthesis for both ground and flight systems. Dr. Buck has also extensively studied the limits of both coherent and direct detection theory as applied to communications, imaging, and remote sensing systems. He earned his Ph.D. in physics from the California Institute of Technology, where he conducted experiments on the physics of individual atom-photon interactions using trapped atoms and high-finesse cavities and carried out research in the areas of quantum information processing as applied to communication protocols. Dr. Buck began his career with the Aerospace Corporation, where he was a member of the team that pioneered some of the early demonstrations of optical aperture synthesis, and he led efforts to combine optical aperture synthesis and laser vibrometry. He then joined Lockheed Martin Coherent Technologies, where he became a principal research scientist leading teams that developed several new remote sensing systems. Dr. Buck is currently serving on the Active Optical Sensing Committee for the Optical Society of America (OSA) Conference on Lasers and Electro-Optics. He is a lifetime member of the OSA; the American Physical Society; and SPIE, the International Society for Optical Engineering; and a member of the Institute of Electrical and Electronics Engineers Photonics Society (formerly LEOS).

MILTON M.T. CHANG is managing director of Incubic Management, LLC. Dr. Chang, who has an exceptional investment track record, founded Incubic to institutionalize this approach in a venture capital and management advisory firm. He personally built Newport Corporation and New Focus, Inc., to successful initial public offering, as chief executive officer, and has provided the first capital to more than a dozen high-tech start-up companies, all of which were successful. Having been an entrepreneur, he is helpful to other entrepreneurs—his operating principle is fairness—he is effective as a sounding board for providing advice to entrepreneurs. Dr. Chang is active in the technical and business community and has received a number of prestigious awards from professional societies. He is a fellow of the Institute of Electrical and Electronics Engineers (IEEE), the Optical Society of America, and the Laser Institute of America (LIA), and a former president of the IEEE Photonics Society and the LIA. He is well known for sharing his experience freely and writes monthly business columns for *Laser Focus World* and contributes articles to *Photonics Spectra.* He earned a B.S. in electrical engineering with highest honors from the University of Illinois and M.S. and Ph.D. degrees in electrical engineering from the California Institute of Technology (Caltech) and completed the Harvard Owner/President Management Program. He has received the Distinguished Alumni Award from both the University of Illinois and Caltech and is a member of the board of trustees of Caltech and an Overseer of the Huntington Library.

CONSTANCE CHANG-HASNAIN is the John R. Whinnery Chair Professor for the Electrical Engineering and Computer Sciences Department, University of California, Berkeley. Professor Chang-Hasnain received her Ph.D. degree from the same department in 1987. Prior to joining the Berkeley faculty, she was a member of the technical staff at Bellcore (1987-1992) and an assistant professor of electrical engineering at Stanford University (1992-1996). She currently serves as chair of the Nanoscale Science and Engineering Graduate Group. She is also an honorary member of the A.F. Loffe Institute (Russia), Chang Jiang Scholar Endowed Chair at Tsinghua University (China), and Visiting Professor of Peking University (China) and National Jiao Tung University (Taiwan). Professor Chang-Hasnain's research interests range from devices to materials and physics, particularly focusing on new optical structures and materials for integrated optoelectronics. Most recently, she and her students achieved groundbreaking results of nanolasers on metal oxide semiconductor-silicon based on their discovery of a brand new nanomaterial growth mode. Professor Chang-Hasnain is recognized by the international scientific community with awards such as the IEEE David Sarnoff Award 2011 for pioneering contributions to vertical cavity surface emitting laser (VCSEL) arrays and tunable VCSELs; the Optical Society of America (OSA) Nick Holonyak, Jr., Award, 2007, for significant contributions to VCSEL arrays, injection locking, and slow light; and the Japan Society of Applied Physics Micro-optics Award, 2009, for distinguished works and contributions to the development and promotion of micro-optics technologies. She received the Guggenheim Memorial Foundation Fellowship, 2009; Humboldt Research Award from the Alexander von Humboldt Stiftung Foundation, 2009; and the Chang Jiang Scholar Endowed Chair Award from the People's Republic of China, 2009. She was also awarded the National Security Science and Engineering Faculty Fellowship, one of the most prestigious faculty fellowships, by the Department of Defense.

CHARLES M. FALCO is the chair of Condensed Matter Physics, professor of optical sciences, and professor of physics at the University of Arizona. Dr. Falco has been a professor of optical sciences since 1992, with a joint appointment in physics. He is a fellow of the American Physical Society (APS), the Institute of Electrical and Electronics Engineers (IEEE), the Optical Society of America, and SPIE, and he served two year terms (1992-1993) as councilor of the APS and member of the Executive Committee of the Division of Condensed Matter Physics of the APS and 4 years (1994-1998) as secretary/treasurer of the Forum on International Physics of the APS. Dr. Falco's principal research interests are the growth (by molecular beam epitaxy and sputtering), structure (using a wide range of probes, including x-ray and electron diffraction, in situ and ex situ surface probes, electron microscopy, scanning probe microscopies, etc.), and studies of the physical properties of metallic superlattices and ultrathin films, including research on magnetism, superconductivity, x-ray optics, elastic properties, and nucleation and epitaxy of thin films, as well as computerized image analysis. He has authored or co-authored more than 250 papers and six book chapters, holds six U.S. patents, and co-edited two books; he has given more than 150 invited talks at conferences in 24 countries and more than 200 seminars at universities and research institutions in 15 countries.

ERICA R.H. FUCHS is an assistant professor in the Department of Engineering and Public Policy at Carnegie Mellon University (CMU). Her research focuses on the role of government in technology development and the effect of location on the competitiveness of new technologies. In 2008 she received the Oak Ridge Associated Universities Junior Faculty Enhancement Fellowship for her research on the impact of offshoring on technology directions, and in 2011 she received a National Science Foundation CAREER award for her research rethinking national innovation systems. During 2011, she played a growing role in national meetings on the future of U.S. advanced manufacturing. Before joining the faculty at CMU, Dr. Fuchs completed her Ph.D. in engineering systems at the Massachusetts Institute of Technology (MIT) in June 2006. She also received her master's and her bachelor's degrees from MIT in technology policy (2003) and materials science and engineering (1999), respectively. Prior to graduate school, Dr. Fuchs spent 1999-2000 as a fellow at the United Nations in Beijing, China. There, she conducted research at state-owned industrial boiler manufacturers on policies to encourage innovation. Her work has been published in *High Temperature Materials and Processes, Journal of Lightwave Technology, Composite Science and Technology, International Journal of Vehicle Design, Issues in Science and Technology, Research Policy*, and *Management Science*. Dr. Fuchs has been an invited speaker at a wide range of venues, including the United Nations Industrial Development Organization, the U.S. Department of Commerce's National Advisory Council on Innovation and Entrepreneurship, and the Council on Foreign Relations.

WAGUIH S. ISHAK, *Alternate Co-Chair*, is the division vice president and director of the Corning West Technology Center, Corning Incorporated. Dr. Ishak's M.Sc. and Ph.D. degrees in electrical engineering were awarded by McMaster University, Ontario, Canada, in 1975 and 1978, respectively. In 1999, Dr. Ishak completed the Stanford Executive Program at Stanford University. In 1987, he became the manager of the Photonics Technology Department of the Instruments and Photonics Laboratory at Hewlett-Packard, which is responsible for research and development (R&D) programs in fiber optics, integrated optics, optoelectronics, micro-optics, and optical interconnects for applications in measurements. In 1995, he was promoted to the position of director of the communications and optics research laboratory. Dr. Ishak led his R&D team in the areas of photonics (fiber optics, integrated optics, optoelectronics, and micro-optics) and integrated electronics. In 2003, he became the director of the Photonics and Electronics Research Lab at Agilent Labs, responsible for the R&D programs in photonics, high-speed electronics, sensors, semiconductor testing, wireless communications, and consumer electronics. In 2005, he became the vice president and chief technology officer at Avago Technologies. Dr. Ishak managed the company's U.S. Advanced R&D Center and was responsible for creating technologies for its Electronic Components Business Unit. In 2007, he joined Corning Incorporated. As the division vice president and director of the Corning West Technology Center, he manages a team of scientists in developing applications for Corning's glass and fiber technologies and conducting state-of-the-art research in the areas of microstructures and nanotechnology. Dr. Ishak has 7 patents, 2 book chapters, and 75 publications and is currently a fellow of the Institute of Electrical and Electronics Engineers and chair of the board-elect of Optoelectronics Industry.

PREM KUMAR is the AT&T Professor of Information Technology for the Electrical Engineering and Computer Science Department at Northwestern University. Professor Kumar is also the director of the Center for Photonic Communication and Computing and a professor of physics and astronomy. He received his Ph.D. in 1980 in physics from the State University of New York, Buffalo, and joined

Northwestern in 1986 after spending 5 years at the Massachusetts Institute of Technology. His publications include 1 edited book, 1 book chapter, 6 patents, 180 peer-reviewed journal papers, 45 proceedings articles, and 300 (90 invited) conference papers. His research focuses on photonic devices and applications utilizing the principles of nonlinear and quantum optics. Current development areas in which he is involved include generation, distribution, and ultrafast processing of quantum entanglement for cryptography and computing; novel optical amplifiers and devices for networked communications; and novel quantum light states for precision measurements and quantum imaging and sensing. He is a fellow of the Optical Society of America (OSA), the American Physical Society (APS), the Institute of Electrical and Electronics Engineers (IEEE), the Institute of Physics (United Kingdom), the American Association for the Advancement of Science (AAAS), and SPIE, the International Society for Optical Engineering. In 2006 he received the Martin E. and Gertrude G. Walder Research Excellence Award from Northwestern University. In 2004 he received the Fifth International Quantum Communication Award from Tamagawa University, Tokyo, Japan. He has been a Distinguished Lecturer for the IEEE Photonics Society (2008-2010). He is active in professional societies (OSA, IEEE, APS, SPIE, and AAAS) in various roles including the following: OSA Long-Term Planning Group (2008-2010); general (program) chair, Quantum Electronics and Laser Science Conference 2008 (2006); chair, Steering Committee, International Conference on Quantum Communication, Measurement, and Computing (Brisbane, Australia, 2010; Calgary, Alberta, Canada, 2008; Tsukuba City, Japan, 2006; Cambridge, Massachusetts, 2002; Evanston, Illinois, 1998, principal organizer). He is the founder and managing partner of NuCrypt, LLC, in Evanston, Illinois.

DAVID A.B. MILLER (NAS, NAE) is the W.M. Keck Professor of Electrical Engineering, a Professor by Courtesy of Applied Physics, and a co-director of the Stanford Photonics Research Center at Stanford University. Dr. Miller received his B.Sc. from St. Andrews University, Scotland, and, in 1979, the Ph.D. from Heriot-Watt University, both in physics. He was with Bell Laboratories from 1981 to 1996, as a department head from 1987, and in latter years of the period as head of the Advanced Photonics Research Department. He has served as director of the Ginzton Laboratory and of the Solid State and Photonics Laboratory at Stanford. His research interests include physics and devices in nanophotonics, nanometallics, and quantum-well optoelectronics, and fundamentals and applications of optics in information sensing, switching, and processing. He has published more than 230 scientific papers, holds 69 patents, and is the author of *Quantum Mechanics for Scientists and Engineers* (Cambridge, 2008). Dr. Miller has served as a board member for both the Optical Society of America (OSA) and the Institute of Electrical and Electronics Engineers (IEEE) Lasers and Electro-Optics Society (LEOS), and on various other society and conference committees. He was president of the IEEE LEOS in 1995. He has also served on boards for various photonics companies and on the Defense Sciences Research Council for the Defense Advanced Research Projects Agency. He was awarded the Adolph Lomb Medal and the R.W. Wood Prize from the OSA, the International Prize in Optics from the International Commission for Optics, and the IEEE Third Millennium Medal. He is a fellow of OSA, IEEE, the American Physical Society, the Royal Society, and the Royal Society of Edinburgh; holds honorary doctorates from the Vrije Universiteit Brussel and Heriot-Watt University; and is a member of the National Academy of Sciences and the National Academy of Engineering.

DUNCAN T. MOORE (NAE) is the Rudolf and Hilda Kingslake Professor of Optical Engineering and Professor of Biomedical Engineering, as well as professor of business administration in the William E. Simon Graduate School of Business Administration at the University of Rochester. In 2006, he was also appointed director for entrepreneurship at the university, and in 2007 he became the vice provost for entrepreneurship. From 2004 until 2009, he was responsible for the $3.6 million Kauffman grant on entrepreneurship with a $7.2 million matching grant from the University of Rochester. The Ph.D. degree in optics was awarded to Dr. Moore in 1974 from the University of Rochester. He had previously earned a master's degree in optics at Rochester and a bachelor's degree in physics from the University of Maine. Dr. Moore has extensive experience in the academic, research, business, and government arenas of science and technology. He is an expert in gradient-index optics, computer-aided design, and the manufacture of optical systems. He has been a thesis advisor for more than 50 graduate students. In 1993, he began a 1-year appointment as science adviser to Senator John D. Rockefeller IV of West Virginia. Dr. Moore was elected to the National Academy of Engineering in February 1998. His major areas of research are in gradient-index materials, computer-aided design (including design for manufacturing methods), the manufacture of optical systems, medical optics (especially optics for minimally invasive surgery), and optical instrumentation. His most recent Ph.D. thesis topics as a thesis adviser for graduate students have been very high efficiency solar cells, polymer gradient index optics, a built-in accommodation system for the eye, terahertz imaging, generalized three-dimensional index gradients, single-point diamond turning of glass, design methods for gradient-index imaging systems, the effect of diffusion chemistry on gradient-index profiles formed by means of sol-gel techniques, quantitative phase imaging in scanning optical microscopy, integration of the design and manufacture of gradient-index optical systems, and interferometric characterization of the chromatic dispersion of gradient-index glasses.

DAVID C. MOWERY is William A. and Betty H. Hasler Professor of New Enterprise Development at the Walter A. Haas School of Business at the University of

California, Berkeley, and a research associate of the National Bureau of Economic Research. He received his undergraduate and Ph.D. degrees in economics from Stanford University and was a postdoctoral fellow at the Harvard Business School. Dr. Mowery taught at Carnegie Mellon University, served as the study director for the Panel on Technology and Employment of the National Academy of Sciences, and served in the Office of the United States Trade Representative as a Council on Foreign Relations International Affairs Fellow. He has been a member of a number of National Research Council committees. In 2003-2004, he was the Marvin Bower Research Fellow at the Harvard Business School. His research deals with the economics of technological innovation and with the effects of public policies on innovation; he has testified before congressional committees and served as an adviser for the Organisation for Economic Cooperation and Development, various federal agencies, and industrial firms. Dr. Mowery has published numerous academic papers and has written or edited a number of books, including the *Oxford Handbook of Innovation; Innovation, Path Dependency, and Policy: The Norwegian Case*; *Innovation in Global Industries*; *Ivory Tower and Industrial Innovation: University-Industry Technology Transfer Before and After the Bayh-Dole Act*; *Paths of Innovation: Technological Change in 20th-Century America*; *The International Computer Software Industry: A Comparative Study of Industry Evolution and Structure*; *U.S. Industry in 2000: Studies in Competitive Performance*; *The Sources of Industrial Leadership: Studies of Seven Industries*; *Science and Technology Policy in Interdependent Economies*; *Technology and the Pursuit of Economic Growth*; *Alliance Politics and Economics: Multinational Joint Ventures in Commercial Aircraft*; *Technology and Employment: Innovation and Growth in the U.S. Economy*; *The Impact of Technological Change on Employment and Economic Growth*; *Technology and the Wealth of Nations*; and *International Collaborative Ventures in U.S. Manufacturing*. His academic awards include the Raymond Vernon Prize from the Association for Public Policy Analysis and Management, the Economic History Association's Fritz Redlich Prize, the *Business History Review*'s Newcomen Prize, and the Cheit Outstanding Teaching Award.

N. DARIUS SANKEY is a currently a portfolio director for Central Portfolio Management at Intellectual Ventures, where he participates in developing intellectual property investment strategies for the firm. Dr. Sankey had recently served as managing director at Zone Ventures, an affiliate venture capital fund of Draper Fisher Jurvetson based in Los Angeles. Dr. Sankey led the Zone Ventures technology assessment efforts and oversaw its portfolio investments for more than 8 years, serving as a board member for several companies including Siimpel Corporation, Lumexis, Inc., and Microfabrica and Neven Vision (acquired by Google). He has led several transactions in the microelectronics, wireless telecommunications, media and entertainment, and business and consumer software sectors. Dr. Sankey has a strong interest in strategizing market applications for basic science research on the university level. This interest has also led him to a position as a visiting professor at the Rady School of Business Management at the University of California, San Diego. Before his tenure at Zone Ventures, Dr. Sankey worked as a management consultant at McKinsey & Company, Inc., and held strategic planning, consulting, and research and development positions at the RAND Corporation and AT&T Bell Laboratories. Dr. Sankey holds a B.S. in physics and electrical engineering from the Massachusetts Institute of Technology and a Ph.D. in optical engineering from the Institute of Optics, University of Rochester.

EDWARD WHITE, the president of Edward White Consulting and a native of New York State, began his career at Kodak after earning a B.S. degree in mechanical engineering from the University of Rochester. He later earned an Executive M.B.A., also from the University of Rochester. At Kodak, Mr. White held a variety of management positions in engineering, research and development, and the business units. As general manager of Kodak's Optical Products Business Unit and vice president of the Commercial Imaging Group, he led an organization responsible for designing and manufacturing optical systems for Kodak products as well as for high-tech customers external to Kodak. His global organization of more than 1,800 people included engineering and manufacturing operations located in the United States, Europe, Latin America, China, Taiwan, and Japan, with sales of $100 million. After retiring from Kodak in 2009, Mr. White took an interim position as president and chief executive officer of JML Optical, a manufacturer of precision optical components and assemblies, and in 2010 he founded Edward White Consulting, LLC. His consulting business specializes in helping engineering and manufacturing companies solve challenging business and operational issues around the world. He is currently engaged in helping companies improve operations in the United States as well as establish new operations in China and India. Mr. White is active in his community and serves on several not-for-profit boards. He currently chairs the Rochester United Way Services Corporation Board, is a member of the Finance Committee of the Greater Rochester United Way, and is a member of Rochester's Children's Success Fund.